£17·40 .

Nature and Origin of Carbohydrates in Soils

Nature and Origin of Carbohydrates in Soils

M. V. CHESHIRE

Macaulay Institute for Soil Research,
Craigiebuckler, Aberdeen

1979

ACADEMIC PRESS

London New York Toronto Sydney San Francisco

A Subsidiary of Harcourt Brace Jovanovich, Publishers

ACADEMIC PRESS INC. (LONDON) LTD.
24/28 Oval Road,
London NW1

United States Edition published by
ACADEMIC PRESS INC.
111 Fifth Avenue
New York, New York 10003

Library of Congress Catalog Card Number 79–40898
Cheshire, M. V.
Nature and origin of carbohydrates in soil.
1. Soils–Carbohydrate content
I. Title
631.4'17   S592.6.C3/   79–40898

ISBN 0-12-171250-8

Printed in Great Britain by Willmer Brothers Limited, Birkenhead, Merseyside.

Preface

This monograph is intended for the soil scientist or student of soil science and subjects in allied fields. Its subject deals with one of the major components of the organic matter of soil, in many areas vital to the maintenance of soil fertility. The nature of the carbohydrate in soil is discussed in relation to carbohydrate in various life forms, in particular in plants and microorganisms, wherein it originates.

A critical examination is made of the methods which have been devised to overcome the unique problems inherent in the analysis of soils for carbohydrate, and of methods for the isolation, purification and characterization of soil polysaccharide. Comparison of the available carbohydrate analyses of soils reveals a remarkable similarity in composition. It is described how early considerations of the origin of this carbohydrate by compositional comparison have been superseded by tracer work using ^{14}C. A description of experimental work over many years shows why it is accepted that soil polysaccharide makes a major contribution to the aggregation of soil, although precise mechanisms await elucidation.

It is hoped that this book will help to further the study of carbohydrate in soil, acting both as a source of reference and as a stimulus to research.

I wish to express my gratitude to Drs George Anderson, Colin Farmer and John Parsons for their helpful criticism and to Walter Bick for translations.

M. V. Cheshire
August 1979

Contents

Chapter 1

Introduction

About 10% of the organic matter in soil occurs as carbohydrate, mostly in the form of polysaccharide containing at least seven neutral, two acidic and two basic sugars. It originates from plant and animal remains, from extracellular gums produced by micro-organisms and from their cellular tissue.

Apart from trace amounts of free sugars which may be leached from soil by water, the carbohydrate is not easily isolated from soil because of its intimate association and chemical bonding with other non-carbohydrate organic matter and with some of the mineral components of soil.

The presence of carbohydrate in soil has been known for almost a century but it is only since the advent of chromatography about 30 years ago that detailed analysis of its composition could be made, revealing its complex nature. Since then investigators have remarked on the surprising similarity in its composition in different soils and the example shown in Table 1.1 is typical of many areas. It has not yet been possible to isolate soluble polysaccharide fractions showing distinctive differences in composition and so it is not known whether the material is predominantly a heterogeneous mixture of polysac-charides, or a single homogeneous but complex polysaccharide containing many different sugars.

From an agricultural viewpoint, the most important property associated with soil polysaccharide is the binding of soil particles into stable aggregates; another function may be to help maintain the water content of soil.

Soil polysaccharide has been the subject of several review articles,

1

TABLE 1.1 *Carbohydrate composition of a Countesswells soil.*

	mg g^{-1} soil
Hexoses	
Galactose	1·4
Glucose	5·2
Mannose	1·5
Pentoses	
Arabinose	1·5
Ribose	0·05
Xylose	1·6
Deoxyhexoses	
Fucose	0·3
Rhamnose	0·8
Uronic acids	
Galacturonic acid	3·0
Glucuronic acid	2·6
Hexosamines	
Galactosamine	0·5
Glucosamine	0·7
Total	19·15

for example by Forsyth (1948), Mehta *et al.* (1961), Gupta (1967), Swincer *et al.* (1969), Finch *et al.* (1971), Greenland and Oades (1975) and Lowe (1978b).

The structures of some of the most common soil sugars are shown below.

THE OCCURRENCE OF POLYSACCHARIDES AND THEIR FUNCTION IN LIVING ORGANISMS

Polysaccharides are found in all living organisms. Table 1.2 lists the major kinds in relation to their sources.

In plants about 75% of the dry weight is polysaccharide, with cellulose, the most abundant of all naturally occurring organic compounds, constituting at least 10% of all vegetable matter. The cellulose has a structural role; in the plant cell wall, linear chains of cellulose molecules occur in cross-linked bundles embedded in a highly branched polysaccharide matrix consisting of the so-called

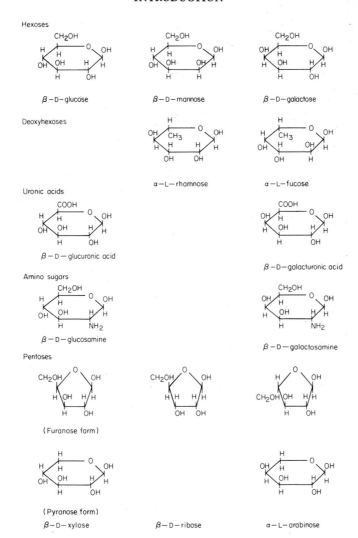

Hexoses

$\beta-D-$glucose $\beta-D-$mannose $\beta-D-$galactose

Deoxyhexoses

$\alpha-L-$rhamnose $\alpha-L-$fucose

Uronic acids

$\beta-D-$glucuronic acid

$\beta-D-$galacturonic acid

Amino sugars

$\beta-D-$glucosamine

$\beta-D-$galactosamine

Pentoses

(Furanose form)

(Pyranose form)

$\beta-D-$xylose $\beta-D-$ribose $\alpha-L-$arabinose

hemicellulose. Hemicelluloses have been defined as the alkali-soluble polysaccharides in plants and are a mixture of homo- and heteropolysaccharides with xylans predominating. They include both neutral and acid polymers, the latter containing uronic acids.

Plants also contain small amounts of water-soluble polysaccharides known as pectic substances, in which polygalacturonic acid is a principal component. Polysaccharides containing acid groups are also found on the surface of plant roots and

TABLE 1.2 *Principal polysaccharides in living organisms (adapted from Greenland and Oades, 1975, Finch et al., 1971 and Gorin and Spencer, 1968).*

Organism	Polysaccharide	Structure or component sugars
Higher Plants	Cellulose	Linear β-(1→4)-D-glucan
	Hemicellulose	Linear β-(1→4)-D-xylan Xyloglucan Arabinogalactan Rhamnogalacturonan Glucomannan Arabinoxylan Galactoarabinoxylan Glucan Glucoarabinoxylan
	Pectin	Linear α-(1→4)-D-galacturonan Arabinogalactan Rhamnogalacturonan Xyloglucan
	Starch amylose	Linear α-(1→4)-D-glucan
	Starch amylopectin	α-(1→4)-D-glucan with α-(1→6) branching
	Inulin	β-(2→1)-D-fructosan
	Fructosans	β-(2→6)-D-fructosan
	Gums	Polymers of uronic acids, hexoses and pentoses
	Mucilages	Galacturonic acid
	Callous	β-(1→3)-D-glucan
Animals	Chitin	Linear β-(1→4)-N-acetylglucosa-minosan
	Glycogen	α-(1→4)-D-glucose with α-(1→6) branching
	Hyaluronic acid	D-glucuronic acid and *N*-acetyl-D-glucosamine alternately in β-(1→3) and β-(1→4) linkages
	Chondroitin	D-glucuronic acid and N-acetyl-D-galactosamine in β-(1→3) and β-(1→4) linkages. The galacto-samine may be sulphated in the 4 or 6-0- position

Organism	Polysaccharide	Structure or component sugars
	Heparin	O-sulphatylglucuronic acid α-(1→4) D-glucosamine-N-sulphate
	Glycoproteins	Mannose, galactose, glucuronic acid, glucosamine, galactosamine, sialic acid
Microorganisms		
Fungi		
	Chitin	Linear β-(1→4) acetylglucosaminosan
	Starch	α-(1→4)-D-glucan α-(1→4)-D-glucan with α-(1→6)-D-glucose branching
	Glycogen	α-(1→4)-D-glucan α-(1→4)-D-glucan with α-(1→6)-D-glucose branching
	Pullulan	Linear α-(1→6) maltotriose
	Mycodextran	Alternate-(1→3) and α-(1→4) linked D-glucan
	Cellulose	Linear β-(1→4)-D-glucan
	Glucans	Linear β-(1→3)-D-glucan β-(1→6)-D-glucan β-(1→3)-D-glucans with β-(1→6) branching
	Mannans	α-(1→6)-D-mannan with α-(1→3) and α-(1→2) side chain β-(1→3) and β-(1→4)-D-mannans
Bacteria		
	Mucopeptide	N-acetylglucosamine glycosidically linked to N-acetylmuramic acid
	Mucopolysaccharide	Glucosamine, galactosamine, galactose, glucose, mannose, rhamnose, fucose

Organism	Polysaccharide	Structure or component sugars
	Teichoic acids	Polymers of glycerol or ribitol phosphate units linked through phosphodiester linkages with either glucose or N-acetylglucosamine linked glycosidically to the polyol
	Cellulose	Linear β-(1→4)-D-glucan
	Dextran	α-(1→6)-D-glucan
	Levan	β-(2→6)-D-fructosan
	Extracellular gums	Glucose, mannose, galactose, fucose, rhamnose, arabinose, xylose, 2-O-methylrhamnose, 3-O-methylrhamnose, glucuronic acid, galacturonic acid
	Lipopolysaccharide	Glucosamine, galactose, glucose, galactosamine, mannose, rhamnose, fucose, 3, 6-dideoxyhexoses

are thought to be involved in the transport of nutrient ions to the plant.

Carbohydrates are also important as a means of storing energy. In the majority of plants this involves the glucose polymer starch, but in a few species fructosan substitutes.

Most animals contain a much smaller proportion of their tissue in the form of carbohydrate than do plants mainly because they do not have a rigid cell-wall structure. Their polysaccharides are predominantly composed of hexoses, hexosamines and uronic acids. Energy storage in the animal utilizes glycogen, a polysaccharide with a structure very similar to that of starch. Lubrication of tissue is effected by hyaluronic acid, which is a chain of alternating glucuronic acid and N-acetylglucosamine units. Some animals have evolved hard exoskeletons which are made up of chitin, a polymer of N-acetylglucosamine.

Microorganisms have some polysaccharides common to plants and animals, for example cellulose, starch, glycogen and chitin, and others which appear more specific. Bacterial cell walls are composed of complexes of amino sugars and peptides called mucopeptides, and polysaccharide, called mucopolysaccharide, which also contains amino sugars. In some bacteria, reducing sugars have been found to account for about 45% of the cell walls. Many microorganisms are encapsulated in polysaccharide. These capsules or slimes are complex heteropolysaccharides whose purpose is not yet known. The polysaccharide is hygroscopic and this may help the organism to avoid or recover from desiccation. It has been suggested that the fixation of nitrogen by microorganisms, a process requiring anaerobic conditions, may be stimulated by the reduction in the rate of diffusion of oxygen to the cells caused by the presence of mucilaginous polysaccharide (Hepper, 1975).

THE OCCURRENCE OF OTHER CARBOHYDRATE SUBSTANCES

A number of carbohydrates, some of which are listed in Table 1.3 can occur in nature either free or in combination with non-carbohydrate material.

TABLE 1.3 *Examples of the occurrence of carbohydrate in forms other than polysaccharide.*

	Source
Ribose	RNA
Deoxyribose	DNA
Glycosides	Plants
Sucrose	Ripe fruit
Fructose	Ripe fruit
Inositol	Seeds
2-ketogluconic acid	Microorganisms

Cell metabolic processes involve a large number of small molecular weight compounds, many of which are phosphate esters of carbohydrates, for example glucose-6-phosphate and glucose-1,6-diphosphate.

Linkage of carbohydrate to other substances in biological material

In glycoproteins there are two types of protein–carbohydrate linkage with the majority involving hexosamine (Montgomery, 1970).
(i) Glycosylaminic bonds occur between N-acetyl glucosamine and L-asparagine:

(ii) Glycosidic bonds occur between the sugars and amino acids listed in Table 1.4. For example galactose and hydroxylysine:

TABLE 1.4 *Sugars and amino acids which are found in glycosidic linkages (from Spiro, 1973).*

Sugar	Amino acid
N-acetylgalactosamine Galactose Xylose Mannose	L-serine or L-threonine
Arabinose Galactose	Hydroxyproline
Galactose	5-hydroxylysine

In glycosides, sugars are linked either to sugar alcohols such as glycerol, arabitol, mannitol and inositol, or to phenols such as the flavones, anthocyanidins and xanthones. Glucose is the most common sugar in glycosides whilst the occurrence of the other hexoses, galactose and mannose, in this form is rare.

THE AMOUNT OF CARBOHYDRATE IN SOIL

The percentage of soil organic matter in the form of carbohydrate shows considerable variation from soil to soil, particularly in uncultivated soils. Folsom *et al.* (1974) found a quadratic relationship between the total monosaccharide (y) in mg $100g^{-1}$ and the total percentage carbon (x) in 43 samples from horizons in 15 prairie and forest soils expressed by the equation,

$$y = 15 + 400\ x - 31\ x^2$$

Most of the samples had carbon contents less than $2 \cdot 0\%$, however, and only a few with high carbon values appeared to distort the linearity of the relationship. I have examined the relationship between the total carbohydrate and the carbon content of the surface horizon of soil more thoroughly by statistical analysis of carbohydrate, determined by anthrone for 113 cultivated and 52 uncultivated soils reported in the literature. For all cultivated soils (Fig. 1(a)), although there are significant ($P < 0 \cdot 01$) second and fourth degree terms in a polynomial regression, the linear regression is not significant showing that carbohydrate carbon does not increase or

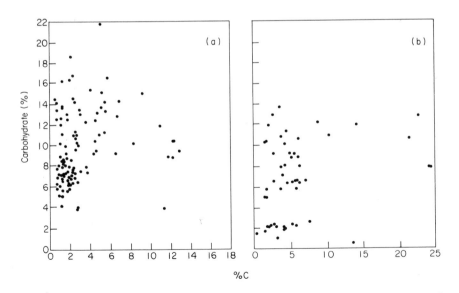

Fig. 1 The relationship between the carbohydrate content and the carbon content of soil for (a) cultivated soils and (b) uncultivated soils.

TABLE 1.5 *Amounts of neutral sugar in uncultivated soils.*

Soil	Class	C (%)	Neutral sugar C as % C	Reference
Uncultivated podzol, Australia	Podzol	1·5	9·4	Oades (1972)
Ando, Australia	Andosol	5·6	9·3	
Black earth, Australia	Vertisol	1·8	14·6	
Rendzina, Australia	Rendzina	3·7	9·2	
Virgin prairie soil developed in loess udollic albaqualf, USA, A_1 0–23 cm	Planosol	2·1	17·3	Folsom et al. (1974)
Soil developed in chertz limestone, typic hapluday, USA, A_1 3–8 cm	Luvisol	4·6	9·4	
Soil developed in sandstone mixed with chertz limestone, typic paleudult, USA, A_1 0–5 cm	Luvisol	2·1	11·1	
Brown forest, Canada upper humus layer	Cambisol	9·2	12·1	Sowden and Ivarson (1962b)
Podzol, Canada, A_2	Podzol	9·6	12·6	
Dark brown, Canada, 0–15·2 cm	Kastanozem	5·7	14·4	
Phragmites peat, Scotland, 130–180 cm	Histosol	52·6	14·4	Cheshire and Mundie (1966)
Coniferous forest soil, Scotland, F layer, 2·5–5 cm	Podzol	51·1	25·7	
Low moor peat, Australia, 0–15 cm	Histosol	47·5	9·1	Oades (1967b)
Krasnozem, Australia, 0–22·5 cm	Nitosol	6·1	10·7	
Podzol, Australia, 95–135 cm	Podzol	2·2	1·8	
Lateritic podzolic, Australia	Luvisol	1·9	16·5	Oades (1972)
Tropical red earth, Australia	Ferralsol	0·76	5·4	
Solodized solonetz, Australia	Solonetz	1·8	16·7	
Low moor peat, Japan, 20–46 cm	Histosol	24·71	8·6	Murayama (1977b)

TABLE 1.6 *Amounts of neutral sugar in cultivated soils*

Soil	Class	C (%)	Neutral sugar C as % C	Reference
Black cultivated, Canada, 0–15·2 cm	Chernozem	6·2	12·0	Sowden and Ivarson (1962b)
Sandy granitic loam, Scotland, 5–10 cm	Podzol	3·6–5·1	9·5–10·1	Cheshire and Mundie (1966)
Red brown earth, Australia, 0–6 cm	Luvisol	1·0–2·5	4·2–5·3	Oades (1967b)
Rendzina, Australia, 0–6 cm	Rendzina	1·6–2·9	5·5–9·0	
Solonized brown soil, Australia, 0–6 cm	Kastanozem	1·7–1·8	11·3–11·7	
Red earth, Australia, 0–10 cm	Ferralsol	1·1–1·6	7·7–7·8	
Terra rossa, Australia, 0–8 cm	Luvisol	1·7–2·7	8·3–8·7	
Grey clay, Australia, 0–10 cm	Vertisol	0·59–1·3	12·6–16·3	
Soil developed in loess, Ap 0–20 cm, typic hapludoll, USA	Phaeozem	1·6	18·3	Folsom et al. (1974)
Soils formed on eocene deposits overlying chalk, England, 0–15 cm	Cambisol:gleysol	0·95–1·73	14·32–14·85	Whitehead et al. (1975)
Greyish brown paddy field soil, Japan, 0–15 cm	Gleysol	1·93	8·6	Murayama (1977)
Greyish paddy field soil, Japan, 0–10 cm	Gleysol	2·13	10·2	

decrease with increasing total carbon. The three main sources of data for cultivated samples, examined separately, did not show a consistent trend. For all uncultivated soils (Fig. 1(b)) there is a significant linear regression ($P < 0.001$) but, when the four soils with about 50% carbon are omitted, the regression is no longer significant and carbohydrate carbon shows no relationship to total carbon. The average percentage of carbon as carbohydrate is 9·9 for cultivated soils, and 7·2 for uncultivated.

A number of values for the percentage of soil carbon present as carbohydrate (neutral sugar) for a wide range of soils are given in Tables 1.5 and 1.6. In all cases the carbohydrate content has been calculated from the sum of individual sugars. Other, far less precise values for total carbohydrate, measured by colorimetric methods, are presented in the Appendix. It may be calculated that, in mineral soils, component sugars usually account for 0·5–2·0% of the soil by weight, but in peats, and horizons rich in organic matter, the values may be as much as 20–30%. The cultivated soils, except for mucks, are generally poorer in organic matter and correspondingly have lower carbohydrate contents, 0·1–2·0%.

TABLE 1.7 *Changes in the amount of carbohydrate with depth.*

Soil	Horizon	Depth (cm)	Carbohydrate C as % soil C	Reference
Loess	A_1	0–23	17·3	Folsom *et al.*
	A_2	25–41	18·2	(1974)
	B_2	46–69	19·6	
Solod	A_1	0–11	7·2	Graveland and Lynch
	A_2	11–15	10·2	(1961)
	B_2	15–22	10·3	
	B_3	22–30	11·7	
Dark grey	L	6·5–5	6·8	Sawyer and Pawluk
wooded soil	F–H	5–0	5·4	(1963)
	A_h	0–4	11·0	
	A_h	4–8	8·6	
	A_h	8–12	6·6	
	A_{he}	16–20	5·4	
	A_e	28–32	5·0	

The carbohydrate contents of soils can also be estimated as reducing sugars, which include uronic acids and hexosamines. These have given values ranging from about 1 to 45%. Scottish arable soils commonly have 0·9–1·5% reducing sugar equivalent to glucose.

The percentage of carbohydrate in soil decreases with depth, as does the percentage of carbon, but the proportion of the carbon present as carbohydrate may increase or decrease (Table 1.7).

Chapter 2

Qualitative Composition of Soil Carbohydrate

The first class of carbohydrate to be identified in soil was pentosan (Chalmot, 1894; Feilitzen and Tollens, 1898) detected by the formation of furfural, on boiling soil or humus fractions with 12% HCl. From a substance in a sulphuric acid hydrolysate of soil, Shorey and Lathrop (1910) synthesized an osazone derivative with a melting point of 160–161°C, corresponding to that of xylose and, by reaction with cadmium carbonate and bromine, they prepared the double salt cadmium xylonate and cadmium bromide, confirming the presence of the pentose. Subsequently, pentosans were shown to be widespread in soil (Schreiner and Lathrop, 1911).

Evidence for hexose-containing polysaccharide was first obtained from peat. Feilitzen and Tollens (1898) hydrolysed peat with dilute sulphuric acid and identified mannose, galactose and fructose in the hydrolysate.

With various soils, in addition to pentosan, Leavitt (1912) found a carbohydrate with properties very similar to starch. Acid hydrolysis released a crystalline sugar considered to be a hexose because of its ease of fermentation. Subsequently most of the sugars found in living organisms have been identified in soil extracts or hydrolysates (Table 2.1).

FREE SUGARS

Sugars in cold water extracts of soils have been identified using paper

14

chromatography. They include glucose, galactose, mannose, arabinose, xylose, ribose, rhamnose, fucose, fructose, sorbose, sucrose and raffinose (Forsyth, 1950; Alvsaker and Michelson, 1957; Griffiths and Dobbs, 1963; Grov, 1963; Kanke and Yamane, 1974 a,b,c; Sparkov and Sparkova, 1970).

TABLE 2.1

Sugars identified in soil extracts or hydrolysates

Hexoses	D-glucose
	D-galactose
	D-mannose
	Fructose
	Sorbose
Pentoses	L-arabinose
	Deoxyribose
	Ribose
	D-xylose
Deoxyhexoses	L-fucose
	L-rhamnose
Uronic acids	Glucuronic acid
	Galacturonic acid
Hexosamines	Glucosamine
	Galactosamine
	N-acetylglucosamine
Methyl sugars	4-O-methyl-D-galactose
	2-O-methyl-L-rhamnose
	2-O-methylxylose
	3-O-methylxylose
Sugar alcohols	Inositol
	Mannitol

Further confirmation of the presence of fructose was obtained by the X-ray diffraction of the osazone derivative (Alvsaker and Michelson, 1957). This sugar was also found in the ammonium oxalate extract of a soil (Lattes *et al.*, 1963).

It is not clear what effect the extraction has on the release of sugars.

Some free sugars must exist in soil, but it is likely that the extraction process enhances the amounts of free sugars found. In contrast to water extraction, 80% ethanol extracted only glucose from four soils (Nagar, 1962). On the other hand ethanol extraction of five incubated soils indicated the presence of free glucose and xylose. With a sixth soil, glucose and arabinose were extracted (Robert, 1964). Cellobiose, cellotriose and gentiobiose were also observed in some of the extracts.

SUGARS RELEASED ON HYDROLYSIS

Neutral sugars

The major neutral sugars occurring in soil polysaccharide are glucose, galactose and mannose (hexoses), arabinose and xylose (pentoses) and rhamnose and fucose (methylpentoses or deoxyhexoses). Minor amounts of ribose, fructose and various methyl sugars are also present. Most of the major sugars were originally characterized by the formation of derivatives and comparison of their properties such as melting points with those of the authentic substance. Identification of the minor sugars has relied on methods involving one or more types of chromatography.

Paper chromatography of sugars in the hydrolysate of a fulvic acid, the alkali- and acid-soluble fraction of soil, indicated the presence of glucose, galactose, mannose, arabinose, xylose and ribose (Forsyth, 1950). Scaling up the procedure by using cellulose column chromatography enabled the glucose and galactose to be identified as the osazone and phenylmethylhydrazone derivatives, respectively (Forsyth, 1950).

The formation of similar types of derivatives from sugars in the hydrolysate of alkali-soluble soil polysaccharide not only allowed identification of the sugars but also enabled their configuration to be determined (Whistler and Kirby, 1956). Galactose, glucose, mannose and xylose were in the D form whereas arabinose was in the L form.

Optical rotations of isolated, carefully purified soil arabinose, shows that at least 90% of the sugar in a Scottish soil is in the L form (Cheshire and Thompson, 1972).

Most of the sugars common to soil have also been identified in peat by means of paper chromatography (Theander, 1954; Black *et al.*, 1955; Strelkow, 1958; Tschaikowa and Rakowski, 1959; Kadner *et al.*, 1962).

Rhamnose in soil was first observed through the isolation of a rhamnose glycoside from which rhamnose was released by hydrolysis with dilute mineral acid. The sugar was characterized by its melting point and that of the phenylhydrazone (Shorey, 1913). The presence of L-rhamnose in the hot-water-soluble fraction of a soil was confirmed by paper chromatography and optical rotation after isolation using cellulose column chromatography (Duff, 1952b). The optical rotation and the melting point of the *p*-nitrophenylhydrazone confirmed that L-rhamnose was present (Whistler and Kirby, 1956).

The occurrence of L-fucose in a soil hydrolysate was shown by paper chromatography and the formation of the phenylmethyl-hydrazone. The derivative was characterized by a mixed melting point with the authentic material, by its optical rotation, and by its X-ray diffraction pattern (Duff, 1952a). Fucose was also one of several sugars identified in peat hydrolysates by the formation of derivatives, namely: fucose as the methylphenylhydrazone; mannose as the phenylhydrazone; xylose as the di-*O*-benzylidene dimethyl acetate; and ribose as the toluene *p*-sulphonylhydrazone (Black *et al.*, 1955).

Deoxyribose, which must be present in all living organisms, is rarely reported but has been observed by paper chromatography of acid hydrolysates of soils (Hardisson and Robert-Gero, 1966).

Substances with the characteristics of nucleotides considered to be derived from RNA and DNA and which contain ribose or deoxyribose have been isolated from alkaline extracts of soils and separated by adsorption chromatography on alumina or by ion exchange chromatography (Shorey, 1913; Adams *et al.*, 1954; Anderson, 1970).

The ketose fructose has been confirmed in peat by paper chromatography (Theander, 1954), in soil hydrolysates by borate sugar complex chromatography on resin columns in combination with thin layer chromatography (TLC) (Murayama, 1977a), and in hydrolysed soil extracts by TLC alone (Kanke and Yamane, 1974b). Another ketose, sorbose, has been detected by paper chromatography in the 6M HCl hydrolysates of humic acid from forest litters (Maksimova, 1973).

Methylated sugars

Several monomethyl sugars occur in soils. 2-*O*-methyl-L-rhamnose and 4-*O*-methyl-D-galactose have been identified in hydrolysates of peats and soils (Duff, 1961). They were separated from other sugars by gradient elution of a charcoal–Celite column with 0–20% ethanol and elution from a Celite column with water-saturated butanol-light petroleum.

The sugar corresponding to 2-*O*-methylrhamnose was characterized initially by degradation studies. Reduction with sodium borohydride followed by sodium periodate gave glyceraldehyde indicating a 2-*O*-methylaldose. Hydrolysis with 13% HCl yielded 5-methylfurfuraldehyde, to be expected from a 6-deoxyhexose, and demethylation with boron trichloride produced rhamnose. The unknown sugar had the same R_F values on paper chromatographs as an authentic sample and it moved at the same rate on electrophoresis. The derivative had the same melting point as the authentic *p*-nitrophenylosazone. Finally an identical infrared spectrum was obtained.

The sugar corresponding to 4-*O*-methylgalactose crystallized more easily, had the expected melting point and optical rotation, and gave galactose on demethylation. In addition to these characteristics the sugar had an infrared spectrum identical to that of an authentic sample.

One of the sugars that Duff isolated from paper chromatograms he believed to be 2-*O*-methylxylose. Subsequent studies on one of the peats using gas–liquid chromatography showed that many sugars were present in that position on the chromatogram but mainly 2-*O*-methylxylose and 3-*O*-methylxylose (Bouhours and Cheshire, 1969).

2-*O*-methylxylose was characterized (Bouhours and Cheshire, 1969) by the retention times on gas–liquid chromatography (GLC) of the alditol acetate, the acetylated nitrile and the methyl glycoside derivatives. The infrared spectrum of the methyl glycoside was identical to that of an authentic sample. Demethylation of the unknown sugar with boron trichloride gave xylose.

3-*O*-methylxylose was identified by similar procedures. In addition to these methyl sugars there is paper chromatographic evidence for the presence of 2-*O*-methylarabinose (Lynch *et al.*, 1957b).

Acidic sugars

Both glucuronic and galacturonic acid are now known to be present in soil hydrolysates but initially the presence of such compounds was assumed on a false premise. This was that CO_2, released on heating carbonate-free soil with acid, derived only from uronic acid (Shorey and Martin, 1930) whereas it is now known that CO_2 is even more readily formed from other humus components (Mehta et al., 1961). Later the presence of uronic acids was confirmed by heating soil extracts with carbazole which reacts to give a green colour (Lynch et al., 1957a).

Glucuronic acid has been identified by paper chromatography (Forsyth, 1950; Whistler and Kirby, 1956; Lynch et al., 1958), by the optical rotation of the barium salt and the formation of a specific red colour with thioglycollic acid (Whistler and Kirby, 1956). Galacturonic acid has also been identified by paper chromatography (Lynch et al., 1958) and the presence of both sugars in hydrolysates of several agricultural soils has been confirmed by gas–liquid chromatography of the silyl ethers of the lactonized aldonic acid derivatives (Mundie, 1976). No derivative corresponding to mannuronic acid was detected by this means although a claim for the presence of this sugar in the humic acid fractions of fresh litter, along with glucuronic, galacturonic and ketogluconic acids, has been made (Maksimova, 1973). Certainly 2-ketogluconic acid has been found in soils incubated with glucose (Duff and Webley, 1959), and in the rhizosphere (Moghimi et al., 1978).

Amino sugars

Two amino sugars, glucosamine and galactosamine, have been identified in soil hydrolysates by paper chromatography (Bremner, 1958) and subsequently isolated as their crystalline hydrochlorides which were readily identified by X-ray analysis. Stevenson (1954) detected glucosamine and galactosamine in soil hydrolysates by ion exchange chromatography and later used the same technique to identify N-acetylglucosamine and possibly N-acetylgalactosamine (Stevenson, 1956).

Evidence for the presence of another amino sugar, muramic acid

(3-*O*-carboxyethyl-D-glucosamine), in soil has been provided by the formation of D-lactic acid on treatment of isolated amino compounds with dilute alkali (Millar and Casida, 1970) and in peat by the identification of the trimethylsilyl derivative by gas–liquid chromatography (Casagrande and Park, 1978).

Sugar alcohols

Two sugar alcohols have been identified in soils. Mannitol has been isolated from the fulvic acid fraction of a soil by precipitation with ammoniacal lead acetate. Mannitol crystals were obtained and the nature of the substance ˋconfirmed by the formation of the tribenzacetol derivative (Shorey, 1913). The hexaacetate has been used to characterize this sugar isolated from a podzol (Sallans *et al.*, 1937). Mannitol has also been identified in soil hydrolysates by paper chromatography (Lynch *et al.*, 1958; Dormaar, 1967).

Inositol has been isolated as a crystalline solid from the alkaline extract of the soil after hydrolysis of the extract by H_2SO_4 (Yoshida, 1940). Treatment of such an extract with hypobromite followed by the addition of iron precipitated a substance with the same composition as ferric phytate (Wrenshall and Dyer, 1941; Bower, 1945). Furthermore acid hydrolysis of a phosphorus-rich alkali-soluble fraction of soil liberated inositol (detected by paper chromatography) and phosphoric acid (Anderson, 1956) and chromatography showed that most of the inositol in the fraction was present as hexaphosphate. The main isomer is *myo*inositol, which together with *scyllo*inositol accounts for about 90% of the esters (McKercher and Anderson, 1968 a,b).

Small amounts of *neo*inositol (Cosgrove and Tate, 1963) and D-*chiro*inositol (Cosgrove, 1970) have been detected.

Inositol has been observed in soil hydrolysates by paper chromatography (Hayashi and Nagai, 1962; Lynch *et al.*, 1958) and by GLC separation of soil sugars as alditol acetates (Oades, 1967a).

Polysaccharides

The presence of cellulose in peats was first indicated by Feilitzen and Tollens (1898) by the presence of residual organic matter after drastic treatment of the peat with hot concentrated alkali.

Identification has also depended on the insolubility of glucose-containing material in ether, water, ethanol and hot 2% HCl, and its solubility in 80% H_2SO_4 (Waksman and Stevens, 1930) or 72% H_2SO_4 (Alvsaker, 1948).

The solubility of glucose-containing material in Schweizer's reagent has been used as a criterion (Oden and Lindberg, 1926). Material extracted from humic acid with Schweizer's reagent yielded 30% octaacetyl cellobiose on acetylation (Scheffer and Kickuth, 1961). The material extracted from seven soils with this reagent has been characterized as cellulose by various chemical and physical means. On hydrolysis it was found to contain between 65 and 85% glucose. Other sugars present, galactose, mannose, arabinose and xylose, accounted for up to 9% (Gupta and Sowden, 1965).

The infrared spectrum of one of these cellulose preparations from an orthic grey-wooded soil was very similar to that of a very pure cellulose but the X-ray diffraction pattern showed that it was amorphous.

Oligosaccharides

An Indian soil rich in plant roots yielded an oligosaccharide from the fulvic acid fraction by precipitation as the barium salt at pH 4·5–4·8 (Majumda *et al.*, 1974). On hydrolysis it gave five sugars: galacturonic acid, mannose, xylose, galactose and arabinose. The order of sugars in the polymer was established by reduction of the end reducing group (in arabinose) and periodate oxidation (which left mannose, xylose and galactose). The two outer sugars of this trio were removed by a second periodate treatment after a Barry's degradation.

The fact that the molecule after etherification and esterification only consumed two molecules of periodate and liberated one molecule of formic acid was taken to mean that the sugars were linked in $(1 \rightarrow 3)$ positions. The material was therefore thought to be: D-galacturonic acid-$(1 \rightarrow 3)$-D-mannose-$(1 \rightarrow 3)$-D-xylose-$(1 \rightarrow 3)$-D-galactose-$(1 \rightarrow 3)$-L-arabinose.

Hydrolysis is best avoided when substances are being examined for the presence of oligosaccharides because small amounts of sugars may recombine during hydrolysis. Nevertheless useful information about polysaccharide structure may be obtained. Partial hydrolysis of mineral soil with 0·5M H_2SO_4 or M acetic acid has given rise to oligosaccharides amongst which were xylobiose and xylotriose suggesting the presence of $(1 \rightarrow 4)$ xylan. An homologous series of $(1 \rightarrow 4)$ linked β-D-xylose oligosaccharides containing up to six xylose units was obtained from a blanket peat (McGrath, 1973).

Chapter 3

Quantitative Analysis of Soil Carbohydrate

DIRECT METHODS

The determination of total carbohydrate in soil has been attempted by heating soil in sulphuric acid with anthrone (Macaulay Annual Report 1953/54) but there was wide variation in the values obtained depending on the conditions of estimation. This may have been related to interference from humified organic matter for which concentrated sulphuric acid is a good solvent. An effect observed with 12M H_2SO_4 extracts of soil is that further dilution of the extract with H_2SO_4 increases the apparent carbohydrate content as measured by anthrone (Cheshire and Mundie, 1966). In contrast the dilution of acid hydrolysates of soil with water, appears to decrease carbohydrate content (Ivarson and Sowden, 1962) and this has been ascribed to the removal of interference by coloured organic matter. Analyses with anthrone are also subject to interference from nitrate (Juo and Stotzky, 1967) which is invariably present in soil.

Methods for the direct determination of uronic acids and pentoses in soil both involve treatment with boiling 12% HCl. Uronic acids are decarboxylated to furfural, the amounts of uronic acid being determined by measuring the amount of CO_2 formed.

$$
\begin{array}{l}
\text{CHO} \\
| \\
\text{H}-\text{C}-\text{OH} \\
| \\
\text{HO}-\text{C}-\text{H} \\
| \\
\text{H}-\text{C}-\text{OH} \\
| \\
\text{H}-\text{C}-\text{OH} \\
| \\
\text{COOH}
\end{array}
\longrightarrow \quad \text{CO}_2 \quad + \quad
\begin{array}{l}
\text{CH}\!-\!\!-\!\text{CH} \\
\quad\| \quad\quad \| \\
\text{CH} \quad \text{C}-\text{CHO} \\
\quad\diagdown\!\!\diagup \\
\quad\quad\text{O}
\end{array}
\quad + \quad 3\text{H}_2\text{O}
$$

23

Interference from inorganic carbonates may be removed by pretreatment of the soil with dilute acid, but more serious is the interference from organic matter. It has been found that some organic matter is decarboxylated even more readily than uronic acids and comparison of values obtained for the uronic acid contents of fulvic acid by this means with those using a colorimetric method with carbazole shows that values based on CO_2 formation are overestimates (Table 3.1).

TABLE 3.1 *Comparison of two methods of analysis for uronic acid in soil fulvic acids (Dubach and Lynch 1959).*

	Uronic acid (p.p.m.)	
	Carbazole	Decarboxylation
Podzol A_0	1160	1165
Podzol A_1	180	470
Podzol A_2	86	2111
Sassafras	127	353
Pocomoke	72	407
Prairie	150	340

Measurement of pentose in soil has been made by distilling off and collecting the furfural formed by the 12% HCl treatment (Ivarson and Sowden, 1962) and reacting it with phloroglucinol or phenylhydrazine to obtain a precipitate which is estimated gravimetrically (Adams and Castagne, 1948). Clearly some of the furfural formed is derived from uronic acids and this must be accounted for. The total yield of furfural from pentoses is in the region of 85–95% of the theoretical value (Mehta and Deuel, 1960) compared with about 40% for uronic acid (Ivarson and Sowden, 1962).

Errors may also arise because substances other than furfural are present in the distillate and the derivatives are not pure. With colorimetric methods of analysis in which the furfural is reacted with orcinol (Mehta and Deuel, 1960) or aniline acetate (Ivarson and Sowden, 1962), interference caused by other aldehydes is negligible.

TABLE 3.2 *Comparison of two methods of analysis for pentose in soils (Ivarson and Sowden, 1962).*

Soil	Pentose ($mg\ g^{-1}$)	
	Adams and Castagne	Column chromatography
Grenville	3·26	3·32
Lennoxville	4·07	4·62
Scott	3·80	3·19
Lacombe	2·82	2·76

Despite the disadvantages, good agreement has been obtained between values for pentose by the distillation method of Adams and Castagne using aniline acetate, and total pentose determined by analysis of separated sugars using column chromotography. This is considered to be the result of compensating errors in the distillation method involving both interference from uronic acids and loss on hydrolysis of pentose.

An attempt has been made to determine water-soluble polysaccharide in soil, commonly referred to as microbial gum, gravimetrically after isolation from the fulvic acid of soil. The soil was extracted with 0·5M NaOH for 3 h and the extract acidified with HCl to pH 2–3 and centrifuged. The soluble fulvic acid fraction was diluted with two volumes of acetone which precipitated polysaccharide. The polysaccharide was purified by redissolving in NaOH, acidifying with HCl and reprecipitating with acetone, and the weight of gum was obtained after drying at 60°C for 48 h (Rennie *et al.*, 1954; see also Forsyth, 1947). These polysaccharides are, however, very impure. In four soils examined, the material initially precipitated had an average of 67·5% ash and a carbohydrate content, equivalent to glucose, of 18·8%, as estimated by a colorimetric procedure using anthrone (Acton *et al.*, 1963a).

Material that had been repeatedly washed with water and reprecipitated still contained as much as 24% ash (Halstead, 1954). Such polysaccharide, even after purification by adsorption on charcoal or treatment with Polyclar, still gives 5% ash and has a sugar content of less than 40%. In addition, more polysaccharide with the same solubility characteristics may be extracted from the humic acid fraction by redissolution and reprecipitation (Forsyth, 1948). Weights of material that have been recorded, can hardly be said to be a quantitative measure of soil gum.

None of the direct methods of analysis discussed here can be commended.

INDIRECT METHODS

Methods of hydrolysis, their effectiveness and the losses involved

There is no universal method of hydrolysis which is satisfactory for all classes of carbohydrate in soil. Furthermore it will be seen that there are often differences between soils when methods are compared for carbohydrates of the same class.

Uronic acids

The presence of uronic acids in polysaccharides increases the resistance of the polysaccharide to hydrolysis by acids (Smith and Montgomery, 1959) so that harsher conditions for complete hydrolysis are needed; but the liberated uronic acids are easily destroyed by acids leading to low recoveries.

Early methods of analysis using carbazole by-passed the problem by converting the uronic acid polymers in fulvic acid to furfural by hydrolysing with concentrated acid. Two ml of the fulvic acid solution was heated with 12·5 ml 18M H_2SO_4 for 30 min. Interference from Fe^{3+} was removed by reducing it to Fe^{2+} with stannous chloride (Lynch *et al.*, 1957a) but nothing was done to remove neutral sugars which also interfere. The use of the fulvic acid fraction probably stemmed from the belief, based on CO_2 formation with 12% HCl, that this fraction contained most of the uronic acid. However, the uronic acid contents of the fulvic acid fractions of four soils measured by carbazole were much less than by the liberation of CO_2 and examination of hydrolysates by paper chromatography suggested that with one of the soils, a podzol, only traces of uronic acid were present (Dubach and Lynch, 1959). Although pretreating soils with M HF:HCl for 2 h increased the amounts of uronic acids in the fulvic acid by 112%, (Graveland and Lynch, 1961) methods of releasing uronic acids from the whole soil were clearly needed.

To this end hydrolysis of soils by 12M H_2SO_4 and formic acid were compared, care being taken to minimize interferences (Ivarson and Sowden, 1962). The hydrolysate was neutralized by $Ba(OH)_2$ to pH 5–6 and the barium sulphate filtered off and washed well with hot water. The filtrate, reduced in volume to 2 ml and made slightly alkaline with NaOH, was added to an anion exchange resin column of Dowex 1 to separate the uronic acids, by absorption, from neutral sugars and iron, which were eluted. The uronic acids were subsequently eluted by 0·25M NH_4Cl at pH 8 and determined using carbozole. It was found that the treatment with 12M H_2SO_4 gave larger yields of uronic acid than those with formic acid or 0·5M H_2SO_4. A prolonged treatment of 22 h with formic acid caused considerable destruction (Table 3.3). Repeated hydrolysis as in (a)

TABLE 3.3 *Comparison of different hydrolysis conditions on the release of uronic acids from litter (Ivarson and Sowden, 1962).*

Hydrolysis conditions	Uronic acid (mg g^{-1} litter)
(a) 12M H_2SO_4 for 15 min	
followed by reflux with 0·5M H_2SO_4 for 3 h	26
(b) 12M H_2SO_4 for 2 h	
followed by reflux with 0·5M H_2SO_4 for 16 h	28·5
(c) 90% formic acid for 8 h	
followed by reflux with 0·5M H_2SO_4 for 3 h	23
(d) 90% formic acid for 22 h	
followed by reflux with 0·5M H_2SO_4 for 16 h	9

released further uronic acid but (b) was the most efficient method. Even so recoveries of glucuronic acid or polygalacturonic acid were less than 50% and a correction factor of two was applied.

The effects of hydrolysis of soils using a large range of HCl and H_2SO_4 concentrations in a single step have been tested (Dormaar and Lynch, 1962). The acid:soil mixtures, 50 ml:10 g soil, were kept at 85°C overnight or autoclaved at 121°C for 1 h. At 85°C the concentration of acid needed to give maximum release of uronic acid increased with increasing organic matter content of the soil. 1·5M

H_2SO_4 was sufficient with three of the soils, but the fourth soil, richest in organic matter (13·6%) required 6M HCl. Hydrolysis at 121°C gave no greater release with H_2SO_4 and results for HCl were erratic and sometimes lower. On autoclaving galacturonic acid with 1·5M H_2SO_4 the recovery was about 85%, and 95% when soil was present. This was taken to indicate a quantitative yield of soil uronic acid.

In another study, hydrolysing soil with 0·5M H_2SO_4 in a sealed tube, in a boiling water bath, gave optimum release of uronic acid after 4 h and the amount in the hydrolysate remained constant thereafter (McGrath, 1971), so a heating period of 10 h was chosen to obtain a greater yield of neutral sugars which were also being determined. In contrast, optimum hydrolysis of uronic acids with 1·5M H_2SO_4 took place over a narrow time range which was different for each soil and considerable destruction occurred with longer heating periods. A variant to the methods of Dormaar and Lynch (1962) and Ivarson and Sowden (1962) was hydrolysis with 12M H_2SO_4 for 1 h followed by autoclaving the diluted acid, presumably 1·5M, at 121°C for 1 h (Lowe and Turnbull, 1968).

On comparison of a three-stage hydrolysis procedure, suggested by Oades et al. (1970) for neutral sugars, with a two-stage procedure, similar values for uronic acid release were obtained (Table 3.4).

TABLE 3.4 *Comparison of different hydrolysis conditions on the release of uronic acids from soil (Mundie, 1976).*

Hydrolysis conditions	Uronic acid $(mg\ g^{-1})$
(a) 20 min reflux with 2·5M H_2SO_4, 16h 12M H_2SO_4, 8h 0·5M H_2SO_4 100°C	2·85
(b) 16h 12M H_2SO_4, 8h 0·5M H_2SO_4 100°C	2·84
(c) 16h 12M H_2SO_4, 8h M H_2SO_4 100°C	2·54
(d) 16h 12M H_2SO_4, 16h M H_2SO_4 100°C	2·37

Using the two-stage procedure, (b), the recovery of glucuronic acid from a mixture with soil was found to be 30%.

Despite the poor recovery, it is concluded that the most suitable hydrolysis procedure for the uronic acid in soil is the two-stage process using cold 12M H_2SO_4 and hot 0·5M H_2SO_4.

Neutral sugars

The complete hydrolysis of many polysaccharides may be accomplished by hot dilute acids, but these can hydrolyse only about 75% of the carbohydrate in most soils. For the complete hydrolysis of polysaccharides such as cellulose, or those in soil, a preliminary treatment with a more concentrated acid is necessary. For example, hydrolysis of soil extracts with 0·3M H_2SO_4 for 6 h at 85°C released sugars equivalent to between 2·3 and 4·9% of the organic matter, compared with the usual value for complete hydrolysis of around 10%, and some of the sugars released had R_F values corresponding to oligosaccharides on paper chromatograms (Lynch *et al.*, 1958). These findings clearly show that only partial hydrolysis had occurred.

In dilute acid hydrolysis of soil, release of hexose increases with concentration of acid. With four soils, 3M H_2SO_4 at 85°C for 24 h gave a slightly greater release of hexoses and deoxyhexoses than 1·5M H_2SO_4 (Brink *et al.*, 1960). Certainly 5·3M HCl was a more effective hydrolysing agent than 0·9 or 1·8 HCl for the hexoses and pentoses of coal humic acid (Havrankova, 1967). After 72-h hydrolysis of soils with 6M HCl at 120–130°C (Singh and Bhandari, 1963) reducing sugar was still being released, despite the destruction of sugars that must have occurred; the recovery of glucose being only 60%. Clearly dilute acid hydrolysis alone is unsuitable for the quantitative release of soil hexoses.

TABLE 3.5 *Effect of H_2SO_4 concentration on the release of pentose from soils (Thomas and Lynch, 1961).*

	Pentose (mg g^{-1})	
	Angus Ridge	Beaverhills
0·25M	3·30	4·03
0·5M	3·64	4·45
1·0M	3·75	3·30

Pentose components of polysaccharide are generally more easily liberated than the hexoses but the released pentose sugar is more liable to destruction during acid hydrolysis. Hydrolysis of soil with 0·5M H_2SO_4 at 121°C for 1 h gave almost the same release of pentose as M H_2SO_4 in one soil and considerably more in another (Table 3.5). Similarly there was little difference in the release of pentose on hydrolysis of soil with 0·5M H_2SO_4 at 100°C for 8 h and 1·5M H_2SO_4 at 80°C for 24 h (Table 3.6).

TABLE 3.6 *Effect of* H_2SO_4 *concentration on the release of pentose from soils (Wen and Chen, 1962).*

	Pentose (mg g^{-1})	
	Brick-red soil	Black soil
0·5M	3·10	1·90
1·5M	3·35 ⎱ 3·20 ⎰	1·75

Methods seeking to obtain maximum hydrolysis of both hexoses and pentoses have usually employed preliminary treatment with quite concentrated acids. One such method involves pretreatment of the soil with cold 12M H_2SO_4 (72% w/w) followed by dilution of the acid to 0·5M and heating at 100°C (Ivarson and Sowden, 1962). With litter samples this caused the release of about three times the amount

TABLE 3.7 *Comparison of hydrolysis conditions on the release of pentoses and hexoses from litter (Ivarson and Sowden, 1962).*

	Hydrolysis conditions	Pentose (mg g^{-1})	Hexose (mg g^{-1})
Coniferous litter	0·5M H_2SO_4 (16h)	52·2	67·4
	12M H_2SO_4 (2h), 0·5M H_2SO_4 (16h)	41·7	170·7
Deciduous litter	0·5M H_2SO_4 (16h)	35·6	50·1
	12M H_2SO_4 (2h), 0·5M H_2SO_4 (16h)	43·0	118·9

TABLE 3.8 *Sugar released by* 0·5M H_2SO_4 *hydrolysis as a percentage of that released by* 12M H_2SO_4 + 0·5M H_2SO_4 *hydrolysis (adapted from Gupta and Sowden, 1965).*

	Orthic podzol L–H	Northern podzol F	Orthic grey wooded L–H	Brown Forest AL
Galactose	38	47	36	52
Glucose	41	42	40	54
Mannose	58	65	63	39
Arabinose	75	75	80	81
Xylose	100	89	107	54
Fucose + ribose	80	125	83	68
Rhamnose	86	100	75	76

of glucose obtained with 0·5M H_2SO_4 hydrolysis, but with coniferous litter less pentose was obtained (Table 3.7). With soil, pretreatment with 12M H_2SO_4 increased the release of most sugars, not just that of glucose (Table 3.8). In choosing this method of hydrolysis, Ivarson and Sowden had compared 12M H_2SO_4 with boiling 90% formic acid but found the formic acid to be less efficient. A 15 min predigestion with 12M H_2SO_4 gave a lower yield of sugars than one of 2 h, which was considered sufficient. There was, however, an indication that a further release of sugars, particularly glucose, occurred with successive treatments of the soil with 12M H_2SO_4.

A more thorough examination of the effect of 12M H_2SO_4 has been made by Cheshire and Mundie, (1966) using a mineral soil. Extraction of carbohydrate into 12M H_2SO_4, as measured by orcinol, reached a maximum within 16 h at 20°C and decreased thereafter, whereas that measured by anthrone continued to increase for at least 40 h (Table 3.9). This shows that some destruction of non-hexose sugar occurs with the extended 12M H_2SO_4 treatment. After treatment with 12M H_2SO_4 a further hydrolysis with 0·5M H_2SO_4 at 100°C was necessary for the maximum release of monosaccharides. Three hours sufficed for pentoses and deoxyhexoses but the period for hexoses depended on the time of treatment with 12M H_2SO_4. The relationship for total reducing sugar is shown in Fig. 2; the longer the period of hydrolysis with 12M H_2SO_4, the shorter need be the subsequent period in 0·5M H_2SO_4. Having taken account of the period in hot dilute acid needed for complete hydrolysis, a greater

TABLE 3.9 *Effect of period of pretreatment with* 12M H_2SO_4 *on the release of glucose (Cheshire and Mundie, 1966).*

Period (h)	Glucose (mg g^{-1})
4	3·1
8	3·3
16	3·4
40	3·7

release of glucose with an extended pretreatment period became apparent.

The amounts of pentose released by a treatment with 12M H_2SO_4:0·5M H_2SO_4 were not significantly less than with 0·5M H_2SO_4 provided the period of pretreatment was not greater than 16 h.

In summary the results suggest that: (1) long periods with 12M H_2SO_4 destroy pentose; (2) hexoses are still not completely released after 40 h in 12M H_2SO_4. It was concluded that there was no variant of this method of hydrolysis which would completely release glucose without destroying pentose or deoxyhexose components. Treatment

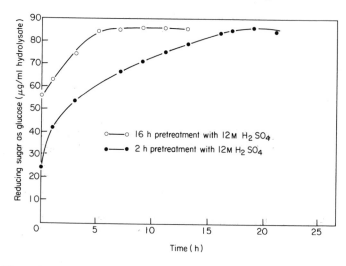

Fig. 2 The effect of time of pretreatment of soil with 12M H_2SO_4 at 20°C on the release of reducing sugar during hydrolysis with 0·5M H_2SO_4 at 100°C.

for 16 h with 12M H_2SO_4 followed by 5 h with 0·5M H_2SO_4 at 100°C was preferred because it appeared to give the best yields of glucose in some cases.

Oades et al. (1970) also became concerned about the different hydrolysis needs of different classes of carbohydrate in soil and investigated the effect of a preliminary short, hot acid hydrolysis. Treatments along these lines had been applied previously in the analysis of soil rich in organic matter by Theander (1952) and Juvvik (1965) using boiling 2% HCl prior to cold 76% H_2SO_4, and 0·25M and M H_2SO_4 prior to 12M H_2SO_4, respectively, but the presence of oligosaccharides in the initial hydrolysates showed that hydrolysis was incomplete. This was avoided by using a more concentrated acid, 2·5M H_2SO_4, in the first stage (Oades et al., 1970). Using a soil from which plant fragments had been removed it was observed that refluxing with 2·5M H_2SO_4 was more efficient than with 6M or 0·5M, or hydrolysis with M H_2SO_4 at 100°C in a sealed tube, and moreover it gave a greater recovery of sugars than treatment with cold 13M H_2SO_4 for 16 h followed by a 5 h hydrolysis with 0·5M H_2SO_4 (Table 3.10). The greatest differences were observed when the period of reflux in 2·5M H_2SO_4 was less than 1 h and it was found that the concentrated acid treatment destroyed at least 33% of the deoxyhexoses and xylose and some of the mannose and galactose. After refluxing with 2·5M H_2SO_4 for 20 min the mixture was filtered through a porous glass filter and the solid washed with acetone and dried, then treated with cold 13M H_2SO_4 (16 h) followed by hot 0·5M

TABLE 3.10 *Comparison of hydrolysis conditions on the release of sugars from soils (Oades et al., 1970).*

| | mg sugar released g^{-1} soil | | |
| | | 13M H_2SO_4 | 2·5MH_2SO_4 followed by |
	2·5M H_2SO_4	0·5M H_2SO_4	13M H_2SO_4:0·5M H_2SO_4
Glucose	0·76	1·17	1·24
Galactose	0·35	0·34	0·37
Mannose	0·36	0·34	0·40
Arabinose	0·33	0·28	0·39
Xylose	0·35	0·17	0·36
Mannose + fucose	0·21	0·12	0·21

H_2SO_4 (15 h). Greater yields of almost all sugars were obtained for the two mineral soils studied. When plant fragments had not been removed however, the single treatment with 13M H_2SO_4:0·5M H_2SO_4 was more efficient.

In experiments in which radioactively labelled plant material was added to soil, support has been obtained (Cheshire et al., 1973) for the suggestion that the amount of glucose unhydrolysed by 2·5M H_2SO_4 gives some measure of cellulose (Oades et al., 1970); 90% of each of the other sugars was released during the 2·5M H_2SO_4 hydrolysis stage. McGrath (1973) examined the possibility of hydrolysing the "hemicellulose" fraction of soil with 0·5M H_2SO_4, but in mineral soils only 70% of the residual sugars was glucose. In nine peat samples a more successful division was observed, the residual sugars being 95% glucose.

An attempt has been made to extend the three-stage procedure of Oades to include hydrolysis of peptide by replacing the H_2SO_4 by HCl (Garyayev et al., 1973). In the first stage 0·7M HCl was used in place of 2·5M H_2SO_4 and a maximum yield of hexoses was obtained by refluxing for 4–5 h. The second stage involved refluxing with 6M HCl for 24 h. No comparisons were reported, but sugars are known to be destroyed by heating with 6M HCl.

Partial hydrolysis of polysaccharides is often useful in structural studies but it has not often been applied intentionally to soil. Hydrolysis of five soils and a peat with 0·5M H_2SO_4 or 1M acetic acid for 4 h released pentoses and oligosaccharides, detected by p-anisidine after paper chromatography (McGrath, 1973). Substances corresponding to xylobiose (in five soils) and xylotriose (in three soils) were present in the hydrolysates. The peat hydrolysate gave spots corresponding to a homologous series of (1–4) linked β-xylose oligosaccharides.

It is concluded that the most effective means of hydrolysis for the release of neutral sugars from soil is the three-stage procedure using 2·5M H_2SO_4, 12M H_2SO_4 and 0·5M H_2SO_4, particularly for soils poor in plant residues. Careful manipulation of the sample is required, however, and for routine work the two-stage procedure using 12M H_2SO_4 and 0·5M H_2SO_4 is quite satisfactory.

Amino sugars

Polymers containing amino sugars are usually hydrolysed with 6M HCl

(Smithies, 1952) and this method has been applied to soil (Bremner and Shaw, 1954; Stevenson, 1957a). To facilitate the hydrolysis the soil was first soaked with concentrated HCl for between 48 and 56 h before diluting the acid to 6M and heating at 100°C. The hexosamine content of the hydrolysate of surface soil as determined colorimetrically reached a maximum after between 6 and 12 h and then declined (Stevenson, 1957a) so a period of 9 h has been recommended. With hydrolysis in a sealed tube at 120° a maximum was released after 4 h both with a surface soil and a claypan but in the former the percentage yield was lower. Soil has also been hydrolysed with 8M HCl for 3 h (Oades and Swincer, 1968). These conditions have been found to give the same degree of hydrolysis of known glycosamine polymers as 6M HCl after 3–6 h but a direct comparison has not been made for soil. Hydrolysis for 1 h with a 1:1 mixture of 6M HCl and 90% formic acid in a sealed tube at 100°C has been found to give less decomposition of amino sugar than 6M HCl alone (Parsons and Tinsley, 1961).

Treatment of hydrolysates to remove hydrolysing agents and other interfering substances prior to analysis

Addition of inorganic salts as precipitants

One of the most commonly used methods of deionizing sulphuric acid hydrolysates is by barium salts. The sulphuric acid reacts with $Ba(OH)_2$ or $BaCO_3$ to form insoluble $BaSO_4$. The $Ba(OH)_2$ may be added in solution whereupon neutralization is very fast. When $BaCO_3$ is used, the excess remains insoluble and the neutral sugars or barium salts of the uronic acids may be recovered by filtration and water-washing of the precipitate. Hot water is sometimes used and soxhlet extraction by ethanol has been applied (Cheshire and Mundie, 1966). It is important to keep the final pH below 6 for although yellow substances which interfere with the colorimetric determination of sugars and uronic acids (Oades, 1967b; Ivarson and Sowden, 1962) are removed by precipitation at pH > 6, with some soil hydrolysates a proportion of the sugars appears to be coprecipitated. The sugars are not completely washed from the barium sulphate by hot water, but a reagent such as 8·5M acetic acid, which is more effective, also releases the yellow substances.

Calcium carbonate has been used to neutralize sulphuric acid hydrolysates by forming insoluble calcium sulphate and removed much of the coloured organic matter (Brink *et al.*, 1960). Repeated analysis of the hydrolysate for sugars by anthrone after neutralization showed no difference in three out of four soils examined. The carbonate of another metal in this group, strontium, has been used to raise the pH of H_2SO_4 hydrolysates to near 7, when complete precipitation of brown iron-organic complexes is achieved by allowing the solution to stand. After the neutralized hydrolysate has been filtered and taken to dryness, sugars are extracted from the solid residue with methanol (Oades *et al.*, 1970). Methanol has also been used to extract and isolate sugars from the dried hydrolysate after neutralization with sodium bicarbonate (Oades, 1967a).

HCl may be removed from HCl hydrolysates by evaporation but alternatively it may be reacted with silver carbonate to form insoluble AgCl (Theander, 1952).

Ion exchange resin

Soil hydrolysates made with H_2SO_4 usually contain large amounts of cations derived from the soil and it is best to remove these by treatment with a cation exchange resin prior to using a strongly basic anion exchange resin to absorb the SO_4^{2-} ion (Black *et al.*, 1955; Finch *et al.*, 1968), otherwise the hydrolysate becomes alkaline and cations compete with the anion exchange resin for the SO_4^{2-}. An initial cation exchange treatment becomes particularly important when exchange resins are used to deionize hydrolysates containing ^{14}C-labelled sugars having very different specific activities. This is because a weakly basic anion exchange resin such as IR-45 must be used (Cheshire *et al.*, 1969) to avoid strongly alkaline conditions which facilitate the transformation of sugars by mutarotation. When soil hydrolysates, which have been partially neutralized by $Ba(OH)_2$ and made slightly alkaline by NaOH to convert lactones to the acid form, are applied to Dowex anion exchange resin, neutral sugars can then be eluted with water, and uronic acids with 0·25M NH_4Cl at pH 8 (Ivarson and Sowden, 1962). This resin treatment removes substances interfering with the colorimetric determination of both sugars and uronic acids. Partially neutralized hydrolysates have also been deionized by successive treatments with anion and cation exchange resins (Duff, 1952b; Acton *et al.*, 1963a; Dormaar, 1967). For the isolation of the uronic acid fraction from soil hydrolysates Lowe and Turnbull (1968) used the ion retardation resin AG-11A8. Sugars were eluted with water, and the uronic acid fraction with 0·25M NH_4Cl at pH 8.

Charcoal

In a comparison of three methods of deionization of soil hydrolysates, using $BaCO_3$, cation and anion resins or charcoal, the recoveries of sugar were similar, within the range 92–95%, but only the charcoal gave a colourless solution (Cheshire *et al.*, 1969).

Extensive use has been made of charcoal columns for deionization of hydrolysates neutralized by NaOH. About 400 g of an equal mixture of charcoal (such as charcoal MFC, Hopkins and Williams) and Celite 535 are mixed together in dilute acetic acid to form a slurry which is allowed to settle in a column 500 mm by 50 mm internal diameter. The column is then washed free of acid by water and the neutralized hydrolysate applied. Salts are washed through with 1 litre of water and the sugars then eluted with 1 litre of 50% ethanol (Cheshire and Mundie, 1966). This is probably the most effective means of purifying sugars from soil hydrolysates, with recoveries of better than 97%. Not all charcoals are suitable, however, and the slowness of the procedure makes it unsuited for routine analysis. It is ideal for handling hydrolysates of radioactively labelled soils.

Absorption on charcoal has been used, prior to analysis by anthrone, with the variant of using 8·5M acetic acid to elute sugars (Oades, 1967b). Of the soil hydrolysates examined, only those of podzol B horizons were not completely clarified by this method (Greenland and Oades, 1975).

It has been surmised that, of at least three components of soil hydrolysates which interfere with the determination of uronic acids by carbazole, two are absorbed on a decolorizing charcoal. The other was thought to be hexose (Lowe and Turnbull, 1968). In a slightly more drastic application, neutralized deionized peat hydrolysates were decolorized by heating the solution with charcoal at 90°C for 15 min (Black *et al.*, 1955).

Filtration

Iron and other transition elements may be precipitated as hydroxides
by neutralizing hydrolysates to pH 7 and so be removed by filtration.

Analysis of hydrolysates for classes of sugar

Most colorimetric methods of sugar analysis depend on one of two
properties; either the reducing power of the sugar, or the formation in
strong acid of compounds based on the furfural structure which react
to give coloured derivatives.

Total reducing sugars

Waksman and Stevens (1930) used alkaline ferricyanide to determine
sugars in HCl hydrolysates of soil. The yellow ferricyanide is reduced
to colourless ferrocyanide. The procedure has since been modified
(Hoffman, 1937) and automated (Cheshire and Mundie, 1966). It has
been suggested that the values obtained will be overestimates because
of the presence in soil hydrolysates of reducing substances other than
sugars (Greenland and Oades, 1975), but the determination of sugars
in soil hydrolysates individually by gas–liquid chromatography
shows that the major neutral sugars account for about 70% of the
reducing power and uronic acids for about 30% (Mundie, 1976).

A method for reducing sugars based on the reduction of alkaline
cupric salt solutions, to give cuprous complexes, has also been
applied to soil (Parsons and Tinsley, 1961). A comparison has been
made between the ferricyanide and copper methods with various soil
samples (Alvsaker, 1948). Although substances such as proteins
cause more reduction of ferricyanide than of copper, the copper has
serious disadvantages. It is very much less sensitive to pentoses than
hexoses, and is more affected by inorganic materials, particularly Fe.

Total sugars by anthrone

Anthrone, 9,10-dihydro-9-oxoanthracene, reacts with furfural
derivatives in sulphuric acid to give a green colour.

CH₂OH—furan—CHO + anthrone ⟶ O=...—CH₂—furan—CH=...=O

Most pentoses give much less colour response than hexoses or deoxyhexoses (Table 3.11) and their response is said to be negligible when the anthrone concentration is greater than $0\cdot05\%$. The reagent was first applied to soil hydrolysates by Brink *et al.* (1960) who used the argument that the polysaccharide in soils of the same type would have a similar composition, and so anthrone would give a relative measure of the total polysaccharide content. Evidence was obtained of interference in the anthrone method from other organic substances in a prairie soil, where treatment with calcium carbonate, which precipitates much organic matter, reduced the apparent carbohydrate content of the soil hydrolysate from $6\cdot75$ to $4\cdot25$ mg g^{-1}. Both nitrate and iron interfere in the anthrone method (Doutre *et al.*, 1978).

Total sugars by phenol

Phenol in sulphuric acid reacts with furfural derivatives from sugars to give yellow colours. The similarity of these colours to that of soil

TABLE 3.11 *Relative absorbance of the reaction products of sugars with various reagents.*

	Relative absorbance		
	Anthrone 625 nm	Orcinol 510 nm	Phenol 492 nm
Galactose	100	100	100
Glucose	183	83	138
Mannose	90	81	136
Arabinose	12	93	72
Ribose	15	84	138
Xylose	17	96	130
Rhamnose	169	55	100
Fucose	172	76	44
Glucuronic acid	11	44	47

TABLE 3.12 *Comparison of anthrone and phenol values for the carbohydrate content of soils (Acton et al., 1963a).*

	Carbohydrate as glucose (mg g^{-1})	
	Anthrone	Phenol
Elslow A$_a$ horizon	4·8	11·1
Melfort A$_a$ horizon	8·8	13·0
Oxbow A$_a$ horizon	2·2	4·4
Oxbow A$_e$ horizon	4·7	6·1

hydrolysates probably partly explains why total carbohydrate values of soil hydrolysates are greater than with anthrone (Yukhnin *et al.*, 1973; McGrath, 1973 and Table 3.12). Fractions rich in non-carbohydrate organic matter show the largest differences. Values with phenol may be lowered by passage of the hydrolysate through cation and anion exchange resin to remove interference by NO_3^- and Fe^{3+} although this treatment might also remove uronic acid (Doutre *et al.* 1978).

Total sugars by orcinol

Another phenol which reacts with furfural derivatives under strongly acid conditions is orcinol, 3,5-dihydroxytoluene. It has the advantage that the various classes of sugar give a more uniform response than with anthrone or phenol (Table 3.11). The reaction product is an orange–brown colour and the method may suffer from some of the same colour interferences as phenol. It has been applied to soil hydrolysates by Bachelier (1966).

Hexoses

From the table of relative responses it is clear that anthrone is a useful reagent for estimating the hexose content of soil hydrolysates. Other reagents which have been used include chromotropic acid (1,8-dihydroxynaphthalene-3,6-disulphonic acid; Ivarson and Sowden, 1962). This compound reacts with the formaldehyde liberated by the breakdown of 5-hydroxymethylfurfural. Only a small amount of interference occurs from pentoses and uronic acids (Deriaz, 1961)

TABLE 3.13 *Comparison of anthrone and chromotropic acid values for the carbohydrate content of soils (Ivarson and Sowden, 1962).*

Soil	Carbohydrate as glucose (mg g^{-1})	
	Anthrone	Chromotropic acid
Grenville	12·42	14·10
Lennoxville	14·70	19·00
Scott	10·91	12·80
Lacombe	7·62	10·90

but the method gives higher values than anthrone with litter samples and soils (Table 3.13).

Pentoses

Acharya (1937) and Alvsaker (1948) criticized the traditional method for estimating pentose in soils, which involves the formation of furfural by heating in HCl, on the grounds that the recovery is not quantitative. The second stage of the method, the determination of the furfural, has been greatly improved by the substitution of a colorimetric method using orcinol:FeCl$_3$ in concentrated HCl in place of phloroglucinol, the orcinol reagent being much more specific for furfural (Mehta and Deuel, 1960). This reagent has also been used

TABLE 3.14 *Relative absorbance of the reaction products of sugars with orcinol: Fe Cl$_3$ (Thomas and Lynch, 1961).*

	Relative absorbance 600 nm
Xylose	100
Ribose	109
Arabinose	101
Glucose	0
Fructose	5
Galacturonic acid	80
Glucuronic acid	39

with H_2SO_4 hydrolysates of soil after the removal of uronic acids by ion exchange chromatography (Thomas and Lynch, 1961). The chromogenic values for various sugars are shown in Table 3.14.

When orcinol:$FeCl_3$ was used directly on soil hydrolysates (Cheshire and Mundie, 1966) in an attempt to judge when maximum release of pentose had occurred on hydrolysis, the relative proportion of pentose determined in the hydrolysate by this method was substantiated by the analysis of individual sugars separated by paper chromatography.

Aniline in acetic acid reacts with pentoses at room temperature (Tracey, 1950) to give a red colour. The reaction does not appear to involve the prior formation of furfural, and hexoses and uronic acids give only slight interference. After treatment of soil hydrolysates by ion exchange resin, there was close agreement between the total pentose content determined by aniline acetate and the sum of the individual pentose sugars obtained by paper chromatography (Ivarson and Sowden, 1962). Wen and Chen (1962) removed interferences caused by coloured organic substances, and sulphate, ferric, ferrous and calcium ions, by treatment with $CaCO_3$ followed by cation exchange resin.

Uronic acids

Lynch *et al.* (1957a) were the first to use a colorimetric method to determine uronic acids in soil. Much lower values were obtained for the uronic acid content of soil hydrolysates by reaction with carbazole than by decarboxylation (Dubach and Lynch, 1959). Carbazole reacts with 5-formylfuroic acid which is formed from the uronic acid by treatment with strong acid. Hexoses show some interference (Table 3.15).

Interference caused by Fe^{3+} was overcome either by the addition of stannous chloride to reduce it to Fe^{2+} or, where excessive amounts were present, by removal using cation exchange resin (Dormaar and Lynch, 1962). No precautions against interference from iron were needed when uronic acids were isolated from the hydrolysate by anion exchange resin (Ivarson and Sowden, 1962). Originally the heating time was 2 h but it was later reduced to 30 min because the colour was found to decrease after 40 min (Dormaar and Lynch, 1962).

TABLE 3.15 *Relative absorbance of the reaction products of sugars with carbazole.*

	Relative absorbance 530 nm
Glucuronic acid	100
Galacturonic acid	116
Mannuronic acid	17
Glucose	10
Galactose	6
Mannose	6
Arabinose	0
Xylose	0
Rhamnose	0
Fucose	0

Lowe and Turnbull (1968) were also able to dispense with the stannous chloride by treating hydrolysates with the ion retardation resin AG-11A8. They also modified the reaction mixture by adding 0·025M sodium tetraborate to the sulphuric acid, which increased the sensitivity, and by heating at 80°C for 45 min. McGrath (1971) used both cation and anion exchange resins.

Deoxypentoses (methylpentoses)

A yellow colour, formed on heating deoxyhexoses with cysteine in sulphuric acid, is thought to be caused by the reaction of 5-methylfurfural with the thiol group of the cysteine. This method, which is said to be specific for deoxyhexoses (Dische and Shettles, 1948), has been applied to soil hydrolysates to test the efficiency of hydrolysis (Cheshire and Mundie, 1966).

Amino sugars

Two types of analysis methods of soil hydrolysates for amino sugars have been described. In one, the sugars are decomposed by heating in borate buffer of pH 8·8 and the liberated ammonia collected and titrated; account being taken of the ammonia initially present in the hydrolysate by distillation with MgO at 40°C under reduced pressure (Bremner and Shaw, 1954). In the other, colorimetric methods involving p-dimethylaminobenzaldehyde are employed. Stevenson (1957a) and Bremmer and Shaw (1954) were the first to apply this reagent to soil hydrolysates for the quantitative analysis of amino sugars. The amino sugars are heated with acetylacetone in alkali and react to form 2-methylpyrrole which gives a

red colour when heated with p-dimethylaminobenzaldehyde. Colour formation in the acetylacetone, caused partly by iron, interferes (Bremner and Shaw, 1954) and there is a small amount of interference from mixtures of hexoses and some amino acids. A refinement of the method has been to apply the hydrolysate to a column of anionic resin in the carbonate form, which removes dark coloured organic substances and precipitates iron and aluminium but allows the amino sugars to be eluted with dilute sodium bicarbonate (Stevenson, 1957a). Where neutral sugars might interfere, the effluent has been applied to a cationic resin column which retains the amino sugars; these are subsequently eluted with dilute HCl. More simply the hydrolysate has been applied to a cationic column which is first eluted with 0.02M EDTA in 0.1M sodium acetate buffer at pH 5.8 to remove iron, and then with 2M HCl to release the hexosamines (Parsons and Tinsley, 1961). Amino sugars have also been determined using ninhydrin after separation from amino acids (Sowden, 1959). The two reagents gave similar results but the ninhydrin method was less time-consuming and gave more reproducible values.

Methods of separating and analysing individual sugars

Uronic acids

(a) Isolation of uronic acids

Various strongly basic ion exchange resins absorb uronic acids quantitatively from soil hydrolysates (Thomas and Lynch, 1961; Ivarson and Sowden, 1962). Ivarson and Sowden (1962) treated the hydrolysate with $Ba(OH)_2$ until the pH was 5–6 and, after concentrating the solution, applied it to the resin. Recoveries of the uronic acids on eluting with $0.25M$ NH_4Cl at pH 8 were 95–108%. Using an ion retardation resin it was possible to apply the acid hydrolysate directly to the resin (Lowe and Turnbull, 1968). Recoveries were good but there was interference from substances derived from neutral sugars. In another procedure hydrolysates have been treated with a cation exchange resin to remove iron, and the uronic acid absorbed from the eluate by an ion exchange resin in the acetate form (McGrath, 1971). After sugars had been eluted with water, uronic acids were eluted with H_2SO_4.

(b) Separation and analysis of uronic acids

Ion exchange column chromatography. Uronic acids isolated from soil hydrolysates by absorption on anion exchange resin have been well separated by subsequent gradient elution with increasing concentration of formic acid, from 0 to $1.5M$. (Mundie, 1976). Mannuronic acid could also be separated by this means and, being absent in soil hydrolysates, was added as an internal standard.

Gas–liquid chromatography. The mixture of uronic acids, isolated by anion exchange, was eluted from the column and dissolved in alkali to convert the lactone forms to the acids (Mundie, 1976). They were then reduced with sodium borohydride to the corresponding aldonic acids, lactonized and silylized.

In this case also, mannuronic acid could be added as an internal standard. The silyl derivatives separated well on gas chromatography and gave reproducible results (Fig. 3). Gas–liquid chromatography was easier and less time consuming than ion exchange column chromatography (Mundie, 1976).

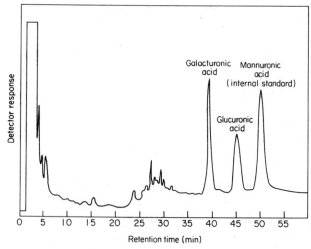

Fig. 3 GLC separation of trimethylsilyl derivatives of uronic acids from soil (Mundie, 1976).

Neutral sugars

(a) Paper chromatography

Paper chromatography was probably the first routine method for the separation of sugars in soil hydrolysates (Forsyth, 1950; Lynch *et al.* 1958).

Paper chromatographic methods have been applied to peat hydrolysates, using solvents based on mixtures of butanol, acetic acid, ethyl acetate and water, to separate and measure the four major sugar components, galactose, glucose, arabinose and xylose (Black *et al.*, 1955). Their efficiency depends upon the choice of solvent: for example butanol separated glucose, galactose, xylose and ribose but failed to separate mannose and arabinose (Forsyth, 1950). More success was achieved using water-saturated *n*-butanol and pyridine, when it was possible to separate and analyse the seven major neutral soil sugars (Lynch, Olney and Wright, 1958). Soil hydrolysates were also analysed qualitatively using the same solvent, and sugars corresponding to glucosamine, *N*-acetylglucosamine, mannitol, inositol, glucuronic acid, galacturonic acid, 2-O-methyl-D-arabinose and 2-O-methyl-D-xylose were detected.

Quantitative paper chromatography using solvents based on butanol–water–pyridine–benzene or butanol–water–acetone mixtures has been compared with other methods of analysis of soil polysaccharide. Paper chromatography was satisfactory for most sugars but not for ribose, rhamnose and sugars of higher R_F because of variation in blank values in paper taken from near the solvent front (Ivarson and Sowden, 1962). Nevertheless paper chromatography was later employed to analyse decomposing litter samples (Sowden and Ivarson, 1962 a).

Paper chromatography with n-butanol saturated with water has been used to assess the specificity of colorimetric methods (Cheshire and Mundie, 1966).

Analysis of a sugar separated by paper chromatography may be made by colorimetric methods appropriate to its class after elution from the paper. This is usually accomplished by soaking the excised area of paper containing the sugar in water for a few hours and analysing the solution by a reducing sugar method, avoiding the use of concentrated acids which could react with small fragments of paper (Cheshire and Mundie, 1966). Another technique has been applied by Parsons and Tinsley (1961) who separated soil sugars by descending paper chromatography with ethyl acetate, water and pyridine or acetic acid mixtures and developed colours on the paper with 2-aminobiphenyl. The amounts of sugars were related to the intensity of colour in the spots, which was determined by reflection densitometry.

Paper chromatography has been used in the separation of radioactively labelled sugars from soil hydrolysates (Cheshire et al., 1969). Although the sugars appeared to be pure as determined by autoradiography of the chromatograms, other methods of separation gave products with different specific activities in some cases. Values for rhamnose and fucose were about 10% of those obtained by paper chromatography (Cheshire et al., 1969; Oades and Wagner, 1970). Paper chromatography is still a useful stage in the separation of labelled sugars, however, particularly when handling hydrolysates rich in strongly labelled glucose, because separation of the glucose on ion exchange resin columns under alkaline conditions leads to the formation of very small amounts of radioactive artifacts, seriously interfering with the determination of the specific activities of other, weakly labelled, sugars. It is therefore an advantage to make a prior separation of glucose from the other sugars by paper chromatography.

(b) Cellulose column chromatography

Cellulose column chromatography has been used as the preliminary stage in the separation of sugars prior to further separation by paper chromatography both for qualitative analysis of sugars present in very small amounts in soil or peat (Duff, 1952a, b) and quantitatively for the major sugars (Sowden and Ivarson, 1962b). In Duff's work, columns of cellulose eluted with water-saturated butanol gave a broad separation into fractions containing relatively large amounts of the sugars, some of which were further separated by paper chromatography into three groups of sugars with R_Fs greater than rhamnose. One of the groups was largely composed of 2-O-methylrhamnose (Duff, 1961). Another was obtained in greater quantity by chromatography of column eluates on 3 mm thick paper and further characterized by gas–liquid chromatography as a mixture mainly of 2- and 3-O-methylxylose (Bouhours and Cheshire, 1969).

In the quantitative analysis of sugars the columns were prepared by settling a slurry of cellulose powder in a mixture of acetone, n-butanol and water (7:2:1 v/v) and, after the addition of the deionized sample, the column was eluted with the same solvent mixture. The order of elution was: rhamnose + ribose, xylose + fucose, arabinose, mannose, glucose and galactose. The rhamnose + ribose fraction also contained high R_G sugars (see p. 57). The fractions were checked for purity by paper chromatography and either analysed directly or after a further separation on paper.

(c) Ion exchange column chromatography

Ion exchange column chromatography in conjunction with other techniques has been particularly useful in the separation of radioactively labelled sugars from soil hydrolysates (Cheshire, Mundie and Shepherd, 1969). Over a period of about 12 h, 1–2 mg of each of the sugars common to soil may be well separated from each other and from substances which do not form borate complexes such as acids and phenols (Fig. 4). More recently a variation has been devised taking 7 h (Murayama, 1977a) but fucose and arabinose were not separated. The cheapness of the resin allows fresh resin to be used

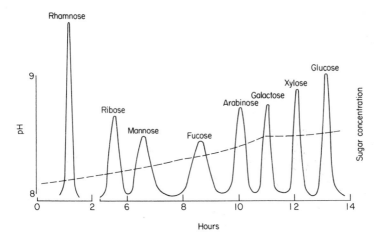

Fig. 4 The separation of sugars common to soil by ion exchange chromatography of the borate complexes. The applied pH gradient is shown by the broken line.

for each column so avoiding possible cross contamination between samples.

Individual sugars from resin column eluates have been analysed in an Auto Analyser using orcinol in 72% sulphuric acid. This procedure involves the acidification of an alkaline borate solution with concentrated acid (Cheshire *et al.*, 1969). A much more useful and less hazardous reagent for alkaline eluates is *p*-hydroxybenzoic acid hydrazide which reacts with sugar in alkaline medium to give a

TABLE 3.16 *Relative absorbance of the reaction products of sugars with* p-*hydroxybenzoic acid hydrazide (Mundie et al., 1976)*.

	Relative absorbance 410 nm
Glucose	100
Galactose	58
Mannose	92
Arabinose	76
Xylose	85
Rhamnose	76
Fucose	66

yellow colour (Table 3.16). Interference from calcium can be made negligible by the addition of calcium to the reagent.

(d) Electrophoresis

Column electrophoresis in phosphate and borate buffers has been employed to separate rhamnose, mannose, galactose and glucose in the hydrolysate of a polysaccharide from a peat soil (Clapp, 1957).

(e) Gas–liquid chromatography

The most useful method for the quantitative analysis of individual sugars is gas–liquid chromatography of the alditol acetates (Oades, 1967a; Cheshire et al., 1973). This is described in the next section. A very rapid method of sugar analysis by gas–liquid chromatography is based on the formation of the aldononitrile derivatives (Easterwood and Huff, 1969).

(f) Enzymic analysis

Glucose may be successfully determined in hydrolysates deionized by charcoal by the use of glucose oxidase (Cheshire et al., 1969). The glucose is oxidised to gluconic acid and H_2O_2

and the peroxide catalysed by peroxidase to form a chromophore derived from O-dianisidine. Neutralized hydrolysates which have not been deionized in this way give very much higher glucose values, probably resulting from the formation of other chromophores in the soil organic matter.

Amino sugars

Paper chromatography with a solvent composed of ethyl acetate–pyridine–water has been used to separate galactosamine and glucosamine in soil polysaccharide hydrolysates (Swincer et al., 1968a). The sugars have also been separated using resin column chromatography in an automatic amino acid analyser, which should also separate muramic acid, although Swincer et al. (1968a) did not observe it in soil hydrolysates.

Recommended analytical procedures

Details are given for a number of methods chosen from those described in the previous sections which are considered most appropriate for soil. The colorimetric methods, though non-specific, are well-tried. The value of those based on GLC, or the enzyme glucose oxidase, is that they measure individual sugars.

Total reducing sugar

(a) Hydrolysis procedure 1 (Cheshire and Mundie, 1966)

Finely ground air-dried soil (1g) is mixed with 2·5 ml 12M H_2SO_4 (72% w/w, specific gravity 1·634) in a 100 ml round bottom flask and left to stand for 16 h. The acid is then diluted to 0·5M by the addition of 57·5 ml water and the flask stoppered with a capillary air leak and heated at 100°C for 5 h. The mixture is cooled and filtered through a sintered glass filter, porosity × 3. The filtrate is neutralized to pH 7 by the addition of NaOH solution, the volume made up to 100 ml, and the solution filtered through a Whatman No. 1 paper.

(b) Hydrolysis procedure 2 (Oades et al., 1970)

Finely ground air-dried soil (2g) is heated under reflux with 25 ml 2·5M H_2SO_4 in a 50 ml round bottom flask. The mixture is filtered through a sintered glass filter and the residue washed with water and dried on the filter over P_2O_5, prior to hydrolysis by 12M and 0·5M H_2SO_4 as described above. The 2·5M H_2SO_4 filtrate and washings are combined and analysed separately or added to the hydrolysate of the residue.

Analysis of hydrolysates for sugars by paper chromatography

Sugars in deionized hydrolysates are spotted onto strips of Whatman No. 1 paper, 150 × 5500 mm, along a line 100 mm from one end and separated by 25 mm. A spot of standard solution of sugars each 1% in isopropanol is also applied. After the spots have dried the chromatogram is developed by descending chromatography with a solvent mixture comprised of butan-1-ol: acetic acid: water, 4:1:5 by volume (upper layer), for between 2 and 7 days or ethyl acetate:pyridine:water, 12:5:4 by volume, for 16 h. The chromatogram is air-dried and dipped in 3% p-anisidine in n-butanol and heated at 80–100°C. Hexoses and deoxyhexoses give yellow–brown spots, pentoses give pink spots and uronic acids, orange–red spots.

In quantitative work, standards are applied to either side of the line of sample spots and after development these side strips are detached and sprayed to locate the position of the sugars.

Carbohydrate by anthrone (Cheshire and Mundie, 1966)

Hydrolysates may conveniently be analysed for total carbohydrate with anthrone using an Auto Analyser (Burt, 1964). The reagent is 0·1% recrystallized anthrone in 76% v/v sulphuric acid.

A stream of anthrone reagent, $1·6$ ml min^{-1}, is joined to the sample stream, $0·32$ ml min^{-1}, segmented with air, $0·42$ ml min^{-1}, in a cooling bath of cold water, and after passing through a heating bath at 95°C and being cooled the absorbance is read at 625 nm. Standard solutions over the range 0–150 μg ml^{-1} are used. The rate of sampling is 40 h^{-1}.

In the manual method, 2 ml of hydrolysate are carefully layered over 10 ml of anthrone reagent in a loosely glass-stoppered 20 ml boiling tube. The layers are mixed and the tube heated in boiling water for 12 min. After cooling, the absorbance of the solutions are measured at 625 nm.

Total reducing sugar by alkaline ferricyanide

This is an automated version (Technicon microglucose method N-9) of the method described by Hoffman (1937). The reagents are alkaline ferricyanide solution containing 0·25 g potassium ferricyanide, 9 g sodium chloride and 20 g sodium carbonate dm^{-3}, and saline solution containing 9 g sodium chloride and 5 g potassium cyanide dm^{-3}.

The sample stream, $0·80$ ml min^{-1}, is mixed with saline, $2·50$ ml min^{-1}, segmented with air, $1·20$ ml min^{-1} and pumped through a dialyser. Sugars diffuse through the membrane into a stream of ferricyanide, $2·90$ ml min^{-1}, segmented by air, $1·20$ ml min^{-1}, which is passed through a heating bath at 95°C and thence to a colorimeter where the absorbance at 420 nm is measured.

Analysis of hydrolysates for sugars by gas–liquid chromatography (Cheshire et al., 1973)

Neutralized hydrolysate equivalent to 0·1 g soil is mixed in a 100 ml round bottom flask with a solution containing 0·5 mg quebrachitol or inositol as an internal standard and 0·1 g $NaBH_4$ is added. The solution is left overnight. One millilitre of glacial acetic acid is added to destroy excess $NaBH_4$ and the solution is evaporated to dryness with 1 ml 10% acetic acid in methanol five times. Acetic anhydride (5 ml) is then added and the flask fitted with an air leak and heated at 100°C for 6 h. The acetic anhydride is removed on a rotary evaporator. Water may be added to facilitate this.

The residue is suspended in water and extracted three times by redistilled diethyl ether in a separating funnel. The sample may be extracted with $CHCl_3$ in place of ether but the product is not so clean. The ether extract is dried by anhydrous sodium sulphate and evaporated to dryness. For gas–liquid chromatography the extract is dissolved in $CHCL_3$ and a sample applied to a column of Gas Chrom Q coated with 3% w/w ECNSS-M at 180°C with 45 ml min^{-1} flow rate of nitrogen carrier gas (Fig. 5). Alternatively a temperature programme may be used. After 5 min at 150°C the temperature is increased to 190°C at the rate of 4°C min^{-1}. An alternative liquid phase is 3% SP-2330.

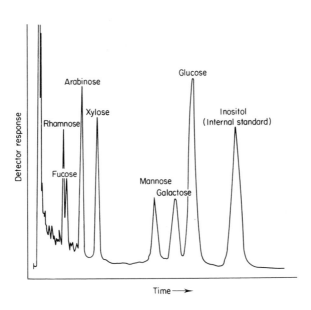

Fig. 5 GLC separation of the alditol acetate derivatives of neutral sugars from soil.

Glucose by glucose oxidase

Glucose may be determined satisfactorily in hydrolysates which have been deionized by charcoal. With most other treatments, hydrolysates appear to contain interfering substances which result in overestimates.

One hundred and fifty units of glucose oxidase and 72 units of peroxidase are dissolved in almost 50 ml of $0.5M$ NaH_2PO_4 buffer at pH 7.0 then 0.5 ml 1% *O*-phenylenediamine in 95% ethanol added and the volume made up to 50 ml with buffer. The solution is filtered through a Whatman No. 1 paper. Three millilitres of this reagent are mixed with 1 ml aliquots of the hydrolysate containing about $20-30\ \mu g$ glucose ($\equiv 0.01$ g soil) in stoppered tubes and incubated at $37°C$ for 1 h together with glucose standards in the range $0-50\ \mu g$. The absorbance of the solution is measured at 420 nm.

Hydrolysis for the release of amino sugars

(a) Method I (Stevenson, 1957a)

Dried soil (2 g) is soaked with 10 ml $11M$ HCl for 48 h, the mixture is diluted with 10 ml water and heated under reflux for 9 h.

(b) Method II (Parsons and Tinsley, 1961)

Dried soil (20 mg) is heated at $105°C$ with 5 ml of a $1:1$ mixture of $6M$ HCl and 90% formic acid in a sealed tube for 1 h.

Colorimetric determination of amino sugars (Parsons and Tinsley, 1961)

The hydrolysate and washings are applied to a 50×10 mm column of cation exchange resin and the column washed with 5 ml $0.02M$ EDTA in $0.1M$ acetate buffer pH 5.8. After washing with a further 5 ml of water the amino sugars are eluted by 10 ml $2M$ HCl.

One to 3 ml of the sample is made up to 4 ml with water after the addition of 1 ml 2% v/v acetylacetone in $0.5M$ Na_2CO_3 and heated to $100°C$ for 10 min. The volume is made up to 9 ml by the addition of ethanol and the solution kept at $75°C$ for 5 min. One millilitre of 1.33% w/v *p*-dimethylaminobenzaldehyde in concentrated HCl is added and the temperature maintained at $75°C$ for 30 min. The volume is adjusted to 10 ml with ethanol and the absorbance read at 512 nm (Belcher *et al.*, 1954).

Uronic acids by GLC

The soil, 2·5 g, is hydrolysed (as in method 1) together with 5 mg mannuronic acid as internal standard. The hydrolysate is titrated to pH 9 and any precipitate filtered off. The filtrate is passed through a column (100 × 15 mm) of Dowex 1 × 8 in the formate form which is subsequently washed free of salts and neutral sugars by 1 litre water at a flow rate of 60 ml h^{-1}. The uronic acids are then eluted with 1 litre 1·5M formic acid. The eluate is evaporated to dryness and taken up in the minimum of water, the pH adjusted to 9, and 150 mg NaBH$_4$ added. After 16 h the solution is treated with 1R-120 H resin to remove sodium ions and evaporated with methanol to remove borate ions.

The aldonic acids are then converted to the 2,3,5,6-tetra-*O*-trimethylsilylaldono-lactones by evaporation with a drop of concentrated HCl and the addition of 0·2 ml trimethylsilane + 0·2 ml hexamethyldisilazane in 2 ml dry pyridine.

The derivatives are separated on a 1·5 m column packed with 3% neopentyl sebacate on 100–200 mesh Diatomite using a flow rate of 20 ml min^{-1} and a temperature programme of 3°C min^{-1} between 110°C and 170°C. An alternative liquid phase is OV1.

THE COMPOSITION OF SOIL CARBOHYDRATE

The overall carbohydrate composition of soil can be determined in more than one way. In particular, (1) by the use of colorimetric methods specific for particular classes of sugars and (2) by analysis of individual sugars.

Analysis by colorimetry

Appendix Table I gives values obtained by colorimetry. The soils are arranged in groups according to the classification devised by the FAO for the soil map of the world.

The soils listed in this table have, on average, 10·1% of their carbon in carbohydrate as determined by anthrone. This value, based on 69 samples, may be compared with one of 9% found for the 165 samples used to examine the relationship between total carbon and the carbohydrate content, described in Chapter 1. Oades (1967) has suggested that total carbohydrate is about 150% of the values obtained using anthrone. Assuming that the anthrone values give a measure of hexose (see pp. 40–41), hexose to pentose ratios range from 2·5 to 4·8. As will be seen from the detailed analysis, a large part of this variation is caused by glucose. Uronic acid C is on average 2·9% of the total C, and amino sugar N 8·7% of the total N.

It is apparent that the variability of the values for the different classes of carbohydrate and for the ratio of hexose to pentose cannot be related to soil type.

Analysis of individual sugars

Neutral sugars

The relative proportions of individual sugars found in numerous soils are presented in Appendix Table II.

TABLE 3.17 *Average relative proportions (%) and amounts of sugars in soils.*

		mg g^{-1}	Relative proportion (%)
Hexoses	Galactose	1·2	14·7
	Glucose	2·8	35·4
	Mannose	1·2	17·1
Pentoses	Arabinose	0·9	11·9
	Ribose	0·1	1·5
	Xylose	0·7	10·4
Deoxyhexoses	Fucose	0·3	3·4
	Rhamnose	0·5	5·9

Glucose is the dominant sugar in every soil, being usually two or three times more abundant than any other sugar. The other hexoses, galactose and mannose, occur in slightly greater proportion than the pentoses, arabinose and xylose, with much smaller amounts of the deoxyhexoses, rhamnose and fucose. Average values are shown in Table 3.17.

The values in the Appendix Table II show that there is a decrease in the proportion of glucose with increasing depth. Another observation, mainly derived from the analyses obtained by Folsom *et al.* (1974) is that mannose increases with depth. Changes observed by Gupta *et al.*, were irregular, however. The mannose content of the polysaccharide isolated from a podzol decreased with depth (Anderson *et al.*, 1977). Contradictory trends reported for changes with depth in the proportion of arabinose (Folsom *et al.*, 1974; Gupta *et al.*, 1963), preclude worthwhile speculation about its source on this basis (see Chapter 5).

Sugar labelled 'high R_G' or 'high R_{Rha}' in Appendix Tables II and III are unidentified sugars which show a large movement during paper chromatography relative to tetramethylglucose or rhamnose. Duff (1952b) estimated that soil contained 0·9% of sugars having R_G 0·37 or 0·48, whereas Bouhours and Cheshire (1969) measured 0·4% methylxyloses (R_G 0·37) in peat. The high R_G sugars determined by Gupta *et al.* accounted for 9·4 of the total sugars on average which seems an unusually large proportion. Further values are given in Appendix Table III.

The amounts of free sugars present in various soils from Canada and Norway are shown in Table 3.18. The lower values obtained by Gupta and Sowden (1963) are probably related to their use of 80% ethanol as extractant. The other values were obtained using water. All the values quoted above were obtained by extracting air-dried soil. More free sugars were released when fresh soil was used (Table 3.19). Robert (1964) found that on incubating some soils at 28°C for 6 weeks there were small changes in the amounts of free sugars extractable with 50% ethanol at 60°C (Table 3.20). The changes were not consistent and probably partially reflect a continuous fluctuation.

Amino sugars

Some total sugar contents of soils are given in Table 3.21. The ratio of glucosamine:galactosamine in soils is usually slightly greater than one. For six West Indian surface horizons the average value was 1·5 (Sowden *et al.*, 1976), for six mediterranean soils 2·1 (Chen *et al.*, 1977) and for horizons in thirty five Canadian soils, 1·8 (Sowden, 1977).

Amounts of free N-acetylglucosamine in soil have been found to be very small (Table 3.22).

C

TABLE 3.18 Amounts of free sugars in the organic matter of air-dried soil.

Soil		Free sugars (mg kg⁻¹ of soil organic matter)								Reference
		Gal-actose	Glu-cose	Man-nose	Fruc-tose	Ara-binose	Xyl-ose	Fuc-ose	Total	
Northern podzol	Podzol	33	100	31		50	16	15	245	Gupta and Sowden (1963)
Well decomposed muck	Histosol	8	10	6		6			30	
Cultivated peat	Histosol		18						18	
Uncultivated peat	Histosol		13						13	
Orthic dark grey gleysolic heavily manured	Gleysol		12	2					14	
Orthic dark grey gleysolic No treatment	Gleysol		10	3					13	
Pine forest soil H layer	Podzol		1890		105	272	42	105	2393	Grov (1963)
Pine forest soil A layer	Podzol		1815			370			2185	
Pine forest soil F layer	Podzol	250	2200		350	400	350		3550	Alvsaka and Michelson (1957)

TABLE 3.19 *Effect of drying soil on the free sugar content (Gupta and Sowden, 1963).*

Soil		Free sugar content (mg kg^{-1})	
		Dried	Undried
Cultivated peat		18	30
Uncultivated peat		13	26
Orthic dark grey gleysolic	Heavily manured	14	23
	No treatment	13	19

TABLE 3.20 *Effect of incubation on the free sugar content of soils (Robert, 1964).*

Soil	Free sugar (mg 100 g^{-1})			
	Hexoses and oligosaccharides	Pentoses	Uronic acid	Total
Rendzina	6·63	3·01	3·04	12·68
Incubated	6·70	2·69	1·72	11·11
Forest Mull A$_2$	11·80	3·62	3·16	18·58
Incubated	11·70	2·96	2·62	17·28
Forest Mor A$_0$	12·41	8·80	13·62	34·83
Incubated	24·90	10·41	8·50	43·01
Natural Prairie	7·90	3·38	0·37	11·65
Incubated	2·89	1·95	0·28	5·12

TABLE 3.21 *Amino sugar content of soils (Chen et al., 1977).*

Soil	N (%)	Glucosamine N as % total N	Galactosamine N as % total N
Brown mediterranean	0·302	6·0	3·3
Brown reddish sand	0·032	6·0	2·8
Sandy alluvium	0·057	4·5	2·2
Brown prairie	0·030	4·7	1·7
Loess	0·046	2·4	1·5
Terra-rossa	0·175	6·3	2·9

TABLE 3.22 *Amounts of free N-acetylglucosamine in soils (Skujins and Pukite 1970).*

Soil	N-acetylglucosamine (mg g^{-1})
Dublin series	5·2
Pygmy series	0
Yolo series	0·5

Polysaccharide gum

A direct measure of the amount of water-soluble polysaccharide in soils is by determination of the weight. The available figures show relatively high values for the gum content in soils rich in organic matter, as might be predicted (Table 3.23). The upper horizons of a silt loam which had numerous grass roots present was much richer in gum than the lower horizons (Table 3.24). The effects of cultivation of a similar soil were to reduce the gum content from 0·25% to 0·12% (Rennie *et al.*, 1954). Fertilization, on the other hand, produced a small increase in the cultivated soil from 0·10 to 0·15%. Levels of gum in a silt loam under various crops formed a relatively constant proportion of the organic matter (Table 3.25) and there were only small differences in the percentage with the different crops.

Cellulose

In the few soils that have been examined a considerable proportion of soil glucose is present as cellulose (Table 3.26). Although the two podzols have higher cellulose contents than the other soils, a smaller proportion of their glucose is in this form.

EFFECT OF FERTILIZATION ON SOIL CARBOHYDRATE

Although soil is sometimes considered as a single natural body (Quastel, 1946) it is far more variable in composition than any one kind of organism. Consequently, meaningful comparisons of the effects of different treatments between soils are very difficult to make.

TABLE 3.23 *Amount of polysaccharide gum in soils.*

Soil	Polysaccharide gum (% of soil)	Reference
Heath raw humus	1·64	Forsyth (1947)
Pine	1·94	
Sphagnum cotton grass peat	1·60	
Gardeners turf soil	1·32	
Silt loam virgin wood	0·25	Rennie *et al.* (1954)
Cultivated	0·12	
Silt loam cultivated	0·10	
Limed	0·14	
Limed and fertilized	0·15	
Silt loams	0·016–0·114	Chesters *et al.* (1957)
Silt loam under Agrostis	0·36	Salomon (1962)
Silt loam under Agrostis	0·34	
Silt loam under red clover	0·35	
Silt loam under continuous potatoes	0·30	
Orthic black	0·14	Acton *et al.* (1963a)
Dark brown solodized solonetz	0·21	
Orthic black	0·78	
Calcareous black	0·78	

(% of soil polysaccharide)

Soil	% of soil polysaccharide	Reference
Lacustrine Chernozem A_h	6	Dormaar (1967)
Lacustrine Chernozem B_m	5	
Alluvial lacustrine Chernozem A_h	5	
Alluvial lacustrine Chernozem B_m	5	
Till Chernozem A_h	5	
Till Chernozem B_m	5	
Eolian Chernozem A_h	7	
Eolian Chernozem B_m	10	

Generally any fertilizer treatment which increases the organic carbon content of the soil is likely to increase the carbohydrate content also. Some fertilizers, such as farmyard manure, themselves contain carbohydrate, but little attention has been paid to the enhancement of the carbohydrate in this way since it is only small in

TABLE 3.24 *Variation of gum content with soil depth (Rennie et al., 1954).*

Soil horizon	Gum (%)
A_1	0·23
A_2	0·02
B_1	0·05

TABLE 3.25 *Effect of various crops on the proportion of the soil carbon present as gum (Salomon, 1962).*

Crop	Carbon as gum (%)
Agrostis	13·2
Agrostis	13·4
Red clover	13·8
Continuous potatoes	12·4

TABLE 3.26 *Cellulose content of soils (Gupta and Sowden, 1964).*

Soil	Horizon	Cellulose ($mg\ g^{-1}$ organic matter)	Cellulose glucose (% of soil glucose)
Orthic podzol	0	19	22
Northern podzol	F	10	21
Orthic grey wooded	L–H	13	18
Brown forest soil	A_1	3	23
Well decomposed muck	0–15·2 cm	8	32
Raw peat	0–15·2 cm	8	40
Cultivated peat	0–15·2 cm	9	33

comparison with that engendered by the crop. There is some evidence that fertilization with nitrogen increases the proportion of organic matter present as carbohydrate. Table 3.27 shows the effect

TABLE 3.27 *Effect of fertilization on the amounts of sugars in soil (Khan, 1969).*

Fertilizer	C (%)	Hexose	Pentose	Uronic acid	Hexosamine
		(as % of total C)			
5-y rotation of grains and legumes					
None	1·42	7·32	3·56	2·46	4·16
Manure	1·73	7·51	3·96	2·66	5·72
NPKS	1·45	8·52	4·42	2·70	4·91
NS	1·44	8·30	4·38	2·57	5·01
Lime	1·40	7·18	3·50	2·72	4·36
Fallow–wheat sequence					
None	1·14	7·16	2·95	2·11	3·73
Manure	1·58	7·40	3·07	2·09	4·88
NPKS	1·25	8·57	3·24	2·32	4·32
NS	1·18	8·56	3·13	2·29	4·41
Lime	1·06	7·28	2·79	2·36	3·97

of various fertilizers on the carbohydrate content of a grey wooded soil which had been consistently cultivated and fertilized over a period of 39 years. Farmyard manure was the only treatment which increased the carbon content of the soil and correspondingly the various classes of carbohydrate. All the treatments with N increased the proportion of hexosamine but only those with NPKS or NS increased the proportion of hexoses and pentoses. Similarly a farmyard manure treatment of a clay soil increased the polysaccharide content of the soil in a corn–oats sequence (Webber, 1965) but this was not the case with other crop sequences which also involved hay, winter wheat or red clover.

When ^{15}N-labelled nitrogen compounds were incubated with soil, 9·9% of the fertilizer-derived organic N was found in the amino sugars whereas only 8% of the native humus was in this form (Allen *et al.*, 1973). This was thought to be because the initial metabolic steps involved the formation of substances such as the amino acids and amino sugars which were subsequently transformed into more stable compounds by reactions involving carbonyl groups (Stevenson, 1973).

Murayama and Inoko (1975) confirmed that addition of moderate amounts of lime caused little change in the proportion of C present as carbohydrate but with larger amounts the proportion was much reduced (Table 3.28). This was explained as the result of increased

TABLE 3.28 *Effect of fertilization and lime on the proportion of soil carbon present as carbohydrate (Murayama and Inoko, 1975).*

Lime only		Complete fertilizer	
Soil pH	Carbohydrate C as% C	Soil pH	Carbohydrate C as %C
5·7	9·0	5·9	8·7
6·1	9·0	6·0	8·5
6·5	8·1	6·3	8·0
7·2	6·9	6·6	6·8
8·1	6·3	7·4	5·7
8·2	6·0	8·0	5·2

biological activity and mobilization and solubilization of the organic matter leading to a decrease in the more easily metabolized organic fractions of soil which include the carbohydrate.

Heavy fertilization of a gleysolic soil by manure or mineral fertilizers was found to cause no significant change in the proportions of different sugars (Table 3.29).

TABLE 3.29 *Effect of fertilization of soil on the relative proportions (%) of sugars (Gupta et al., 1963).*

	Untreated	Heavily manured	Heavily fertilized
Galactose	15	14	20
Glucose	31	28	31
Mannose	15	20	15
Arabinose	15	15	13
Xylose	11	10	10
Rhamnose	8	8	7
Fucose + ribose	5	5	4

EFFECT OF VEGETATION AND CROPPING ON SOIL CARBOHYDRATE

In the long term the carbohydrate content of a soil is very dependent on the kinds of plant grown on it (Table 3.30) because these affect the

TABLE 3.30 *Effect of various crops on the carbohydrate content of soil (adapted from Webber, 1965).*

Crop	Carbohydrate (% of soil)
Continuous bluegrass	0·724 ± 0·041
Continuous corn	0·422 ± 0·023
Corn, oats, winter wheat and red clover	0·553 ± 0·019

level of organic matter present. The figures in Table 3.31 show that the carbohydrate is a fairly constant proportion of the total organic

TABLE 3.31 *Effect of rotation on the carbohydrate content of soils (Oades, 1967b).*

Cropping history[a]								C (%)	Carbohydrate C as % C	
									Whole soil	Heavy fraction
P	P	P	P	P	P	P	P	2·50	6·9	4·8
P	P	W	W	P	P	P	P	1·76	6·7	5·0
P	P	W	P	P	W	P	P	2·28	6·4	4·3
P	W	P	P	W	P	P	W	1·92	5·9	4·2
W	W	W	W	W	W	W	W	1·76	5·7	4·3
F	W	F	W	F	W	F	W	1·04	6·1	5·3
W	F	W	F	W	F	W	F	1·03	5·2	4·9
Belalie clay loam, 0–6 cm from various farms,										
P	P	P	P	P	P	P	P	1·24	11·1	9·7
W	P	P	P	P	P	F	W	2·40	11·0	10·4
L	L	L	L	L	P	F	O	2·64	11·0	9·5
P	W	P	P	W	P	P	W	1·13	12·1	11·8
F	W	F	W	F	W	F	B	1·18	7·2	7·3

[a] P = pasture
W = wheat
F = fallow
L = lucerne
O = oats
B = barley

matter of a soil whatever cropping practice is applied and this has been the experience of several workers (Webber, 1965; Salomon, 1962). Differences between soils are considered to relate to soil types.

Seasonal changes in the amount of carbohydrate in soil can sometimes be accounted to the increased growth of plants (Table 3.33; Webber, 1965), but fluctuations from season to season, such as

TABLE 3.32 *Effect of fallowing for 10 y on the carbohydrate content of soil (Cheshire and Anderson, 1975).*

Soil	Carbohydrate $(mg\ g^{-1})$	
	Beans	Fallow
Highfield	13·5	6·8
Fosters	5·2	4·2

those indicated in Table 3.30 have been related to soil moisture content (Webber, 1965; Guckert, 1975), carbohydrate production by the microflora being stimulated by periods of high rainfall.

A period of fallow in a cropping sequence may considerably reduce the organic matter content, and consequently, the carbohydrate content. Thus Toogood and Lynch (1959) showed that, given the same fertilizer treatments, soil under a fallow–wheat sequence always has a lower polysaccharide content than under a cereal and legume

TABLE 3.33 *Variation of the carbohydrate content of soil under wheat with time (Oades and Swincer, 1968).*

Period (months)	Carbohydrate	
	Season	(% of soil)
1	Autumn	0·34
3	Winter	0·33
5	Spring	0·34
7	Summer	0·40
9	Summer	0·42
12	Autumn	0·41

rotation. An effect of fallowing may also be seen from the findings presented in Table 3.31. A more extreme example is provided by an extended period of fallowing which considerably reduced the carbohydrate content of two Rothamsted soils (Table 3.32). Usually the carbohydrate content of a soil will be greater under pasture than under other crops (Table 3.30; Oades and Swincer, 1968), but not always. In Urrbrae fine sandy loam, for example, the carbohydrate increased during the growth of a single wheat crop after 4 years under grass (Table 3.33).

In a soil carrying wheat a greater proportion of the carbohydrate was extractable in NaOH in the Australian winter months, the period of active plant growth, and the extract contained a larger proportion of material with a molecular weight $< 100\,000$ (Oades and Swincer, 1968). This is consistent with the observation that polysaccharide extractable by NaOH from old pastures has a higher proportion of high molecular weight material than soils supporting a continuous wheat crop.

Permanent pasture is very effective in increasing the carbohydrate content of the soil. One reason for this is that the total amount of plant residues is greater than with other crops, partly because the grass plant is present in the soil all year round. Huntjens and Albers (1978) have suggested that another contributing factor is nitrogen deficiency induced by a continuous supply of carbonaceous material to the microflora from the grass roots.

Chapter 4

Soil Polysaccharide

METHODS OF EXTRACTION FROM SOIL

Many different methods of extraction have been used and these are listed in Tables 4.2–4.7 (soils) and 4.10 (peats). Although some of the milder methods are very inefficient they do not cause chemical changes such as deacetylation which may, for example, occur with alkaline extractants. The choice of extractant depends on the purpose for which the polysaccharide is required. Some indication of the range of extractant efficiencies may be obtained from Table 4.1 where the effect of different reagents has been examined for an arable sandy loam soil. The polysaccharide in each fraction and the whole soil was determined as reducing sugar, with ferricyanide, after hydrolysis. In the other tables, also, the efficiency of extraction is indicated by the

TABLE 4.1 *Comparisons of various methods for the extraction of polysaccharide from a Countesswells sandy loam (Cheshire, 1977).*

Method	Polysaccharide extracted (% of total)
Water at 20°C	0·47
Water at 20°C with sonification	8·0
Water at 80°C	2·9
0·05M H_2SO_4	0·97
0·2M NaOH	42
0·2M NaOH with sonification	69
Chelex 100 resin Na form	37
0·1M EDTA at pH 7	66
0·2M acetylacetone at pH 8	64

TABLE 4.2 *The proportion of total soil carbohydrate extracted by water.*

Soil	Temperature	Period	Carbohydrate extracted (% of soil carbohydrate)	Reference
Garden loam developed on granitic till	85°C	2 × 4 h	21	Duff (1952b)
Red–brown earth	70°C	16 h	14	Swincer et al. (1968a)
	80°C	10 h	2·9	Cheshire (1977)
Countesswells sandy loam developed on granitic till	20°C	10 h	0·47	
	20°C Sonificaton	10 h	8·0	
14 cultivated soils	100°C	1 h	15	Jenkinson (1968)

percentage of the total polysaccharide that has been extracted. Where values for the total soil polysaccharide are not available they have been calculated on the assumption that 10% of the organic carbon is present as carbohydrate.

Cold water extracts only a very small proportion of the total soil polysaccharide, $<1\%$, unless sonification is used (Table 4.2). Grinding the soil has been found to increase the polysaccharide extracted with cold water by a factor of two to four (Murayama and Inoko, 1975), the optimum grinding period being about 30 min. In this study sodium sulphate was added at the end of the 2h period of extraction with water, just before separation of the extract, in order to flocculate the clay. Hot water at 70–85°C extracts at least five-fold more polysaccharide than cold water. The amount of soil polysaccharide extracted with water probably depends on the pH of the soil and the cationic species present in so far as these factors determine whether or not the polysaccharide is present in a soluble salt form.

Cold buffer solutions at pH 7 such as sodium and potassium phosphate, with potassium chloride to flocculate clay, are much more efficient than cold water (Bernier, 1956). These extractants are thought to be the least damaging of all because the isolated product has the highest viscosity indicating the highest molecular weight.

The value of dilute acids as extractants (Table 4.3) is in preventing the soil humic acid fraction from dissolving, although very little carbohydrate may be extracted (Wright et al., 1958). From 6–18% of the polysaccharide in an arable soil was extracted with various dilute acids by Swincer et al. (1968a). Barker et al. (1967) chose 0.3M H_2SO_4 to extract the polysaccharide from a fen peat because the isolated material contained a much higher proportion of polysaccharide (29%) than those isolated with other reagents ($<13\%$).

Alkali is a very effective extractant, bringing into solution 30–50% of the soil polysaccharide. Sonification increases the value to almost 70% (Table 4.4). Amongst the disadvantages of alkali are the large quantities of non-carbohydrate organic matter also extracted, the possible oxidation of the polysaccharide that may occur and the interaction induced with other soil organic matter compounds. Alkali has been used to extract soil gum for studies on the aggregating properties of soil polysaccharide (Rennie et al., 1954; Dubach et al., 1955; Whistler and Kirkby, 1956) based on the initial studies made

TABLE 4.3 *The proportion of soil carbohydrate extracted with acid.*

Soil	Extraction conditions			Carbohydrate extracted (% of soil carbohydrate)	Reference
	Acid	Time	Temperature		
Urrbrae sandy loam (red–brown earth)	0.5M H_2SO_4	16h	20°C	6	Swincer et al. (1968a)
	M HCl	16h	20°C	9	
	M HF	16h	20°C	18	
	0.05M H_2SO_4	1h	20°C	3	
Countesswells sandy loam developed on granitic till	0.05M H_2SO_4	1h	20°C	1	Cheshire (1977)
Lateritic podzolic	M HCl	16h	25°C	26	Oades (1972)
Tropical red earth				10	
Solodized solonetz				12	
Grey clay				10	
Terra rossa				12	
Sandy podzol				22	
Ando				27	
Black earth				10	
Krasnozem				8	
Solonized brown soil				14	
Rendzina				20	
Fen peat				5	

TABLE 4.4 *The proportion of soil carbohydrate extracted with alkali.*

Soil	Extraction conditions			Carbohydrate extracted as % of soil carbohydrate	Reference
Melfort A_a	0·5M NaOH	3h	RT	46·4	Acton *et al.* (1963a)
Urrbrae sandy loam	0·5M NaOH	2×16h	20°C	33·4 ⎫	Swincer *et al.* (1968a)
(red–brown earth)	0·5M NaOH	1h	100°C	37 ⎬	
Countesswells					
sandy loam	0·2M NaOH	4h	20°C	42 ⎫	Cheshire (1977)
	with sonification	4h	20°C	69 ⎬	
14 soils	0·05M Ba(OH)	3 min	RT	5·5 ⎫	Jenkinson (1968)
	0·01M NaHCO$_3$	15 min	RT	6·5 ⎬	
Podzol Taintrux A_1	1% pyrophosphate			45·9 ⎫	
Podzol Taintrux B_h	followed by 0·1M NaOH			34·1 ⎪	
Podzol Taintrux B_s	after bromoform in			26·3 ⎬	Gallali *et al.* (1972)
Sol brun acid A_1	ethanol (d=1·8)			34·8 ⎪	
Sol brun acid B				22·1 ⎪	
Chernozem A_1				9·0 ⎭	
Belalie clay 4-y pasture, loam 0–6 cm	0·2M NaOH			30·9	Oades and Swincer (1968)

by Forsyth (1947). Usually less than half the polysaccharide is recoverable from the extract by adsorption on charcoal, or precipitation by acetone.

Juvvik (1965) treated peats with chlorite solution (Table 4.5) to destroy lignin-type compounds prior to extracting polysaccharide with alkali, but the yield was disappointing because of losses on dialysis. Nevertheless the unextracted carbohydrate, regarded as cellulose, only amounted to 12% showing that the extraction was effective.

The success achieved with complexing reagents such as EDTA (Table 4.5) (see also Dubach et al., 1964) indicates that the polysaccharide is present in soil largely as a water-insoluble salt. Barker et al. (1967) suggested that, in the fen peat they investigated, it was in Ca or Mg form, or bonded to humic acid salts of these elements. Sodium pyrophosphate solutions at pH 7 (Bernier, 1958; Mehta et al., 1960; Mortenson and Schwendinger, 1963; McKenzie and Dawson, 1962) probably owe their effectiveness to the simultaneous extraction of Fe.

Some reagents such as 8M urea have been tried because of their ability to break hydrogen bonds (Table 4.6) but with disappointing results. The effectiveness of hot anhydrous formic acid as an extractant (Tables 4.5 and 4.6) may be related to hydrolysis of the polysaccharide. Kibblewhite (1977) used a 4:1 mixture of formic acid and dimethyl sulphoxide as extractant because this mixture had a higher dielectric constant and he found that it was more effective than either solvent used on its own.

Flotation (Table 4.7) removes some plant fragments but others are not removed until the soil has been treated with acid and alkali and it has been suggested that the heavier plant fragments are initially encrusted with inorganic materials. The most commonly used reagent for densimetric separation is Nemagon (1,2-dibromo-3-chloropropane, sp. gr. 2·06) containing 0·1% w/w Aerosol OT-100 (sodium dioctylsulphosuccinate) as a wetting agent (Ford et al., 1969). Previously, mixtures of bromoform with light petroleum (Greenland and Ford, 1964) or ethanol (Monnier et al., 1962) have been used. The dry soil is suspended in the reagent and the mixture subjected to ultrasonics for 3 min at < 35°C. The mixture is then centrifuged and the light fraction obtained by filtering the supernatant suspension. Both fractions are washed free of Nemagon

TABLE 4.5 *The proportion of soil carbohydrate extracted by miscellaneous methods.*

Soil	Extracted conditions		Carbohydrate extracted as % of soil carbohydrate	Reference
Meadow soil	Anhydrous formic acid + 0·2M LiBr	Refluxed 2 × 30 min	41 ⎫	Parsons and Tinsley (1961)
Podzol A₀			45 ⎬	
Podzol B			31 ⎭	
Compost 11 month			75	
Countesswells sandy loam	EDTA pH 7	20°C 2 × 2 h	66	Cheshire (1977)
Countesswells sandy loam	Na Chelex 100	20°C 2 × 24 h	37 ⎫	Juvvik (1965)
Forest soil F layer	Chlorite	4 h	18·75	
Urrbrae sandy loam	Dimethylformamide 0·4M boric acid, 0·4M oxalic, 0·2M LiCl	Refluxed 1 h	29 ⎫	Swincer et al. (1968a)
	Acetic anhydride 1:1 v/v pyridine	Refluxed 16 h	12 ⎬	
	Acetic anhydride: $(CH_3)_2SO_4$ 20:1 v/v	60°C 10 min	76 ⎭	
Brown forest podzol	Schweizer's reagent	6 h	1 ⎫	Gupta and Sowden (1964)
A₁ Orthic podzol 0			12 ⎬	
Well decomposed muck			12 ⎭	
Sassafras loam	0·5M pyrophosphate pH7	24 h	65–70	Lynch et al. (1958)
Cultivated and uncultivated pocomoke				

TABLE 4.6 *The amounts of carbohydrate extracted from peats by various reagents.*

	Extraction conditions		Carbohydrate extracted as % of peat carbohydrate	Reference
Sphagnum eriphorum peat				
	Extraction with water			
		100°C 8h	38	Duff (1961)
Sphagnum peat		100°C 3h	31	Theander (1952)
Sphagnum peat		100°C 3h	5	Theander (1952)
Sphagnum peat		60–65°C 3 × 3h	0·8–1·7	Theander (1954)
Sphagnum peat		100°C 4h	24[a]	Black et al. (1955)
	Extraction with acid			
Sphagnum peat	0·09M HCl	60–70°C 1h	3·5	Black et al. (1955)
Fen peat	0·3M H$_2$SO$_4$	RT 24h	7·0	Barker et al. (1967)
	Extraction with alkali			
Sphagnum cotton grass peat	0·5M NaOH	2 × 48h	16	Forsyth (1947)
	Miscellaneous extraction reagents			
Fen peat	7% Na$_2$ EDTA	RT 24h	1·1	
	Amberlite IRC-50 (H)	RT 24h	0·6	
	N,N-dimethylformamide	RT 24h	3·0	Barker et al. (1967)
	N-methyl-2-pyrrolidone	RT 24h	0·13	
	8M urea	RT 24h	1·7	
Peat	Anhydrous formic acid + 0·2M LiBr	Reflux 2 × 30 min	30	Parsons and Tinsley (1961)
Raw peat	Schweizer's reagent		14 }	Gupta and Sowden (1964)
Cultivated peat	Schweizer's reagent		17 }	
Woody peat	Amberlite LA-2 in hexane and n-butanol	RT	—	Mingelgrin and Dawson (1973)

[a]After extraction with methanol and benzene.

TABLE 4.7 *Proportion of total soil neutral sugars in the light fraction of soils.*

Soil	Neutral sugars in the light fraction (%)	Reference
Urrbrae old pasture 0–6 cm	43	Oades and Swincer (1968)
Urrbrae old pasture 9–15 cm	26	
Continuous wheat 0–6 cm	36	
Continuous wheat 9–15 cm	17	
Wheat after pasture 0–6 cm	34–40	
Belalie clay loam		
Wheat after pasture 0–6 cm	29	
Wheat after pasture 9–15 cm	9	
Frilsham series		
arable 0–15 cm	11	Whitehead *et al.* (1975)
Frilsham series		
17-y pasture after arable 0–15 cm	38	
Frilsham:Windsor series		
old pasture 0–15 cm	29	

with acetone. Light fractions contain the same kinds of sugars as the whole soil, as well as siliceous matter which may include opaline phytoliths (Greenland and Ford, 1964). Analysis of the ash indicates a high content of amorphous oxides. From Table 4.7 it can be seen that the upper horizons of soils, which are expected to be richer in plant remains than the lower, have a greater proportion of their carbohydrate in the light fraction. Experiments using ^{14}C-labelled materials in a sandy loam soil have shown that, with a liquid of sp. gr. 2, plant and microbial matter fractionate in a fairly similar way to each other (Murayama *et al.*, 1979). Presumably the behaviour of the organic matter depends on the extent of its bonding with mineral matter, particularly in the case of microbial material.

Schweizer's reagent, cuprammonium hydroxide, has been applied to soils and peats to extract cellulose (Tables 4.5 and 4.6). The isolated material accounted for 20–40% of the total glucose in the soil hydrolysate. It contained 65–82% glucose and other sugars included galactose, mannose, arabinose and xylose.

Two examples of the sequential extraction of soil with acids and alkali are shown in Tables 4.8 and 4.9. Sequential extraction usually removes more carbohydrate than repeated extraction with the same reagent. The efficiency of extraction is also dependent on the type of acid used. Table 4.10 shows that HCl or HF releases more carbohydrate from an Urrbrae sandy loam than H_2SO_4. Although HF extracts more carbohydrate than does HCl, the subsequent alkali treatment extracts less carbohydrate after HF than after HCl, so in this instance the total extracted is almost the same. The distribution of carbohydrate between fractions derived by Tyurin's fractionation scheme, using alternate acid and alkali treatments of soil, varied with different soil types (Sadovnikova and Orlov, 1976), with a few soils showing considerable deviation from the average (Table 4.9). Soils described as grey and light grey forest and those of arid steppe, steppe desert and subtropic zones were found to have a smaller proportion of the carbohydrates in the humic acid fraction than podzols, humus gleys and chernozems. Considerable hydrolysis probably occurs with the extraction by acetic anhydride: H_2SO_4 in the sequential

TABLE 4.8 *The proportion of the soil carbohydrate obtained by sequential extraction of soils.*

	Carbohydrate as % of soil carbohydrate			
Soil Extractant	1MHCl	0·5M NaOH	Acetic anhydride:H_2SO_4	Sum total
Lateritic podzolic	20	42	32	94
Tropical red earth	9	32	21	63
Solodized solonetz	11	42	26	79
Grey soil of heavy texture	13	38	30	81
Terra rossa	12	42	33	87
Ando soil	20	18	19	57
Groundwater rendzina	21	31	35	87
Red–brown earth pasture	12	46	23	81
Red–brown earth cultivated	13	42	24	79
Krasnozem	12	45	36	94
Solonized brown soil	12	38	30	80
	14	38	28	80

extraction procedure devised by Swincer *et al.* (1968a) (Table 4.8) but this insoluble carbohydrate would otherwise remain uncharacterized.

TABLE 4.9 *Proportion of soil carbohydrate (%) in sequentially extracted fractions (adapted from Sadovnikova and Orlov, 1976).*

	Range	Mean
Ethanol benzene extract	0–4·9	1·5
Dilute HCl and H_2SO_4 extract	4·6–19·9	8·4
0·1M NaOH soluble, acid insoluble (humic acid)	0·3–45·0	16·9
acid soluble (fulvic acid)	15·9–72·4	39·9
0·05M H_2SO_4 soluble	0–2·4	1·4
0·1M NaOH soluble, acid insoluble (humic acid)	0–2·0	1·0
acid soluble (fulvic acid)	0–6·2	3·9
Residue	4·2–65·1	34·9

TABLE 4.10 *The amounts of carbohydrate obtained by sequential extraction of Urrbrae sandy loam (Swincer et al., 1968a).*

Extraction conditions			Carbohydrate extracted as % of the soil
M HCL	16h	20°C	9%
0·5M NaOH	2 × 16h	20°C	38%
			47%
M HF	16h	20°C	18%
0·5M NaOH	2 × 16h	20°C	32%
			50%
0·5M H_2SO_4	16h	20°C	6%
0·5M NaOH	16h	20°C	28%
			34%
0·05M H_2SO_4	16h	20°C	3%
0·5M NaOH	2 × 16h	20°C	29%
			32%

Conclusion

Of all the common reagents dilute alkali is clearly the most efficient extractant for soil polysaccharide. To minimize the possibility of chemical changes occurring during extraction, cold buffer solutions at neutral pH should be used.

Complexing reagents such as EDTA and pyrophosphate at pH 7 have been quite effective in the few instances they have been used. Further studies using a greater range of soils are necessary to evaluate these reagents fully. Fractionations based on density may be useful although, in those reported so far, considerable mineral matter is present in the light fraction.

ISOLATION OF POLYSACCHARIDE FROM EXTRACTS

Precipitation of humic acid

Acidification of alkaline extracts of soils and peats precipitates humic acid, which may be separated by centrifugation. Some polysaccharide, sometimes as much as 12% of the total extracted, is coprecipitated with the humic acid (Acton *et al.*, 1963b). It is probable that this polysaccharide is linked to the humic acid by cation bridges because treatment of the alkaline extract with cation exchange resin largely prevents the coprecipitation (Swincer *et al.*, 1968a). The proportion of carbohydrate precipitated from an extract has been reduced from 8 to 1% by this means.

Adsorption procedures

Fractionation on charcoal

Acidification precipitates the humic acid and leaves fulvic acid in solution. One of the most effective ways of isolating polysaccharide from fulvic acid, developed by Forsyth (1947), has been by the use of charcoal. The fulvic acid is applied to a pad of charcoal on a paper filter in a Buchner funnel under slight suction, followed by 0·1M HCl.

The two eluates, combined, constitute fraction A and contain most of the salts and simple organic compounds. 90% acetone elutes a wine-red coloured fraction B which was originally thought to be a phenolic glycoside. Although it contains some sugars and phenols it is essentially aliphatic in nature (Anderson and Russell, 1976). The colour is possibly caused by conjugated carbonyl groups. Schlicting (1953) has argued that the presence of reducing sugars in the hydrolysate could equally well show the presence of polysaccharides as of glycosides. Fraction C, which is eluted by water, contains polysaccharide very largely free of brown humic material and salts. Finally 0·2M NaOH elutes fraction D, which is a brown humic material. Using this technique with a garden soil Forsyth obtained yields of fractions B, C and D corresponding to 2·13, 1·32 and 2·60%, respectively, of the soil organic matter. Polysaccharide has been recovered from B and D fractions by ion exchange chromatography on DEAE cellulose (Linehan, 1977). Stevenson (1960) detected amino sugars in all the fractions.

Cheshire et al. (1977) fractionated the fulvic acid from a sandy loam of the Countesswells series in a similar manner, using a column of charcoal and Celite, a diatomaceous earth, and obtained a good recovery of organic matter. Carbohydrate was found in all the fractions but its recovery appeared to be less satisfactory, the distribution between fractions being A, 12, B, 18·4, C, 27·8 and D, 7·7%, a total of only 55% of the total in the fulvic acid.

The apparently poor recovery has led to considerable criticism of this method. Although 8·5M acetic acid was a more efficient eluant than water, Swincer et al. (1968a) were only able to recover 30% of the fulvic carbohydrate by its use. Treatment of the charcoal with stearic acid, which reduces the number of adsorption sites, increased the yield to 65%.

But it is important to distinguish here between carbohydrate and polysaccharide. In the author's experience the yield of alcohol-precipitable polysaccharide obtained by the charcoal adsorption method is about 90% of that isolated by precipitation with ethanol after treatment of the fulvic acid with Polyclar, which is considered to give a recovery of polysaccharide of over 95%. Bernier (1958) eluted buffer-extracted polysaccharide almost quantitatively from a charcoal pad using water and sodium tetraborate solution. Furthermore, soil polysaccharide purified by other means, such as treatment with Polyclar, is usually more highly coloured than charcoal derived material suggesting that it has a greater content of humic substances.

Fractionation on Polyclar AT

Polymeric organic materials which contain amino groups that react specifically in acid conditions with phenolic substances have been used to free polysaccharide from some other components of fulvic acid. Polyclar AT, a cross-linked polyvinyl pyrrolidone polymer, retains most of the coloured materials from the fulvic acid leaving the polysaccharide unadsorbed and elutable with acid. Celite may be mixed with the Polyclar to speed up the flow through the column. The recovery of polysaccharide measured as carbohydrate is greater than 95%. Most of the coloured material may easily be eluted subsequently by dilute alkali although there is a small amount of irreversible adsorption.

Fractionation on Amberlite XAD-1

Components of fulvic acid have also been separated by adsorption chromatography using the polymeric adsorbent Amberlite XAD-1. By elution with acid, water, methanol and alkali, four fractions were obtained which contained different proportions of carbohydrate (Table 4.11), the first having the greatest.

TABLE 4.11 *The fractionation of fulvic acid on Amberlite XAD-1 (Inoko and Murayama, 1973).*

Fraction	eluent	Absorbance at 400 nm (%)	Carbohydrate by anthrone (%)
1	$0.05M$ H_2SO_4, then $0.005M$ H_2SO_4	18·7	63·2
2	Water [a]	54·3	14·3
3	Methanol	11·6	12·3
4	$0.01M$ NaOH	1·3	2·7

[a]The water was preceded by a small amount of alkali to neutralize the remaining acid.

Precipitation procedures

Precipitation by organic solvents

A very common treatment is the addition of excess ethanol or acetone to the alkali extract of the soil or the fulvic acid solution to precipitate the polysaccharide. The addition of acid is sometimes necessary to initiate precipitation. In one variation Garyayev *et al.* (1973) precipitated polysaccharide by adding two volumes of isopropanol to hot water extracts of soil. The disadvantages are that humic substances present in the salt form are also precipitated and much polysaccharide remains in the solution which, from its sugar composition, is not distinguishable from the precipitate. Repeated precipitation with ethanol and acetic acid has been a useful method for reducing the content of humic material (Duff, 1952b).

Precipitation of alkali-insoluble substances

When soil extracts made with acids are neutralized with alkali such as NaOH or $NaHCO_3$ some polysaccharide is precipitated (Black *et al.*, 1955; Barker *et al.*, 1965). This does not seem to involve coprecipitation with metal ions, for the simultaneous precipitation of metal hydroxides can be prevented by the addition of EDTA (Swincer *et al.*, 1969).

Gel filtration

Filtration through gel columns has been used to separate carbohydrate-rich organic fractions of high molecular weight from other soil organic materials. Roulet *et al.* (1963b) fractionated M HCl extracts of three different soil types on Sephadex G-75 eluting with pH 10 buffer but there was an improvement in the separation by using 0·2M acetic acid as eluant which retarded the humic fractions without affecting the elution of the polysaccharide. Similarly, brown polymeric fulvic substances in acid extracts of a fen peat have been separated from high molecular weight polysaccharides by gel filtration using G-100 (Barker *et al.*, 1965; 1967). It was not possible to separate the polysaccharides present in extracts made with other reagents in this way unless they had been treated with acid, so it was concluded that the acid was necessary to break a linkage between the polysaccharide and the humic substances. It is most unlikely that all the polysaccharide is separated from the brown fulvic material by gel filtration; for example Swincer *et al.* (1965a) using various gels, Sephadex G-25, G-75, G-100, G-200 and Biogel P-20 and P-200, obtained polysaccharide fractions with a range of nominal molecular weights including some <4000. Considering the separation from another viewpoint, Martin (1976) reduced the polysaccharide content of fulvic acid preparations to 3.7% using Sephadex gel filtration. Treatment with Polyclar, charcoal, acetone, or hot water were each less effective.

Dialysis

Low molecular weight substances and salts have been removed from soil extracts by dialysis in Visking tubing (Duff, 1952a, b; Whistler and Kirkby, 1956) or cellophane with an average pore diameter of 4·8 nm (Mortensen and Schwendinger, 1963) but considerable losses can occur. It has been claimed that 20% of a fulvic acid carbohydrate was adsorbed onto the Visking tubing and could not be washed off (Swincer *et al.*, 1968a). Some carbohydrate is lost by dialysis through the membrane; Swincer *et al.* found a loss in this way of 20% of an alkali-extracted polysaccharide whilst Muller *et al.* (1960) retained, within the membrane, only about 8% of the carbohydrate extracted from a forest soil.

In my experience dialysis in the presence of sodium EDTA prevents the formation of precipitates and, after subsequent dialysis against water, gives polysaccharide which, after precipitation with ethanol or after freeze-drying, is far more readily soluble in water than material dialysed only in water. Analysis indicates that EDTA removes iron and aluminium.

Treatment with ion exchange resins

Polysaccharide extracted from soil and hot water and precipitated by acetone has been treated with the cation and anion exchange resins

IR-120 (H) and IR-410 (acetate) (Mortensen, 1960; Keefer and Mortensen, 1963) resulting in a considerable reduction in ash content.

Treatment with metal salts

The nitrogen content of soil polysaccharides has been diminished by adding excess $CuSO_4$ to an alkaline solution to precipitate a copper complex of the polysaccharide which is separated and decomposed by ethanol containing HCl (Forsyth, 1950). Polysaccharides from several Scottish soils contained only 0·1–0·5% after this treatment. When this technique was applied by the author to polysaccharide from a Countesswells sandy loam, only a small decrease in the N content was observed but a small amount of material rich in N remained in solution:

%N:	original polysaccharide	precipitated	not precipitated
	3·58	2·96	5·81

In another procedure soil polysaccharide has been stirred with $CdSO_4$ in M NaOH at 60°C for 1 min and the mixture centrifuged to precipitate the protein complex, leaving polysaccharide in solution (Bernier, 1958). An attempt to remove more protein by treatment with $Co(NH_3)_6Cl_3$ was unsuccessful.

Other methods

Insoluble gels are formed when chloroform is shaken with a solution of protein in water and, when this technique of Sevag was applied to soil polysaccharide solutions in an attempt to remove protein, precipitable material was formed (Bernier, 1958; Mortensen, 1960; Thomas et al., 1967). In one of the studies at least, amino acids were still present in the residual unprecipitated polysaccharide.

Bernier also shook a solution of soil polysaccharide in M acetic acid with Fuller's earth to remove protein by adsorption, but no indication was given of the effect.

Polysaccharide material extracted from soil by anhydrous formic acid + LiBr could be precipitated by the addition of di-isopropylether containing 1% CH_3COCl which maintained anhydrous conditions by reacting with water but kept inorganic cations in solution (Parsons and Tinsley, 1961). Zirconium hydroxide has been used to precipitate a carbohydrate-rich fraction from solution in aqueous acetone (Kibblewhite, 1977). Subsequent removal of zirconium required treatment with fluoride and anion exchange resin.

Scheffer and Kickuth (1961) extracted polysaccharide from humic acid with Schweizers reagent. It was thought to be a form of oxycellulose because of the ease with which it could be hydrolysed by 2M HCl.

Conclusions

Much of the polysaccharide which may be extracted from soil appears to be chemically linked in proteinaceous complexes. Although some of these complexes may be separated from others by treatments designed to precipitate or otherwise remove proteins from solution, leaving behind complexes with smaller proportions of protein, fractionation or purification techniques aimed at isolating N-free polysaccharide would appear to be fruitless. Separation of the polysaccharide from humified materials has been more successful, the humified substances appearing to be associated with the complex through ionic bonding. Of the various techniques described, the most useful would appear to be the use of Polyclar for the removal of humic materials and precipitation of the polysaccharide with ethanol, or dialysis for the removal of low molecular weight substances, salts and metallic cations.

FRACTIONATION OF SOIL POLYSACCHARIDE

In discussing the fractionation of soil polysaccharide a distinction is made between methods which fractionate the whole of the soil polysaccharide (pp. 86–92) and those which fractionate the isolated soluble soil polysaccharide (pp. 92–105).

Fractionation of the whole soil

Physical separation

One first stage of fractionation has been flotation to separate plant debris and other unhumified components from the bulk of the soil organic matter. Ford *et al.* (1969) treated a suspension of soil in Nemagon (1,2-dibromo-3-chloropropane, sp. gr. 2·06), containing 0·1% w/w Aerosol OT-100 (sodium dioctyl sulphosuccinate) as a wetting agent with ultrasonics. With twelve varied soils it was found that the light fraction had glucose as the predominant sugar (Table 4.12) but showed considerable variation in the amounts of other sugars. In some soils this fraction was rich in arabinose and xylose, consistent with the sugars being mostly present in plant remains. In other soils, mannose and galactose were more abundant than the pentoses and there appeared to be an inverse relationship between these two hexoses and the arabinose and xylose. Possibly the hexoses were present in fungal remains which had replaced the original polysaccharides in plant structures. With many soils a considerable proportion of the polysaccharide is present in the light fraction. The mean value for the 12 Australian soils examined by Ford *et al.* was 52%.

The light fraction of three English pasture soils examined were richer in glucose and particularly in xylose than the whole soil (Table 4.13). Similarly the separation of ten New Zealand tussock grassland soils gave light fractions with very much greater proportions of xylose than the heavy fractions (Molloy *et al.*, 1977). The proportions of arabinose were not very different, however.

After the residual heavy fraction of soil has been extracted with acid, alkali and acetic anhydride, it may be resubjected to separation by density, to yield a further light fraction. This has been considered to originate from the release of encrusted plant fragments indicating an incomplete separation of plant material in the initial separation. The average composition of this further light fraction material isolated from eight soils was distinctive only in the small proportion of galactose present (Table 4.14). Separation of soil into several fractions on a density basis has been reported to give a concentration of polysaccharide of microbial origin in fractions with intermediate densities (Turchenek and Oades, 1974/5).

TABLE 4.12 *Relative proportions of sugars in the light fraction of soils (adapted from Oades, 1972).*

	Lateritic podzolic	Solodized solonetz	Solonized brown soil	Rendzina	Sandy podzol	Black earth	Terra Rossa	Tropical red earth	Krasnozem	Grey clay	Fen peat	Ando
Glucose	100	100	100	100	100	100	100	100	100	100	100	100
Galactose	13	23	17	18	23	13	21	21	23	29	30	44
Mannose	10	14	14	16	24	24	29	33	33	37	43	60
Arabinose	71	50	22	20	24	14	23	15	18	26	22	17
Xylose	100	51	24	19	16	20	19	12	16	37	18	11
Rhamnose	10	6	2	8	5	2	6	5	2	9	13	11
Fucose	5	2	3	2	2	<1	3	1	1	5	7	5
Ribose	4	<1	<1	<1	<1	<1	1	<1	<1	2	1	<1

TABLE 4.13 *Comparison of the relative proportions of sugars in the light fraction and the whole soil (Whitehead et al., 1975).*

	Soil 1		Soil 2		Soil 3	
	Whole soil	Light fraction	Whole soil	Light fraction	Whole soil	Light fraction
Glucose	100	100	100	100	100	100
Galactose	35	15	38	28	36	28
Mannose	26	10	23	11	27	14
Arabinose	31	19	30	27	29	30
Xylose	36	54	48	59	43	60
Rhamnose	12	3	11	4	15	7
Fucose	3	1	5	4	5	3
% of the soil sugars in the light fraction		11		38		29

In my experience physical separation into light and heavy fractions using liquid density 2 cannot distinguish plant and microbial products. This has been clearly shown by fractionation of soil containing ^{14}C-labelled plant and microbial material (Murayama *et al.*, 1979).

Extraction of the light fraction by water

Only very small amounts of sugars, 0·6–0·18% of the dry weight, of the light fractions from tussock grassland soils are soluble in water (Ross and Molloy, 1977). Glucose, mannose and arabinose were the predominant sugars.

Sequential extraction by acid, alkali and acetylation

In Oades' study, (Table 4.14) after removal of the light fraction of the soils, the heavy fraction was successively extracted with M HCl, 0·2M NaOH and acetic anhydride:sulphuric acid. The fraction soluble in alkali contained most sugar (48%). Each of the other fractions contained 15% of the sugars and 22% remained in the residue. No great distinction between the sugar composition of the extracted

TABLE 4.14 *Relative proportions of sugars in various soil fractions averaged for 12 soils (adapted from Oades, 1972).*

	Light fraction	Heavy fraction						
		M HCl			0·2M NaOH	Acetic anhydride sulphuric acid	Residue	Light fraction from residue (eight soils)
		Molecular fraction obtained by gel filtration						
		<4000 (<G-25)	4000–100 000 (>G-25 <P-100)	>100 000 (>P-100)				
Glucose	100	100	100	100	100	100	100	100
Galactose	23	21	57	54	46	29	23	5
Mannose	28	28	61	109	59	38	34	46
Arabinose	27	13	15	12	36	46	25	32
Xylose	29	11	23	29	32	21	13	16
Rhamnose	7	7	24	46	19	11	7	—
Fucose	3	1	18	25	7	4	3	—

D

materials was found; glucose was the dominant sugar in all the fractions except the high molecular weight material extracted by HCl where mannose was predominant in 11 of the soils studied. Rhamnose and fucose were also comparatively abundant in the HCl soluble fraction.

In a rather similar fractionation scheme applied to humus layers of forest soils, using sequential extraction with $0.05M$ H_2SO_4 and $0.1M$ NaOH but without prior separation of the light fraction, Lowe (1978a) found that the alkali-soluble fraction only accounted for 15% of the soil polysaccharide, the major fraction being present in the humin (39%). As with the heavy fractions of the Australian soils, glucose was the predominant sugar in each of the fractions. The polysaccharide extracted by dilute acid had smaller proportions of mannose and xylose than had the other polysaccharide fractions which contrasts with the distribution observed by Oades with the heavy fractions. In fact analysis of the largest of the four fulvic acid fractions subsequently obtained shows a composition with almost equal amounts of the three hexoses, characteristic of the composition of polysaccharide obtained by direct extraction of soil by alkali described in the next section.

Extraction by alkali

Direct treatment of soil with dilute alkali extracts a large proportion of the polysaccharide although the major part sometimes remains in the humin fraction (Gallali et al., 1972). Acidification of the alkaline extract precipitates the humic acid but most of the polysaccharide remains in the acid-soluble fulvic acid fraction. It is recovered by adsorption on charcoal or by precipitation with ethanol (pp. 79–82). Analyses for a number of soils appeared to show that this polysaccharide is rich in pentose (Forsyth, 1950; Lynch et al., 1958; Johnston, 1961) but later studies using more concentrated acids for hydrolysis suggest that some of the earlier analyses were biased in favour of pentoses because the mild hydrolysis conditions used would give incomplete hydrolysis of hexosans. Where complete hydrolysis has been achieved (Cheshire et al., 1974a) the isolated polysaccharide from soil is found to contain very similar amounts of the three hexoses glucose, galactose and mannose, and smaller amounts of pentoses and methyl pentoses, which nevertheless are

TABLE 4.15 *Relative proportion of sugars in soil polysaccharide isolated from fulvic acids and in whole soil (after Cheshire et al., 1974a).*

	Countesswells series soil		Insch series soil	
Glucose	100	100[a]	100	100[a]
Galactose	86	27	81	29
Mannose	84	29	109	31
Arabinose	35	29	35	36
Xylose	49	31	56	34
Rhamnose	29	15	37	20
Fucose	27	6	33	8

[a] Whole soil.

present in higher proportions than in the total soil polysaccharide (Table 4.15).

Analysis of the alkali-soluble holocellulose of sphagnum peats showed it to be a polysaccharide, rich in glucose and xylose, with no variation in composition with age over a period of 200–4800 years (Theander, 1954).

The alkali-soluble, acid-insoluble organic matter, the humic acid fraction, also contains sugars with a composition not very different from that of whole soil (Lynch *et al.*, 1957b). Corresponding values for the fulvic acid are not available so that it is not possible to judge whether there is specific adsorption of polysaccharide on the humic acid or the precipitation of polysaccharide which is not soluble in acid. It has been suggested that polysaccharide may be combined in humic acid through linkage with amino acids as occurs in glycoprotein (Sadovnikova and Orlov, 1976) but most of the polysaccharide appears to be too easily released by resin treatment to be held in that way.

Glucosamine and galactosamine have both been observed in humic acid hydrolysates (Piper and Posner, 1968). A greater proportion of glucosamine to galactosamine, about 3:1, was found in the humin fraction of two soils than in the alkali-soluble humic or fulvic fractions (about 1·3:1) (Sowden *et al.*, 1976). Only about 3–5% of the

N of the humic acid and 2–3% of the N of the fulvic acid was present as amino sugar N.

Extraction by acid

Dilute acid has been used directly on soils without prior separation of light fractions. Peat extracted with 0·09M HCl at 60–70° for 1 h yielded a polysaccharide, precipitable with ethanol, which was rich in glucose and had galactose, mannose and xylose as minor components (Black *et al.*, 1955).

Organic matter isolated from soils by extraction with hot anhydrous formic acid containing 0·2M LiBr and precipitated by di-isopropyl ether, had reducing sugar contents of between 10 and 30%. The sugar composition was not greatly different from that of the residual unextracted organic matter although it was richer in the deoxysugar rhamnose (Parsons and Tinsley, 1961).

Conclusion

It is hard to assess the success of the methods of fractionation as we do not yet know the composition of the individual polysaccharides contributing to the whole soil polysaccharide. The composition of the isolated materials varies depending on the methods of fractionation and the fractions appear to relate to the way in which the polysaccharide is associated with other soil components rather than to the type or origin of the polysaccharide.

Methods of fractionation of isolated soil polysaccharide and the composition of the fractions

Fractional precipitation

(a) Alcohol

Attempts have been made to fractionate soil polysaccharide by precipitation from aqueous solution on the addition of ethanol. Bernier (1958) obtained three fractions from a forest soil polysaccharide in this way but their sugar composition appeared to be similar.

The precipitation obtained by the gradual addition of ethanol to a 1% solution of a soil polysaccharide at pH 1·85 is illustrated in Fig. 6 (Whistler and Kirkby, 1956). All the common sugars were present in each fraction though in differing amounts, with no consistent trends. At higher pH a smooth curve was obtained relating the weight of material precipitated to ethanol concentration.

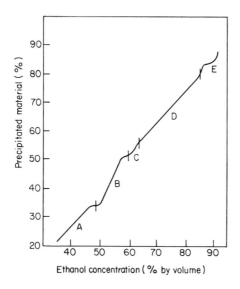

Fig. 6 The effect of ethanol concentration on the precipitation of soil polysaccharide from aqueous solution (Whistler and Kirby, 1956).

(b) Acetone

A fractional precipitation of the polysaccharide from three different forest soils occurred by the drop-wise addition of acetone to an aqueous solution (Saini and Salonius, 1969). One soil gave three fractions and the other only two. No account of their sugar composition was reported.

(c) Cetyl pyridinium chloride (cetavalon)

Long-chain quaternary ammonium salts from insoluble complexes with polysaccharide polyanions, which become soluble in various salt solutions. Fractionation of a forest soil polysaccharide was unsuccessful with cetavalon (Bernier, 1958), but a fractionation has been achieved with other soil polysaccharides (Streuli *et al.*, 1958; Parsons and Tinsley, 1961). On adding cetavalon to a solution of the polysaccharide in water a precipitate formed which was partially soluble in NaCl and LiCl (Parsons and Tinsley, 1961). The unprecipitated "neutral" polysaccharide had a much greater reducing sugar content than the insoluble "acidic" precipitate and proportionally a much lower uronic acid content. It was not possible to distinguish the "neutral" polysaccharide fraction on the basis of neutral sugar composition except possibly in the lower proportion of galactose it contained.

The sugars in this "neutral" polysaccharide-rich fraction still made up less than 16% of the ash-free weight of the fraction; a proportion little different from that of the sugars in the initial isolate.

(d) Light petroleum

Two methylated soil polysaccharides, dissolved in chloroform, were fractionally precipitated by the addition of light petroleum, b.p. 60–80° (Bernier, 1958). One soil polysaccharide gave six fractions, the other, three. There were small differences in the proportion of the sugars in the fractions and some of them contained only traces of arabinose and fucose.

(e) Copper salts

The polysaccharide from a forest soil has been separated into two fractions by adding an equal volume of Benedicts solution, an alkaline citric acid–copper complex, to a solution of the polysaccharide in 4% NaOH and shaking. The precipitate which formed was washed with 0·01M NaOH and redissolved by adding M HCl to decompose the copper complex. The polysaccharide in the precipitate and that remaining in solution, could also be distinguished by their different behaviour on electrophoresis (Bernier, 1958), the unprecipitated soluble material being more mobile.

(f) Sodium salts

Part of the polysaccharide extracted by 0·3M H_2SO_4 from a fenland soil was precipitated on neutralization with sodium bicarbonate (Finch *et al.*, 1966). The soluble fraction appeared to have mannose as the predominant sugar whereas the precipitate had a more uniform

TABLE 4.16 *Composition of the formic acid extracts of soils and their neutral polysaccharide fractions unprecipitated by cetavalon (adapted from Parsons and Tinsley, 1961).*

	Sandy loam		Meadow Soil		Peat		Podzol A_0	
	Formic acid extract	Neutral polysaccharide fraction	Formic acid extract	Neutral polysaccharide fraction	Formic acid extract	Neutral polysaccharide fraction	Formic acid extract	Neutral polysaccharide fraction
Galactose	92	57	82	53	93	78	92	69
Glucose	100	100	100	100	100	100	100	100
Mannose	64	53	56	48	57	59	61	53
Arabinose	28	23	33	20	36	30	19	20
Xylose	26	23	36	22	43	38	22	22
Rhamnose	26	40	31	22	57	70	19	33

sugar composition with greater amounts of pentoses than the soluble polysaccharide.

Gel filtration

Gels, made from dextran or agarose, have the ability to exlude solute molecules which exceed certain sizes and this forms the basis of gel filtration. Filtration through Sephadex G-100 gel with water as solvent, has been used to obtain four fractions from each of the alkali soluble and insoluble polysaccharides extracted from a fenland peat soil by 0·3M H_2SO_4 (Barker et al., 1965; Finch et al., 1968). Only the lowest molecular weight material was coloured. No sugar analyses were given for these fractions because they were subsequently further subdivided using anion exchange chromatography (see Fig. 7).

Ethylenediamine extracts of the same soil were acidified by HCl and the fulvic acid fraction remaining in solution separated into two fractions on Biogel P-150 (Hayes et al., 1975). These fractions were

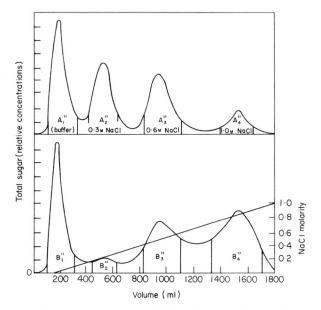

Fig. 7 Separation by stepwise elution (A″ fractions) and by gradient elution (B″ fractions) with sodium chloride solutions, of soil polysaccharides on DEAE Sephadex A-50 (Barker et al., 1965).

TABLE 4.17 *Relative sugar composition of Sephadex G-100 fractions of Biogel P-150 fraction 2 (adapted from Hayes et al., 1975).*

	Sephadex G-100 fraction	
	1	2
Galactose	8	114
Glucose	100	100
Mannose	8	79
Arabinose	7	39
Xylose	11	53
Rhamnose	3	43
Fucose	3	35

then further separated on Sephadex G-100, eluting the column first with water then with 0·5M NaCl. Analyses of the second Biogel fraction only are given and show (Table 4.17) an apparently higher molecular weight component very rich in glucose and a lower weight one in which galactose is dominant. The similarity of the distribution of sugars to that obtained using DEAE cellulose (see p. 102) suggests that charged sites in the Sephadex are causing it to behave as an anion exchanger.

The polysaccharide extracted from an acid pretreated Urrbrae loam by 0·5M NaOH was divided into several fractions by gel chromatography; there being a greater proportion of low molecular weight species. Subsequently, polysaccharides obtained by the sequential extraction of soil from the same site by HCl and NaOH were separated on Sephadex G-25 and Biogel P-100 into three fractions with nominal molecular weights of <4000, 4000–100 000 and >100 000 (Table 4.18). With each extract the composition of the <4000 fraction was very different from the other fractions in having a much smaller uronic acid content and being richer in amino acids in relation to the neutral sugar content. The amino sugar content was similar in all fractions. The uronic acid content was also much smaller in the lower molecular weight fractions of the fulvic acids of three soils obtained using Sephadex G-75 by Roulet *et al.* (1963b). Ribose and glycerol occurred in relatively high proportion in the low molecular weight fractions from the

TABLE 4.18 *Relative proportions of sugars in the* HCl *and* NaOH *extracts of Urrbrae loam (Swincer et al., 1968b).*

	Nominal molecular weight					
	<4000		4000–100 000		>100 000	
	HCl	NaOH	HCl	NaOH	HCl	NaOH
Galactose	15	28	50	42	34	37
Glucose	100	100	100	100	100	100
Mannose	12	67	49	57	76	63
Arabinose	20	25	4	25	17	12
Ribose	2	9	1	3	2	1
Xylose	5	23	7	35	35	24
Fucose	2	9	7	5	29	12
Rhamnose	2	11	8	8	39	25
Uronic acids[a]	<1	4	69	37	40	25
Amino sugars[a]	8	14	19	16	24	16
Amino acids[a]	194	106	135	25	70	12

[a] In relation to neutral sugars = 100.

Urrbrae sandy loam, and the deoxysugars rhamnose and fucose, were concentrated in the high molecular weight fraction (Swincer *et al.*, 1968b). Glucose was the dominant sugar in each of the molecular weight fractions.

In a study of twelve different Australian soils about 50% of the polysaccharide in the HCl extract was of molecular weight <4000 and this fraction was rich in glucose (Table 4.14). In the high molecular weight fraction, however, mannose was the predominant sugar in 11 of the soils.

All the material obtained by acetylation of soil residues after acid and alkali extraction was found to be excluded from Sephadex LH-20, a chemically treated G-25 Sephadex (Swincer *et al.*, 1968b). However, this might have been expected whatever the molecular size because retention depends on the presence of free hydroxyl groups.

Ion exchange chromatography

Separation by means of anion exchange materials depends on the nature and quantity of carboxyl groups in the polysaccharide. The

TABLE 4.19 *Relative proportions of sugars in fractions of soil polysaccharides separated by ion exchange chromatography (adapted from Chahal, 1968).*

	Acid soil fraction				Neutral soil fraction			Alkaline soil fraction		
Peak number	1	2	3	4	1	3	4	1	3	5
Galactose	93	75	36	48	83	163	115	83	95	76
Glucose	100	100	100	100	100	100	100	100	100	100
Mannose	58	44	38	10	40	66	64	291	41	49
Arabinose	70	24	48	13	19	38	40	160	11	3
Xylose	33	66	53	206	41	31	51	58	8	10
Fucose	13	95	57	93	155	100	210	112	11	60
Rhamnose	177	434	320	382	302	527	195	167	69	84

acidic nature of soil polysaccharide has enabled successful fractionation to be made on cellulose or Sephadex substituted with diethylaminoethyl groups (DEAE) (Thomas *et al.*, 1967; Mehta *et al.*, 1961; Barker *et al.*, 1965; Chahal, 1968). Despite this, there is rarely any consistency in changes in neutral composition of fractions eluted with increasing ionic strength. For example the composition of fractions obtained using DEAE cellulose with an applied NaCl and pH gradient are shown in Tables 4.19 and 4.20.

TABLE 4.20 *Relative proportions of sugars in the first three fractions in the separation of the acid-soluble alkali-insoluble polysaccharide on DEAE cellulose (Finch et al., 1968).*

Sugar	Fraction		
	1	2	3
Galactose	28	62	77
Glucose	100	100	100
Mannose	47	90	55
Arabinose	16	9	9
Xylose	22	40	29
Rhamnose	39	86	50
Fucose	18	62	59
Uronic acid	6	14	64

TABLE 4.21 *Uronic acid content and electrophoretic mobility of fractions of soil polysaccharide separated by anion exchange chromatography (from Thomas et al., 1967).*

Fraction	Uronic acid (%)	Mobility $\times$ 10^5 (cm^2 V^{-1}s^{-1})
1	6·0	2·54
2	6·1	4·18
3	7·8	6·25
4	6·6	7·95
5	6·6	6·17

DEAE cellulose is considered to be better than DEAE Sephadex A-50 gel; because some of the polysaccharide is excluded from the gel not all the active sites in it are accessible. Five or six fractions from the gel have been obtained with DEAE cellulose (Muller *et al.*, 1960; Thomas *et al.*, 1967) by eluting columns first with phosphate buffers and then with dilute alkali. There was only a small difference in the proportion of uronic acids present in each fraction (Table 4.21). Recoveries ranged from 60% (Thomas *et al.*, 1967) to 100% (Muller *et al.*, 1960).

A considerable effort has gone into the separation of the acid-soluble polysaccharide from a fenland soil using anion exchangers (Barker *et al.*, 1965, 1967; Finch *et al.*, 1967, 1968). A summary of the scheme is shown opposite.

Using DEAE Sephadex A-50, Barker *et al.* (1965) obtained four fractions from an isolated polysaccharide (A in Fig. 7), by eluting with increasing concentrations of NaCl in phosphate buffer, pH 6·0. Neutral sugars were similar in each fraction but the third and fourth fractions contained higher amounts of uronic acid. In a later study (1967) they found that the uronic acid contents of corresponding fractions were, in order of elution, 3·2, 5·0, 14·7 and 23·8%, assuming that the acid groups were in uronic acids. By a colorimetric method for uronic acid the contents were 2·3, 5·8, 13·1 and 5·1% respectively, showing that acidic compounds other than uronic acid were present in the polysaccharide. Colorimetric methods for different types of carbohydrate showed that the fractions varied in composition, with the least acidic fraction being richest in hexose (Finch *et al.*, 1967).

Fractions with greater differences in composition were obtained from the polysaccharide precipitable on neutralization with Na$_2$CO$_3$

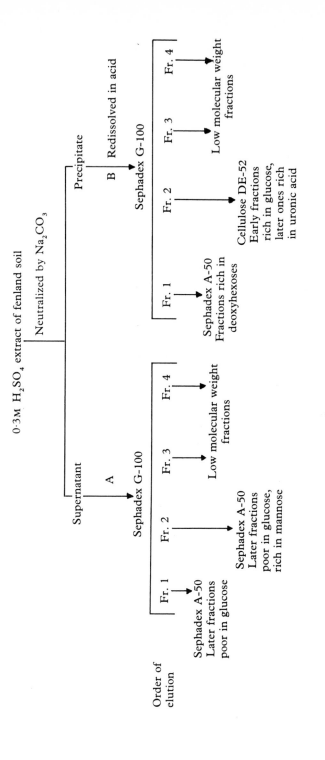

0·3M H₂SO₄ extract of fenland soil

Neutralized by Na₂CO₃

Supernatant

A

Sephadex G-100

Order of
elution

| Fr. 1 | Fr. 2 | Fr. 3 | Fr. 4 |

Sephadex A-50
Later fractions
poor in glucose

Sephadex A-50
Later fractions
poor in glucose,
rich in mannose

Low molecular weight
fractions

Precipitate

B   Redissolved in acid

Sephadex G-100

| Fr. 1 | Fr. 2 | Fr. 3 | Fr. 4 |

Sephadex A-50
Fractions rich in
deoxyhexoses

Cellulose DE-52
Early fractions
rich in glucose,
later ones rich
in uronic acid

Low molecular weight
fractions

using DEAE cellulose DE-52 (Finch *et al.*, 1968). Echelon elution with buffers of increasing ionic strength gave four fractions, the first of which was richer in glucose than the others, and the third much richer in uronic acid (Table 4.20). The fourth was too small for analysis. Reseparation of the first fraction showed that several components were still present.

With polysaccharide extracted by ethylenediamine from the same peat, Hayes *et al.* (1975), using DEAE cellulose and a chloride gradient at pH 9, obtained two fractions from each of the two molecular weight fractions given by Biogel P-150. The less charged of these, eluted early from the column, was rich in glucose, whereas the other had galactose as the dominant sugar (Table 4.22).

Diafiltration

Diafiltration is filtration through membranes with a defined pore size, usually under pressure. Diafiltration at a pressure of 25 psi (1·72 Mbars) has been applied to fulvic acid polysaccharide from a peaty soil and has yielded fractions with nominal molecular weight of <1000, 1000–30 000 and >30 000 (Lowe, 1975). The two latter fractions accounted for 2·3 and 15·2% of the organic matter, respectively, and contained 24·2 and 31·2% carbohydrate determined by anthrone. They had very similar sugar and uronic compositions

TABLE 4.22 *Relative proportions of sugars in fractions of the ethylenediamine-soluble polysaccharide of fen peat separated by anion exchange chromatography and gel filtration (adapted from Hayes et al., 1975).*

	Smaller molecular weight fraction		Larger molecular weight fraction	
	Neutral	Acidic	Neutral	Acidic
Galactose	67	183	13	163
Glucose	100	100	100	100
Mannose	27	92	13	89
Arabinose	32	26	3	21
Ribose	—	4	—	8
Xylose	58	49	5	41
Rhamnose	14	64	8	65
Fucose	10	29	3	18

(Table 4.23). Subsequently the method was applied to polysaccharides from forest humus soils, using an additional filter to distinguish material > 300 000 (Lowe, 1978a). The lower molecular weight fractions were richer in glucose and arabinose, and the higher molecular weight, in galactose and xylose.

TABLE 4.23 *Relative sugar composition of fractions of soil polysaccharide obtained by diafiltration (adapted from Lowe, 1975).*

	Nominal molecular weight	
	1000–30 000	> 30 000
Galactose	100	93
Glucose	100	100
Mannose	53	67
Arabinose	51	46
Ribose	6	14
Xylose	169	160
Rhamnose	31	51
Uronic acid	29	18

Adsorption on Polyamide

Polyamide has been used by Sequi et al. (1975) to fractionate the polysaccharide in fulvic acid extracts, eluting the column successively with 0.25M H_2SO_4 and 0.5M NaOH.

From 46 to 64% of the carbohydrate in the fulvic acid from five soils was eluted with the acid, 23–34% with the alkali and 12–13% was not recovered.

Adsorption on Polyclar

Fulvic acid from a muck soil has been separated into two fractions by application of the solution to Polyclar (Lowe, 1975). Elution with acid removed 82% of the recoverable polysaccharide and a further 18% was recovered with alkali: altogether 95% of the fulvic acid carbon was recovered. The acid eluted polysaccharide, further purified by precipitation and diafiltration was unusual in having xylose as the predominant sugar (see Table 4.23). The good recovery of polysaccharide shows that this high xylose content did not result from the fractionation.

Electrophoresis

Soil polysaccharides have been investigated by (a) paper (b) column and (c) free boundary electrophoresis. The findings have been very variable.

(a) Paper electrophoresis

Paper electrophoresis of the polysaccharide extracted from two forest soils with phosphate buffer separated it into two components, one stationary and one fast moving (Bernier, 1958) with similar neutral sugar compositions.

Continuous flow paper electrophoresis, of water-extracted poly-saccharide in phosphate buffer of pH 5·25 or 0·02M borate, was more successful (Mortensen, 1960; Mortensen and Schwendinger, 1963). Scanning the electrophoretogram for uronic acid revealed four maxima, although scanning for total sugar content, using α-naphthol reagent, showed only one maximum which coincided with most of the organic matter. Material extracted from soil by pyrophosphate was separated into two judging by analysis of the sugar content of the fractions, one of which was rich in brown polymeric material. The pattern of separation resembles that obtained by Clapp (1957) using water-soluble polysaccharide from a muck soil. In phosphate buffer the polysaccharide material was completely resolved from the coloured materials. McKenzie and Dawson (1962), on the other hand, observed three fractions in separations of material extracted from peats and muck soils by pyrophosphate, of which the least charged was predominant.

(b) Column electrophoresis

Water-soluble soil polysaccharide moved as a single substance during column electrophoresis in both borate and phosphate buffers (Clapp, 1957).

(c) Free-boundary electrophoresis

The buffer-extracted polysaccharide from forest soils moved as a single substance during free-boundary electrophoresis in borate or phosphate buffers (Bernier, 1958). However, fractionation of the polysaccharide by alkaline copper solutions gave three components with different mobilities.

The hot-water-soluble polysaccharide from a silty clay loam examined by Whistler and Kirkby (1956) showed a single peak on electrophoresis in borate buffer at pH 9·9, but it was broad, indicating that it was heterogeneous.

Polysaccharide extracted by 0·3M H_2SO_4 from a fen peat soil, and not precipitated on neutralizing with Na_2CO_3, appeared to contain one neutral and five negatively charged components on free boundary electrophoresis in phosphate buffer pH 7·0 (Barker et al., 1965). In later studies on this material Finch et al. (1967, 1968) used electrophoresis to test the homogeneity of fractions isolated by DEAE cellulose. All of the fractions were shown to contain more than one component.

A direct relationship between the uronic acid content and the electrophoretic mobility has been observed in fractions obtained from the polysaccharide of Brookston soil (Table 4.21) (Thomas et al., 1967). Neutral sugar components were not estimated but qualitatively the fractions appeared similar.

Conclusion

The general conclusion is the same as for the fractionation of the whole soil. When, with different methods, several fractions are obtained with similar sugar compositions the impression is gained that the fractionation depends on non-carbohydrate matter associated with the polysaccharide.

The most successful fractionation of isolated soil polysaccharide has been achieved by methods such as electrophoresis and use of anion exchange columns which depend upon the separation of charged species. Even so, few fractions are obtained, and they have usually only small differences in composition. Separations based on molecular weight differences are useful in distinguishing fractions rich in glucose.

Methods of hydrolysis of extracted polysaccharide

Milder conditions suffice for the complete hydrolysis of isolated soluble soil polysaccharide than for the whole of the polysaccharide in soil. Indeed a solution of polysaccharide obtained from a humus-rich soil underwent some autohydrolysis on heating in boiling water for 24 h (Forsyth, 1950). When the polymer was subsequently precipitated from solution with ethanol, arabinose remained in solution and could not be detected in the precipitate (Table 4.24). The ease of hydrolysis was taken to indicate that the arabinose was present in the polysaccharide in the furanoside form. Dilute sulphuric acid has been the most frequently used agent for complete hydrolysis but other acids such as formic, nitric, hydrochloric and trifluoracetic have also been used. One advantage of hydrochloric, formic and trifluoracetic acids is that they may be completely removed from the hydrolysate by evaporation. Heating with 6M hydrochloric acid has been used to give a more universal hydrolysis of soil constituents but it is known that sugars form brown polymers under these conditions. Table 4.25 summarizes the hydrolysis procedures that have been applied.

Few critical comparisons of the different methods have been made. The author found that a polysaccharide from a Countesswells series soil gave 33% reducing sugar on hydrolysis with trifluoracetic acid compared with 30% with 0·5M H_2SO_4 and 30·4% with cold 12M H_2SO_4 followed by 0·5M H_2SO_4 at 100°C. It seems most probable that

TABLE 4.24 *Relative proportions of sugars in soil polysaccharide before and after autohydrolysis (Forsyth, 1950).*

	Before autohydrolysis	After autohydrolysis
Glucose	100	100
Galactose	97	86
Mannose	105	96
Arabinose	59	0
Xylose	116	114

TABLE 4.25 *Procedures used to hydrolyse isolated soil polysaccharide.*

Hydrolysing agent	Conditions	Period	Reference
Sulphuric acid			
pH 0·5	Under N_2 sealed tube 100°C	8h	Sequi et al. (1975)
0·25M	Under N_2 sealed tube 100°C		Sinha (1972)
0·5M	Steam bath, 100°, or 100–105°C (sealed tube)	3–24h	Forsyth (1950); Duff (1952b); Theander (1954); Whistler and Kirkby (1956); Clapp (1957); Wright et al. (1958); Bernier (1958); Mortensen (1960); Duff (1961); Hayashi and Nagai (1962); Keefer and Mortensen (1963); Barker et al. (1965); Barker et al. (1967); Chahal (1968)
1·0M	100°C	7h	Wright et al. (1958)
1·5M	80°C 85°C or steam bath	24h	Acton et al. (1963a); Dormaar (1967); Gallali et al. (1972)
2·5M	Under reflux	20 min	Oades (1972)
3M	Boiling water bath	18h	Lynch et al. (1958); Hayashi and Nagai (1962)
12M	Cold	16h	Cheshire et al. (1974a)
followed by			
0·5M	100°C	5h	
2M trifluoracetic acid	Sealed tube 116°C	1h	Hayes et al. (1975)
0·5M HNO_3	Sealed tube 110°C	4h	Muller et al. (1960)
1M HCl	Sealed tube	7h	Bernier (1958)
2M HCl		4h	Forsyth (1947)
6M HCl	Sealed tube 105°C	6h	Maksimova (1973)
8M HCl	Refluxed	1h	Swincer et al. (1968 b)
98% formic acid	Sealed tube 100°C	12h	Parsons and Tinsley (1961)
0·5M H_2SO_4	Sealed tube 100°C	6h	

hydrolysis occurs very readily and that the differences observed are caused by subsequent destruction of sugars.

Composition and structure of isolated soil polysaccharide

Composition

Most isolated polysaccharide contains a relatively low proportion of carbohydrate on analysis; 24–33% by reducing sugar methods (Cheshire et al., 1974); 19–31% by colorimetric methods such as anthrone (Swincer et al., 1968a; Lowe, 1975; Otsuka, 1975). Indeed a soil organic fraction described as an organic phosphorus component contained similar amounts of carbohydrate (Anderson and Hance, 1963). Table 4.28 shows a typical analysis.

The neutral sugar compositions of various soil polysaccharides have already been presented in pp. 95–103.

(a) Amino sugars

Glucosamine, galactosamine and N-acetylglucosamine have been identified in the H_2SO_4 hydrolysates of isolated soil polysaccharides (Thomas, 1964). Such hexosamines account for about 4–6% of the alkali-soluble polysaccharide of soils (Swincer et al., 1968; Cheshire et al., 1974a) and 3–5% of the polysaccharide obtained from soil by anhydrous formic acid extraction (Parsons and Tinsley, 1961).

Of the polysaccharides extracted sequentially from soil by Swincer et al., (1968b), the high molecular weight fraction of the material soluble in HCl was richest in amino sugars. Very small amounts were present in the fraction extracted by acetylation.

Swincer et al. observed a ratio of 1·2:1 for glucosamine to galactosamine with an Urrbrae soil (South Australia) whereas Cheshire found 2·2% galactosamine and 1·7% glucosamine in the polysaccharide of a Countesswells series soil (Scotland).

(b) Uronic acids

Polysaccharide from a Countesswells soil contains 2·0% galacturonic acid and 1·3% glucuronic acid (Mundie, 1976), values comparable to those of the amino sugars.

TABLE 4.26 *Elemental composition (%) of soil polysaccharide.*

Source	C	H	O	N	S	P Inorganic	P Organic	Reference
Countesswells series soil	39·4	5·6	—	3·8	0·05	0·55	0·43	Cheshire
Insch series soil	38·0	5·6	—	3·3	—	0·47	0·49	
Lulu muck	41·15	6·29	48·72[a]	2·95	0·61	0·28		Lowe, 1975
Florida peat	39·1	6·6	49·8[a]	2·4	0·3	—	—	Lucas, 1970

[a] By difference.

(c) Elemental analysis

Elemental analysis gives values of C, H and O expected for a carbohydrate, but also shows the presence of small amounts of N, P and S (Table 4.26).

(d) Nitrogen

The nitrogen in soil polysaccharide preparations is present in amino sugars or amino acids and ranges from about 0·25 to 3 or 4%. Shaking an aqueous solution with chloroform has been used to remove N-rich material, presumably protein, but only lowered the N content to 1·78 in Brookston clay soil polysaccharide (Thomas, 1964). Rigorous purification of polysaccharide using dialysis, charcoal adsorption, precipitation with ethanol, treatment with $CdSO_4$, shaking with $CHCl_3$ and adsorption on Fuller's earth yielded materials with 0·26–0·60% N (Bernier, 1958).

(e) Sulphur

The carbohydrate-rich fraction of fulvic acid separated by Polyclar contains a low proportion of the fulvic sulphur (Lowe, 1975). Nevertheless it is considered that polysaccharides isolated from grey wooded and chernozemic soils are sulphated (Lowe, 1965). They account for no more than 2% of the total sulphur. Sulphur contents of 0·06–0·41% found for nine Alberta soil polysaccharides are equivalent to one S atom every 40–250 hexose units (Lowe, 1968). Weak infrared adsorption peaks at 1240 and 820 cm^{-1} in soil polysaccharide (Lowe, 1968) have been interpreted as ester sulphate groups.

(f) Phosphorus

Soil polysaccharide usually contains 1–2% phosphorus. The polysaccharide from two Scottish arable soils each contained about 0·5% in organic form and 0·5% in inorganic (Cheshire et al., 1974a). The organic form appears to be resistant to hydrolysis by acid and alkaline phosphatases.

Hydrolysis of polysaccharide-containing soil fractions has yielded phosphate esters of monomers with some of the properties of sugars or uronic acids but these have not been fully characterized (Cheshire and Anderson, 1975).

(g) Non-carbohydrate components of the polysaccharide

Amino acids. Protein-like material has been detected in the water-soluble polysaccharide of a soil (Duff, 1952b) and amino acids were observed in hydrolysates of various other soil polysaccharide preparations (Whistler and Kirkby, 1956; Mortensen, 1960; Parsons and Tinsley, 1961). Muller *et al.*, (1960) identified glycine, alanine, valine, tyrosine, cystine, aspartic and glutamic acids and lysine in alkali-soluble polysaccharide isolated by adsorption on charcoal.

Swincer *et al.*, (1968a) found that amino acids accounted for 6% of the alkali-soluble polysaccharide from an Urrbrae fine sandy loam. Alanine, glycine, aspartic and glutamic acids were the most abundant (column 1, Table 4.27) and it has been suggested that this composition indicates a microbial cell-wall origin. Separation of the polysaccharide by gel filtration showed that generally there were

TABLE 4.27 *Amino acids present in isolated soil polysaccharides.*

	(mol 100 mol^{-1} amino acid) Urrbrae soil	(μmol g^{-1}) Insch soil	(μmol g^{-1}) Countesswells soil
Alanine	11·7	46	100
Arginine	0·4	1	2
Aspartic acid	19·1	24	4
Cysteic acid[a]	2·5	6	6
Glutamic acid	17·4	17	34
Glycine	14·9	48	81
Histidine	0·5	4	4
Isoleucine	2·7	1	5
Leucine	4·4	1	4
Lysine	2·0	13	21
Phenlalanine	1·1	1	3
Proline	4·9	4	9
Serine	5·7	19	33
Threonine	6·2	16	36
Tyrosine	0·6	1	1
Valine	6·1	16	29
Diaminopimelic	—	2	4
Reference	Swincer *et al.* (1968a)	Cheshire (unpublished)	

[a] Derived from cysteine and cystine.

TABLE 4.28 *Sugar composition (%) of polysaccharide isolated from Countesswells soil.*

Galactose	Glucose	Mannose	Arabinose	Xylose	Fucose	Rhamnose
5·0	9·7	3·7	2·5	5·1	1·2	2·2

larger proportions of the peptide-like substance in the low molecular weight fractions. Polysaccharides extracted by alkali from Scottish arable soils and isolated by adsorption on charcoal have been found to contain 2–3% of amino acids in which alanine and glycine are the predominant components. The presence of a small proportion of diaminopimelic acid also indicates a microbial origin (Cheshire *et al.*, 1974a).

The inability to remove all the protein from soil polysaccharide preparations leads to the suggestion that glycoproteins or proteoglycans are present. The amino sugars may form bridging units between the glycan and the peptide components.

Structure

A preliminary study of the structure of soil polysaccharide has been made by analysis of the hydrolysis products by GC-MS after complete methylation (Cheshire *et al.*, 1979a). The results of this analysis are shown in Table 4.29. The data show that hexoses are predominantly linked by (1→3) and (1→4) bonds and xylose by (1→4) (Table 4.30). Sugar derivatives representing branching points account for about 20% of the residues with a far greater proportion in hexose and deoxyhexose than pentose residues. The average molecular weight of the methylated polysaccharide was calculated from end group analysis to be in the region of 1500. This order of value was confirmed by vapour pressure osmometry, and contrasts with the far larger values proposed for unmethylated polysaccharide presented in Chapter 4. It is probable that part of the reason for this difference is that much of the polysaccharide is linked to proteinaceous material and is in the form of a glycoprotein or peptidoglycan.

TABLE 4.29 *Relative molar proportions of sugars obtained on hydrolysis of methylated Countesswells soil polysaccharide (Cheshire* et al., *1979a).*

Relative retention time OV225	Designation of methoxy derivatives	Relative molar proportion	Possible origin structure
55	2,3,5-Ara.	2·7	Terminal Ara. (f)[a]
64	2,3,5-Xyl.	3·7	Terminal Xyl. (f)
	2,3,4-Rha.	3·1	Terminal Rha.
	2,3,4-Ara.	0·6	Terminal Ara. (p)[b]
	2,3,4-Xyl.	4·3	Terminal Xyl. (p)
69	2,3,4-Fuc.	1·3	Terminal Fuc.
	3,5-Ara./Xyl.	3·3	(1→2) Ara. (f)/Xyl. (f)
74	2,5-Ara./Xyl.	0·4	(1→3) Ara. (f)/Xyl. (f)
	3,4-Rha./Fuc.	0·9	(1→2) Rha./Fuc.
87	2,3-Rha.	0·8	(1→4) Rha.
	2,4-Rha.	1·4	(1→3) Rha.
	2,4-Ara./Xyl.	1·0	(1→3) Ara. (p)/Xyl. (p)
	2,3-Ara.	1·9	(1→4) Ara. (p) or (1→5) Ara. (f)
	2,3-Fuc.	0·4	(1→4) Fuc.
	2,3-Xyl.	12·0	(1→4) Xyl. (p)
100	2,3,4,6-Glc./Man.	7·0	Terminal Glc./Man.
106	2,3,4,6-Gal.	7·0	Terminal Gal.
114	2-Rha./Fuc.	1·4	(1→3) and (1→4) branched Rha./Fuc.
	4-Rha./Fuc.	0·5	(1→2) and (1→3) branched Rha./Fuc.
	3-Rha./Fuc.	0·6	(1→2) and (1→4) branched Rha./Fuc.
	2-Ara./Xyl.	3·0	(1→3) and (1→4) branched Xyl. (p)/Ara. (p) (1→3) and (1→5) branched Ara. (f)
	3-Ara./Xyl.	0·7	(1→2) and (1→4) branched Xyl. (p)/Ara. (p) (1→2) and (1→5) branched Ara. (f)
137	2,4,6-Glc.	6·7	(1→3) Glc.
	3,4,6-Glc. and/or Man.	6·6	(1→2) Glc. and/or Man.
	2,4,6-Man.	1·0	(1→3) Man.
143	2,3,6-Man.	2·5	(1→4) Man.
	2,4,6-Gal.	2·5	(1→3) Gal.
	2,3,6-Gal.	1·2	(1→4) Gal.
	2,3,4-Glc.	1·2	(1→6) Glc./Man.
	2,3,6-Glc.	6·4	(1→7) Glc.
154	2,3,4-Gal.	1·6	(1→6) Gal.
176	4,6-Man./Gal.	1·4	(1→2) and (1→3) branched Man./Gal.
	2,6-Gal./Man.	1·4	(1→3) and (1→4) branched Gal./Man.
	2,6-Glc./Man.	1·4	(1→3) and (1→4) branched Glc./Man.
188	3,6-Glc./Gal./Man.	2·5	(1→2) and (1→4) branched Glc./Gal./Man.
204	2,3-Man.	1·5	(1→4) and (1→6) branched Man.
215	2,4-Glc.	1·9	(1→3) and (1→6) branched Glc.
222	2,4-Man.	0·3	(1→3) and (1→6) branched Man.
	2,3-Glc.	0·2	(1→4) and (1→6) branched Glc.
	2,3-Gal.	0·9	(1→4) and (1→6) branched Gal.
236	2,4-Gal.	0·5	(1→3) and (1→6) branched Gal.

[a] (f) furanose.
[b] (p) pyranose.

TABLE 4.30 *Glycosidic linkages (%) present in soil polysaccharide.*

Linkage	%
Hexose	
(1→2)	6·6
(1→3)	10·2
(1→4)	10·1
(1→6)	2·8
(1→2) and (1→3)	1·4
(1→2) and (1→4)	2·5
(1→3) and (1→4)	2·8
(1→3) and (1→6)	2·7
(1→4) and (1→6)	2·6
Pentose	
(1→2)	3·3
(1→3)	1·4
(1→4)	13·9
(1→3) and (1→4) or (1→5)	3·0
(1→2) and (1→4) or (1→5)	0·7
Deoxyhexose	
(1→2)	0·9
(1→3)	1·4
(1→4)	1·2
(1→2) and (1→3)	0·5
(1→2) and (1→4)	0·6
(1→3) and (1→4)	1·4

It was concluded that the variety of linkages present supports the concept of soil polysaccharide as a mixture of diverse composition and structure.

Physical and chemical properties

Physical characteristics

(a) Viscosity

Viscosity in solutions can be regarded as a type of internal friction. For non-spherical molecules such as polysaccharides the viscosity of solutions increases with increase in molecular weight. Bernier (1958) favoured extraction of polysaccharide from soil with phosphate buffer at pH 7 over extraction with alkali because the buffer-extracted polysaccharide had a higher viscosity which was interpreted to mean that less breakdown of the structure had occurred.

The viscosity, η, of solutions of such buffer-extracted polysaccharide was measured using a Fenske-type viscometer at 25°C. From the specific viscosity, η_{sp}, for a given concentration c of solute, defined as $(\eta-\eta_0)/\eta_0$ where η_0 is the viscosity of the solvent, the intrinsic viscosity $[\eta]$ may be obtained, defined as $(\eta_{sp}/c)\ c\rightarrow 0$. The intrinsic viscosities of polysaccharides from two mull soils were 0·16 and 0·42, compared with about 0·04 for the polysaccharide from two mor soils (Bernier, 1958). An intrinsic viscosity of 0·21 was obtained for a peat polysaccharide in 0·5M cupriethylene hydroxide (Lucas, 1970) and values obtained by Saini and Salonius (1969) ranged from 0·70 to 1·50 for the alkali-soluble polysaccharide of three forest soils. These values should be compared with values of 1·0–2·0 quoted for plant starches.

The viscosity of solutions of the polysaccharide extracted from Paulding clay by hot water was found to be affected by change in pH and salt concentration (Mortensen, 1960). The polysaccharide behaved as a polyelectrolyte, the viscosity being decreased by increasing concentrations of electrolytes such as NaCl and $CuCl_2$. These changes possibly relate to the shape of the molecule. Variation of the pH of the polysaccharide solution gave a viscosity minimum at a pH near 4·9 suggesting the presence of protein.

(b) Molecular weight

The molecular weight of soil polysaccharide has been determined in three ways; directly by sedimentation using the ultracentrifuge and by gel filtration, and indirectly by viscosity.

Ultracentrifuge sedimentation. Ultracentrifuge sedimentation of a soil polysaccharide prepared by phosphate buffer extraction (Bernier, 1958) showed an approximate molecular weight of 130 000 (Ogston, 1958) compared with 124 000 from viscosity measurements. The measurements suggested that the average polysaccharide molecule was a rod with axial ratio of 80, but the breadth of the sedimentation boundary showed that the polysaccharide was very polydisperse. Similar single symmetrical peaks were observed with the water-

TABLE 4.31 *Proportions of extracted polysaccharides in three molecular weight ranges (Swincer et al., 1968a).*

	Nominal molecular weight		
Extractant	<4000	4000–100 000	>100 000
M HCl	55	35	10
0·5M NaOH	22	54	24

soluble polysaccharides from a muck soil (Clapp, 1957) and a woody peat soil (Mingelgrin and Dawson, 1973).

Sedimentation of the polysaccharide fractions from an Urrbrae soil, previously separated by gel filtration, showed the materials to be so polydisperse that sedimentation coefficients could not be obtained (Swincer *et al.*, 1968a).

A value for the molecular weight of a fen peat polysaccharide in

TABLE 4.32 *Optical rotation of soil polysaccharides.*

Source of polysaccharide	Extractant	Optical rotation $[\alpha]_D^{20}$ in water (C, 1·0)	Reference
Mull on calcareous grit ⎫		−18°	Bernier (1958)
Mull on calcareous clay ⎪		−12	
Mull on non-calcareous ⎪ gravel ⎬	Phosphate buffer, pH 7	−19	
Non-calcareous gravel ⎪		−11	
Non-calcareous gravel ⎭		−7	
Humus-rich agricultural soil	NaOH	−5	Forsyth (1947)
Heather raw humus, ⎫ Garden soil ⎪			
Pine raw humus, ⎬ Deciduous woodland ⎪	NaOH	−2–10	
Permanent pasture ⎭			
Brookston silty clay loam	Water	+36·8	Whistler and Kirby (1956)
Dunmore peat	0·09 MHCl	+27[a]	Black *et al.* (1955)
Florida peat	NaOH	+31	Lucas (1970)

[a] $[\alpha]_D^{16}$

phosphate buffer has been calculated as 51 400 $\pm$ 3900 (Finch et al., 1967).

Viscometry. From the viscosity of the soil polysaccharides in solution (Bernier, 1958), the molecular weight was obtained by applying Staudingers equation.

$$(\eta_{sp}/c)\ c \rightarrow 0 = K_m P$$

where c is the concentration in g litre^{-1}, $K_m/5 \times 10^{-4}$ and P is the degree of polymerization. Assuming a monomeric composition with equal numbers of pentose and hexose molecules, the polysaccharides extracted with phosphate buffer from four soils had molecular weights ranging from 10 300 to 124 000. Hot-water-soluble polysaccharide, in 0·1M NaCl, had an apparent $(\eta_{sp}/c)\ c \rightarrow 0$ of 0·85 (Mortensen, 1960) and using the same assumptions, this corresponded to a molecular weight of 450 000. Saini and Salonius (1969), however, suggested that the modified equation for the Staudinger index,

$$[\eta] = KM^a$$

where M is the molecular weight of the polymer and K and a are constants, should be used. They related M to the precipitability, γ, by the equation $\gamma = a + b/m$, where a and b are constants. γ was defined as $V/(V + V_0)$ where V is the volume added to the original volume V_0 to cause precipitation. Precipitabilities ranged from 0·48 to 0·74 and were significantly inversely related to the Staudinger indices, but with the modified equation, insufficient constants were known to allow calculation of molecular weights.

Optical rotation

Most values are laevo-rotatory (Table 4.32), some strongly so (Haworth et al., 1946), indicating a predominance of β-glycosidic linkages.

The reported behaviour of soil polysaccharides on sedimentation and gel filtration is confusing and contradictory. It seems much more likely that polymers of a wide range of molecular weights are present in each of the polysaccharide preparations investigated rather than the narrow range implied by the finding of single symmetrical sedimentation peaks. Inevitably viscosity measurements can only give an

TABLE 4.33 *Equivalent age (y) for the carbon in soil fractions.*

| Soil | Soil fraction | | | | |
	Whole soil	Humin	Humic acid	Fulvic acid	Reference
Melfort (Canada)	—	1140 ± 50	1235 ± 60	470 ± 90	Campbell *et al.* (1967)
Broadbalk 1881 (England)	1450 ± 95	2395 ± 90	750 ± 90	420 ± 85	Jenkinson (1970)

average value for the molecular weight, and because of ignorance about the shapes of the molecules this can only be very approximate.

(c) Age

The polysaccharide has been shown by carbon dating to be one of the youngest fractions of the organic matter of soil (Table 4.33).

The higher ratios of $^{13}C:^{12}C$ in the fulvic acid and polysaccharide than in the humic acid fraction of a soil has been considered to show that they are isotopically closer in composition to living plant material than the humic acid fraction (Nissenbaum and Schallinger, 1974) and from this it has been argued that the fulvic acids represent an intermediate stage in the formation of humic substances.

(d) Complexing of metal ions

Soil polysaccharide fractions obtained by precipitation with acetone or alcohol may contain inorganic matter giving as much as 50% ash (Mortensen, 1960). Most of this inorganic matter represents coprecipitated salts which are easily removed by treatment with ion exchange resins, ethanol containing acetic acid and dialysis, yielding polysaccharide with about 5% ash. Soil polysaccharide isolated by the Forsyth method of adsorption on charcoal also contains about 5% ash; further purification by dialysis and treatment with strongly acidic cation exchange resin reduces the ash content to about 1% (Muller *et al.* 1960; Thomas, 1964).

Alkali-soluble polysaccharide, isolated from a soil derived from granitic till, contained 0·23% and 0·006% of aluminium and iron respectively (Cheshire *et al.*, 1977), possibly combined with the COOH groups of the uronic acid constituents (Gaponenkov and Shatsman, 1964). Other major inorganic substances in the ash were

TABLE 4.34 *Effect of adding iron or aluminium chloride on the pH of a soil polysaccharide solution (Saini, 1966).*

Addition to 1 mg polysaccharide and 20 ml water	pH
1 ml water	5·6
1 ml 0·0001M $AlCl_3$	5·2
1 ml 0·0001M $FeCl_3$	4·8

silica and phosphate. Examination by electron resonance spectroscopy confirmed that few transition elements were present.

Other evidence for the complexing of metal ions has usually come from experiments involving the addition of metal salts to the isolated polysaccharide. Thus Mortensen (1960), on adding copper salts, observed perturbations in the infrared absorption bands assigned to carboxyl and amide groups. Complex formation with the radioactive isomers ^{90}Sr and ^{90}Y was shown by the separation by gel filtration and paper electrophoresis of the two negatively charged ^{90}Y complexes and one positively charged ^{90}Sr complex (De Datta et al., 1967). When the Sr and Y salts were added to soil, although only about 1·5% of the ^{90}Sr was extractable with hot water, about 75% of the radioactivity was in the polysaccharide fraction.

Complex formation between soil polysaccharide and iron and aluminium has been shown by a decrease in the pH of a solution of the polysaccharide on the addition of the metal ion (Table 4.34), showing a release of protons. The greater decrease of pH with iron addition is taken to indicate that it forms a stronger complex than aluminium. The metal addition caused a slight increase in the u.v. absorption, which was taken as confirmation that complexes had formed. When Florjanczyk (1965) applied fulvic acid to anion exchange resin and eluted with $NaHCO_3$ solution he observed minute red flecks in the eluate which he interpreted from analysis to be iron complexed with a polyglycose, resembling complexes which form with cellulose. Aluminium appeared to form a similar complex.

(e) Iso-electric point

An iso-electric point at a pH of about 4·9 was attributed by Mortensen (1960) to contamination of the polysaccharide by protein. The iso-electric point was lowered by adding NaCl and $CuCl_2$

suggesting to Mortensen that the anion was more strongly bound to the polysaccharide than the cation.

(f) Infrared absorption

Infrared absorption of soil polysaccharide commonly shows the presence of carboxyl and amide groups. Typical infrared spectra of the acid and salt forms are shown in Fig. 8. Weak bands at 895, 805 and 800–750 cm^{-1} are not inconsistent with the presence of β-anomers in the polysaccharide.

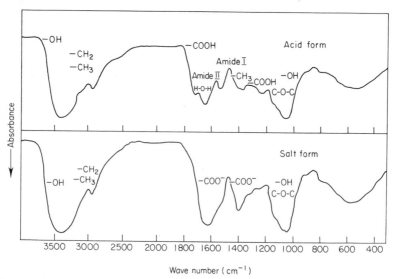

Fig. 8 Infrared spectrum of soil polysaccharide in the free acid and salt forms.

In the salt form there should be no absorption at ≈ 1250 from COOH but absorption in this region is still observed. Possibly this is from carboxylic acid esters but deesterification by boiling the polysaccharide with 0·1M NaOH only partially reduces the absorption. Part may be related to a sulphate ester group.

Clark and Tan (1969) claim to have discerned ester linkages between soil polysaccharide and hymatomelanic acid, the ethanol soluble fraction of the humic acid. Their spectra, however, show gross contamination with bicarbonate and silica gel, and allow no such conclusion to be drawn (Russell and Anderson, 1974).

The amide bands are probably caused by peptide because amino acids are obtained on chemical analysis.

(g) Electrophoretic mobility

The electrophoretic mobility of soil polysaccharides depends on the charges on the molecule. In neutral or alkaline conditions these are mainly negative charges associated with carboxyl groups in uronic acids and amino acids. The mobility is increased in borate buffer showing that complex formation between the borate and the polysaccharide occurs. For example, polysaccharide extracted from a muck soil by hot water was apparently homogeneous on electrophoresis (Clapp, 1957) with an average mobility of $-2\cdot8 \times 10^5$ cm^2 V^{-1}s^{-1} in phosphate buffer at pH 7·05 and $-7\cdot7 \times 10^5$ cm^2 V^{-1} s^{-1} in borate buffer at pH 9·05.

A direct relationship between the uronic acid content and the mobility has been observed for five fractions run at pH 7·65, at which pH the carboxyl groups are the main source of charge (Table 4.21).

Polysaccharide, extracted from a fenland peat soil by 0·3M H$_2$SO$_4$, separated, in phosphate buffer at pH 7, into about ten fractions with electrophoretic mobilities ranging between $-0\cdot14$ and $-12\cdot6 \times 10^5$ cm^2 V^{-1} S^{-1} (Finch et al., 1967, 1968).

The mobilities of the electrophoretically homogeneous polysaccharides extracted from four soils by phosphate buffer pH 7 lay between $-5\cdot1$ and $-7\cdot9 \times 10^5$ cm^2 V^{-1}s^{-1} in borate buffer at pH 9·1 (Bernier, 1958). With a fifth soil the polysaccharide was heterogeneous with mobilities of $-0\cdot9$ and $-2\cdot8 \times 10^5$ cm^2 V^{-1}s^{-1} in pH 7·8 phosphate buffer. In view of the variability in electrophoretic mobility reported for different soil polysaccharides it is surprising that relatively few components in each preparation can be separated by electrophoresis.

Chemical characteristics

(a) Susceptibility to enzyme attack

Several attempts have been made to split chemical linkages in soil polysaccharides with enzymes.

A snail-crop fluid prepared from *Helix aspersa* which contained cellulase was applied to soil polysaccharide which had been extracted by phosphate buffer at pH 7·0 (Bernier, 1958). Reducing sugars were released amounting to about 12% of the polysaccharide and the relative viscosity was decreased by about 20%, indicating that the molecular weight had become smaller.

E

Barker *et al.* (1967) were able to release between 25 and 50% of the rhamnose of the polysaccharide extracted from a peat with dilute sulphuric acid by using an α-rhamnosidase, and so concluded that this rhamnose was in terminal groups.

Other studies have not been so successful. Thomas (1964) applied the culture filtrate of *Pseudomonas solonarum*, which contained a cellulase, a pectin methylesterase and a polygalacturonase, to a soil polysaccharide solution. Neither this filtrate nor commercial preparations of cellulase and polygalacturonase had any measurable effect on the viscosity of the solution and this was taken as evidence for the absence in soil polysaccharide of β-(1→4) glucan, (1→4) polygalacturonic acid or the methyl ester linkage of pectin.

Cheshire and Anderson (1975) obtained an insignificant release of sugars from alkali-extracted soil polysaccharide with (1→4) polygalacturonase, β-(1→3), β-(1→4) and α-(1→6) glucanase, α-(1→6) mannosidase and β-gluconidase. The inability of proteases to release amino acids, considered to be present in peptide linkage, suggests that the enzymes may be inhibited by the structure of the substrate.

Interaction of polysaccharide and clay

Various plant polysaccharides such as starch, cellulose, dextrin, inulin and water-soluble corn polysaccharide, or animal poly-saccharides such as glycogen have been found to be strongly adsorbed by kaolinite and montmorillonite clays (Lynch *et al.*, 1956; Parfitt and Greenland, 1970).

Adsorption occurs on two types of sites; surfaces and interlamellar spaces. With kaolinites, only the surface site is available, but with montmorillonites, polysaccharide is also adsorbed into the interlamellar spaces. Using X-ray diffraction Greenland (1956), by observing changes in the basal spacings of Ca montmorillonite during complex formation (Table 4.35), showed that soil polysaccharide could enter these spaces. Much greater amounts of polysaccharide may be adsorbed by clays which possess internal adsorption sites. Finch *et al.* (1967) observed that 40 mg of a peat polysaccharide was adsorbed g^{-1} of sodium kaolinite compared with >180 mg g^{-1} of montmorillonite. With the montmorillonite the adsorption isotherm showed inflection points corresponding to about

TABLE 4.35 *Effect of polysaccharide on the inter-*
lamellar spacing of montmorillonite (Greenland,
1956).

Polysaccharide content of clay (%)	d (001) nm
1	1·46
4	1·46
10	1·6–1·8

61 mg g^{-1} clay at which stage sites other than those on the surface were thought to become available for adsorption. To estimate the adsorption surface area, Finch et al. (1967) assumed the area of projection of a hexapyranose ring to be 48×10^{-20} m^2, that the average monomer in the polysaccharide had a molecular weight of 147 and that the polysaccharide would be fully collapsed onto the surface of the clay. The surface area corresponding to 61 mg g^{-1} adsorption was then calculated to be 118 m^{2}g^{-1}. With the same polysaccharide on kaolinite a surface area of 83 m^{2}g^{-1} was calculated. With some polysaccharides of known structure, for example β-linked dextrans, the amounts adsorbed are greater than could be accommodated on the surface (Olness and Clapp, 1973, 1975), and it is suggested that linkage occurs to only some of the sugar units so forming free loops of polymer chain (Hayes and Swift, 1978).

The reaction of montmorillonite with particular soil polysacc-haride fractions has been followed by comparing the fractionation of the polysaccharide on DEAE cellulose before and after removal of the clay-adsorbed fraction by centrifugation (Finch et al., 1967). Some high molecular weight polysaccharides were adsorbed. The much greater adsorption of the polysaccharide with montmorillonite in the H form than the Ca form (Finch et al., 1967) was taken as support for the idea that the polysaccharide is hydrogen-bonded to the clay (Kohl and Taylor, 1961), since an ionic linkage would be weaker at lower pH. The H bonding is envisaged to occur between the OH groups of the polysaccharides and the oxygen atoms of the clay.

Other investigations have indicated that metal ions may be involved in the adsorption. Using a labelled soil polysaccharide made by incubating soil with ^{14}C-glucose, Saini and Maclean (1966) reported that different amounts of polysaccharide were adsorbed with different forms of kaolin following the order $Fe^{3+} > Al^{3+} > H^+ > Ca^{2+} > Mg^{2+} > Na^+$. A similar relationship was observed for montmorillonite with the $Fe^{3+} > H > Ca$ forms, the order being unrelated to the size of the outer surface areas which were 29, 78 and 135 m^2g^{-1}, respectively (Guckert *et al.*, 1975b).

Parfitt and Greenland (1970) found that polygalacturonic acid was adsorbed on Al montmorillonite but not, unlike the neutral sugar polymers of soil, on the Na form. Parfitt (1972) concluded that uronic acids were adsorbed mainly by complex formation.

Despite the attractiveness of hydrogen bonding to explain the adsorption of polysaccharide by clays, Greenland argues that there is little evidence for it and that multiple dispersion forces and entropy changes, arising from the replacement of water molecules from the clay surface, are sufficient to account for the phenomenon.

Chapter 5

The Metabolism of Carbohydrate in Soil

Organic substances added to soil are gradually broken down by microorganisms to small molecules such as glyceraldehyde and monosaccharide phosphates which are either oxidized to CO_2 to provide energy or used to synthesize microbial tissue. The as-yet undecomposed part remains as a residue, although this may have been modified by the microorganisms. It is rarely possible to determine exactly how much of a natural material added to soil has been decomposed after a period of incubation because of the similarities in the biochemical make-up of all living things. The agents of decomposition, the microflora, may resynthesize compounds similar in composition to those they are attacking and may also be stimulated by the fresh substrate to attack other organic components of soil. Thus neither the CO_2 evolved from the soil nor the residual carbon provides an accurate measurement of the degree of decomposition of the substrate. The use of ^{14}C in uniformly labelled substrates has greatly helped incubation studies, because the carbon evolved as CO_2 can be identified as coming from the substrate and gives a measure of the minimum decomposition that has occurred.

The biochemical decomposition of polysaccharide starts by enzymatic hydrolysis of the glycosidic linkages resulting in a depolymerization to mono- and oligosaccharides and in most instances, only after monosaccharides are formed, does oxidation or transformation take place. Polysaccharides therefore show a slower rate of breakdown than monosaccharides and oligosaccharides, and this is illustrated for glucose and one of its polymers, a dextran, in

TABLE 5.1 *Percentage of glucose and dextran remaining in the soil after incubation as measured by residual glucose (adapted from Oades and Wagner, 1971).*

	2 d	5 d	12 d	28 d
Glucose	7·3	2·4	1·4	1·1
Dextran	43·3	22·8	6·2	3·8

Table 5.1. Two classes of enzyme take part in this hydrolysis. Exoenzymes attack the linkages to the terminal sugar residues at the non-reducing end of the chain, whereas endoenzymes attack all bonds within the chain. The rate of decomposition of carbohydrates also depends upon other intrinsic factors such as solubility and structure, and association with non-carbohydrate components.

In the soil, biological attack on a carbohydrate occurs mainly in two ways: in one, water-soluble components are assimilated by diffusion into microbial cells where they may be attacked by the enzymes of the cytoplasm. In the other, both water-soluble and water-insoluble polysaccharides are attacked by exocellular enzymes synthesized by microorganisms. The mono- and oligosaccharides liberated in this way diffuse throughout the soil solution and so become generally available. Thus, because enzymes function much more effectively in solution, with plant material a general increasing order of stability has been found as follows: monosaccharides < oligosaccharides < starch < pectin < mannan < xylan < cellulose (Mehta *et al.*, 1961), very largely in order of decreasing solubility. A similar order of stability has been given by Burges (1967): water soluble components (sugars, starch, pectins) < hemicellulose < cellulose. Using CO_2 formation as a measure of decomposition, Visser (1968) came to the surprising conclusion that hexoses and pentoses were much more stable in soil than oligosaccharides and polysaccharides such as glycogen, inulin and starch. It is probable that the monosaccharides underwent transformation rather than oxidation to CO_2 and only in that sense were more stable.

Differences in the biological stability of polysaccharides can also be related to sugar composition and structure. It has been found, for example, that galactans are more resistant to microbial degradation in soil than mannans and xylans (Waksman and Diehm, 1931a,b,c). It seems unlikely that there are significant differences in the rate of hydrolysis by the appropriate enzymes in this instance, but rather that the amounts of enzyme may be limited, and this would depend on how commonly the polysaccharide occurs as a substrate. Chitin, although very resistant to chemical hydrolysis, does not persist for long in the soil and this is possibly because it is of such common occurrence that a large number of organisms have developed enzymes to hydrolyse it (Greenland and Oades, 1975). Some microbial polysaccharides are very persistent in soil. These appear to contain uronic acids as well as neutral sugar components (Martin *et al.*, 1965) and are probably stable because of their structure, although, as will be discussed later, association with soil components may also be a factor. Many microbial polysaccharides are branched. Their enzymic decomposition is therefore more complicated than that of linear polysaccharides and this will contribute to their persistence.

With complex heteropolysaccharides, the glycosidic linkages would require many different specific polysaccharases in a particular order for complete hydrolysis. Martin has implicated complexity of structure as a reason for the apparent stability of soil polysaccharide (Martin, 1971). Incubation experiments which show a rapid decomposition of isolated soil polysaccharide do not support this idea, however.

Substitution of hydroxyl groups in sugars by methoxyl or acetyl groups may increase a polysaccharide's resistance to enzyme decomposition. Indeed it has been claimed that, when more than every alternate sugar in a chain is substituted, a polymer cannot be degraded by microorganisms (Reese, 1968). Duff (1961) suggested that the stability of soil polysaccharide may be related to the methylated sugars present. Some polysaccharides in grasses appear to be quite heavily substituted with acetyl groups (Bacon *et al*, 1975) and this must contribute to the persistence of plant material in soil.

The stability of polysaccharide in soil is also very dependent upon its association with other cell components. Thus polysaccharide which has been isolated from the organism in which it occurs is much more easily decomposed than when part of the original structure. For example only about 1–2% of the xylan in an isolated hemicellulose fraction remained after 448 d incubation compared with 23–28% in

TABLE 5.2 *Persistence of labelled carbohydrates in soil incubations.*

Substrate	Incubation period (d)	% remaining in the soil		Reference
		Soil A	Soil B	
Cereal rye straw				
Glucose in cellulose	56	44		Cheshire et al. (1973)
	224	48		
	448	22		
Xylose	56	43	37	
	112	—	19	
	224	36	26	
	448	28	23	
Isolated hemicellulose				
Hemicellulose residual carbon	14	68		Cheshire et al. (1974b)
	112	51		
	448	30		
Hemicellulose xylose	14	19	—	
	56	5·1	—	
	112	4·4	—	
	224	—	2·7	
	448	2·3	1·4	
Tobacco starch				
Glucose	42	2·3	1·4	Cheshire et al. (1969)
	84	2·3	1·4	

the whole unfractionated plant tissue (Table 5.2). In this instance the difference may have been caused by the greater solubility of the hemicellulose brought about by the extraction, but there are possibly additional or alternative causes. For example, the rate of decomposition of the polysaccharide in plant residues depends on the nitrogen and the lignin contents (Lueken *et al.*, 1962; Herman *et al.*, 1977). Generally, materials with high nitrogen content decompose more easily than those with low nitrogen, whereas lignin has a protective effect and delays decomposition.

In addition to the constraints on decomposition inherent in the carbohydrate material, the soil itself provides mechanisms for protection. Perhaps the most effective of these is the physical state of soil. The agents of decomposition, hydrolysing enzymes, require a liquid phase in which to react with polysaccharide molecules. The water in soil mostly occurs as a film of liquid on the surface of soil particles and free water may not penetrate to their interiors. Even when it can penetrate, the substrate may not be accessible to organisms or their enzymes. Increased microbial activity in soil as a result of shearing soil particles (Rovira and Greacen, 1957) has been explained as the result of breaking micropores too small for microorganisms to enter.

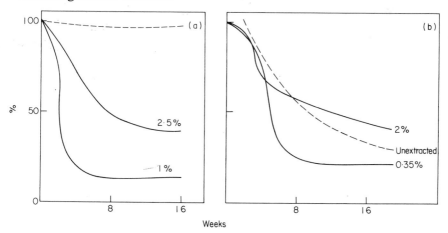

Fig. 9 (a) The effect of incubation on the reducing sugar present in hydrolysates of soil (– – –) in relation to that present originally and in soil polysaccharide–soil mixtures (——) in relation to the added polysaccharide. (b) The percentage loss of radioactivity from the sugars of ¹⁴C-labelled soil polysaccharide added to the soil (——) and from polysaccharide synthesized *in situ* in 14 d using ¹⁴C glucose as substrate (– – –).

TABLE 5.3 *Decomposition of carbohydrate during incubation in soil. Residual substrate expressed as a percentage of the original (Cheshire et al., 1974a).*

Substrate	Residue (%)		
	2 weeks	4 weeks	8 weeks
Rye grass roots	75	71	48
Soil polysaccharide	70	18	11
Wheat starch	62	34	10
Cellulose	44	24	2
Aspergillus mycelium	39	31	0

In its native state soil polysaccharide appears to be very stable with a slow rate of turnover, but, as shown in Fig. 9, after extraction and isolation, the material is rapidly decomposed on reintroduction to soil (Cheshire *et al.*, 1974a; 1975) and, contrary to the claim made by Kilburn (1967), the rate of decomposition is not apparently slower than that of plant polysaccharides (Table 5.3). The effect is therefore very similar to that observed with hemicellulose when isolation after alkaline extraction greatly increases the rate of decomposition, and it may have similar causes, in particular increased solubility and the disruption of chemical bonds with other soil components.

Another factor affecting decomposition, at least partly physical, is the formation of faecal pellets. It has been observed that plant material incorporated in faecal pellets resists decomposition, the pellets remaining intact over a 10-month incubation period (Grossbard, 1969). This direct observation using ^{14}C-labelled substances contrasts with the general opinion that the overall rate of decomposition of litter is increased by transformation to faeces (Lodha, 1974).

In soil many different organic molecules are present as a result of microbial activity, and chemical interaction of carbohydrate with these is to be expected, particularly with those of a phenolic nature. As a result of this a considerable degree of protection may be afforded. For example, in decomposing mor litter, Handley (1954) obtained evidence that cellulose in cell walls is protected by tannin-like polyphenolic substances. He observed that cell walls became coated with organic material which had to be removed with sodium hypochlorite solution before a colour reaction with zinc and iodine, characteristic for cellulose, could be obtained. Precipitation of gelatin occurred with water extracts of fresh leaves showing that tannins were present and it was suggested that the cytoplasmic protein in leaves is tanned and so stabilized to biological attack. The effect of wattle tannin on the decomposition of various polysaccharides has been studied experimentally (Benoit and Starkey, 1968; Lewis and Starkey, 1968). Inhibition of decomposition of over 50% was obtained with some substrates during the 3-week incubation period (Table 5.4). Polysaccharides which contain amino groups, such as chitin, have been shown to be partially protected against oxidation by synthetic phenolic polymers (Bondietti *et al.*, 1972). After 8 weeks incubation 31% of the carbon had been lost from a chitosan–phenol complex compared with 53% from chitosan. There are three ways in which tannins may effect decomposition (Benoit and Starkey, 1968).

1. By tanning the proteins of plants and animals to produce resistant complexes.

2. By tanning the enzymes which decompose the carbohydrate.

3. By tanning the polysaccharide itself.

Leaf leachates have been examined for tannins and their presence confirmed, but tannin-like properties are also possessed by soil organic matter fractions (Davies, 1971). Thus humic substances are adsorbed by Polyclar (polyvinylpyrollidone) and by powdered nylon, both of which strongly adsorb tannins. Furthermore, some of the soil polysaccharide appears to be physically if not chemically bound to humic substances in the soil as exemplified by the association of carbohydrate and humic acid (Mehta *et al.*, 1961; Swincer *et al.*, 1969). Martin *et al.* (1978) think that protection of polysaccharide by

TABLE 5.4 *Inhibition of decomposition of carbohydrates by tannin (Benoit and Starkey, 1968).*

Inhibition of decomposition by 2% tannin (%)			
Soil pH	Pectin	Rye hemicellulose	Cellulose
4·0	16	62	35
7·0	25	59	51

humic substances depends on chemical bonding because merely mixing various humic substances with polysaccharides has no effect on the rate of decomposition. An indirect form of protection possibly provided by organic compounds must be mentioned. The rate of decomposition of carbohydrates in a sphagnum peat is slow compared with deciduous and coniferous leaf litter (Ivarson, 1977). Phenols are present in the peat and the possibility that they have a specific inhibitory effect on microbial activity has been suggested. An alternative explanation involves nutritional factors, such as a limitation on nitrogen supply (Given and Dickenson, 1975). Similarly, cellulose was found to be more rapidly destroyed in pure culture than when mixed with soil unless the soil had been sterilized (Went and Jong, 1966). It was suggested that inhibiting substances may be present which are heat-sensitive.

At least two other components of soil have been found to affect the rate of decomposition of polysaccharides in soil, namely clays and metal cations. When cellulose dextrin was shaken with calcium montmorillonite in a mineral-rich medium and incubated for periods of up to 7 weeks, at one stage 38% less CO_2 was evolved than with the clay omitted (Lynch and Cotnoir, 1956). Similar reductions in the rate of decomposition of glucose and microbial products caused by montmorillonite occur in solid media (Sørensen, 1965; Guckert et al., 1977). The reduced rate of oxidation could have been caused by the effect of the clay in reducing enzyme activity; both cellulase and hemicellulase show decreased activity in mixtures with clay. Alternatively it may have been because of adsorption of the substrate by the clay as described in Chapter 4 (Dubach et al., 1955; Lynch et al., 1956; Greenland, 1956; Emerson, 1963; Clapp et al., 1968).

The form in which polysaccharides occur in soil is not yet known, but it is considered they will be present as metal salts or complexes (Martin, 1971). Many of them contain carboxyl groups and also mannose, which is capable of forming metal complexes involving the cis-OH groups (Martin et al., 1972). Good evidence for a salt form is found in the ash content, often as high as 50% in polysaccharides isolated from soil by mild methods such as extraction by buffers. This association with metal ions may easily be overlooked when polysaccharide is isolated by other methods. For example, by extraction by alkali and adsorption on charcoal, the polysaccharide is one of the organic matter fractions with the highest loss on ignition and appears poor in those transition elements which form the strongest complexes with organic compounds (Cheshire *et al.*, 1977). Polysaccharides synthetically complexed with metals usually show retarded decomposition rates. With several plant and microbial polysaccharides Martin *et al.* (1966, 1972) found a general order of effectiveness Cu > Zn, Fe > Al. The effects are difficult to explain because the order is not related to solubility or the inhibition of enzymes. Possibly more than one factor is operative. A difference in the rate of decomposition of two bacterial lipopolysaccharides was attributed to the formation of complexes between the more stable material and both metal ions and clays (Cortez, 1977).

The accumulation of organic matter that generally occurs in soils developed on granitic parent materials, which usually have a high concentration of free aluminium ions, has been accounted to immobilization of the organic matter by reaction with the aluminium (Carballas *et al.*, 1978). In laboratory experiments using such soil, the addition to the soil of aluminium as a hydroxide gel has been reported to decrease the decomposition of glucose and maize straw during the first 22 d incubation in soil by 30 and 42% respectively.

In conclusion it is envisaged that many factors contribute to the stability of polysaccharide in soil but particularly chemical linkage to biologically stable humic substances.

DECOMPOSITION AND TRANSFORMATION OF SACCHARIDES

Glucose

The decomposition of glucose in soil has been studied more than that of any other substrate because glucose is such a common biological metabolite.

Decomposition is very rapid; at the 1% addition level almost all the glucose disappears within 5–7 days at normal temperatures in the temperate regions and the initial rate of decomposition is governed by first order kinetics, that is to say it is proportional to the amount of glucose present. (Mayaudon, 1971). By measuring changes in respiration rate Drobnik (1960 a,b) observed that the decomposition of glucose, added to soil batchwise, occurred in two stages. In the first stage, which lasted while glucose was still present, he distinguished a primary oxidation which had two components. In the first, glucose was oxidized to CO_2 and in the second, glucose or its partially oxidized products were assimilated. In the second stage, when glucose was no longer present, a secondary oxidation occurred. A thorough investigation of the way glucose is catabolized in soil has been carried out using glucose labelled in the 1, 2, 3, or 4, or 6 positions or uniformly (Mayaudon, 1968). This showed that glucose is oxidized to CO_2 predominantly by the pentose phosphate cycle and the Embden-Meyerhof-Parnas cycle (glucose→fructose diphosphate→glyceraldehyde phosphate→acetyl CoA) (Table 5.5). The

TABLE 5.5 *The proportions (%) of glucose metabolized by various pathways in soil (Mayaudon, 1968).*

	Unfertilized soil	Fertilized soil
Pentose phosphate	48	42
Embden–Meyerhof–Parnas	30	33
Entner–Doudoroff	0	14
Krebs	22	10

Embden-Meyerhof-Parnas cycle and Entner-Doudoroff cycle (6-gluconate→2-oxo-3-deoxy-6-gluconate) are mutually exclusive in microorganisms. The decomposition of glucose in soil has also been studied by continuous addition in a flow of water (Macura and Kunc, 1961). The proportion of the glucose mineralized to CO_2 was found to be higher when NH_4^+ and PO_4^- were added, as has also been observed by Stotzky and Norman (1961) using batch additions of glucose. Macura and Kunc (1961) concluded that when glucose was continuously added alone, its conversion into a reserve material stored in the microbial cells predominated over oxidation to CO_2. Huntjens and Albers (1978) reported that even less glucose carbon was oxidized to CO_2 during continuous addition of glucose when the soil was nitrogen deficient.

Other studies have shown that glucose carbon which is not oxidized to CO_2 is well distributed amongst the various components of the soil organic matter and reflects the way in which microbial matter fractionates. The distribution of radioactivity remaining in the soil after a 60-d incubation is shown opposite.

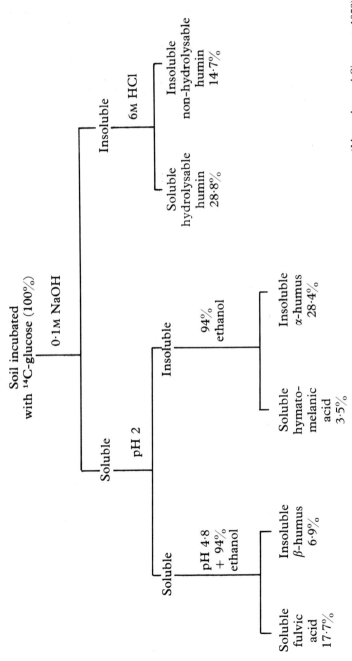

Soil incubated
with ^{14}C-glucose (100%)

0.1M NaOH

Soluble

pH 2

Soluble

pH 4.8
+ 94%
ethanol

Soluble
fulvic
acid
17.7%

Insoluble
β-humus
6.9%

Insoluble

94%
ethanol

Soluble
hymato-
melanic
acid
3.5%

Insoluble
α-humus
28.4%

Insoluble

6M HCl

Soluble
hydrolysable
humin
28.8%

Insoluble
non-hydrolysable
humin
14.7%

(Mayaudon and Simonart, 1958)

Taking into account the total carbon in each fraction, the material which could be dissolved by hydrolysis of the humin had the highest specific activity. Martin *et al.* (1974) found that on incubating glucose in a garden soil for 42 d, 63% of its carbon had been evolved as CO_2. Of the remainder, 2·8% was soluble in 0·1M HCl, 13·2% was in the humic acid, 7·0% in the fulvic acid and 15·3% in the humin. The ^{14}C material in three of these fractions was further characterized by chemical hydrolysis and separation into carbohydrate and amino acid fractions (Table 5.6). In a soil incubation conducted over a similar period by Guckert *et al.* (1970; 1971), 18·7% of residual ^{14}C activity in the humin fraction was present as sugar but, in contrast to the data of Martin, about 66% of the carbohydrate derived from the glucose was in this fraction. Persson (1968) argued that none of the chemical methods of fractionating humus differentiate between materials of different biological character. In his experiments, after the initial rapid mineralization of glucose, the residual labelled carbon was mineralized very slowly, 15–20% remaining after 3 years. He compared the relative specific activity of various fractions after 28 d incubation and found that generally the more soluble fractions had higher specific activities (Table 5.7).

Others have shown that a large proportion of the sugars labelled after incubation of soil with glucose is present in polysaccharide which can be extracted with hot water, alkali or acid (Keefer and Mortenson, 1963; Cheshire *et al.*, 1975; Wagner and Tang, 1976). The individual neutral monosaccharides obtained by complete hydrolysis of a Scottish soil accounted for about 10% of the radioactivity of the glucose substrate after 56 d and about 3% after 448 d incubation (Cheshire *et al.*, 1971). The transformation of

TABLE 5.6 *Distribution of* [14]C *in organic matter fractions after incubation with radioactively labelled glucose for 42 d (Martin et al., 1974).*

Fraction	[14]C as % of original[a]		
	Humic acid	Fulvic acid	Humin
Carbohydrate	2·3 (17·4)	5·1 (72·9)	1·9 (12·4)
Amino acid	8·5 (64·4)	0·4 (5·7)	0·6 (3·9)

[a] Figures in parenthesis are % of residual activity.

TABLE 5.7 *Specific activity of the carbon in three sequential fractionation treatments of soil incubated with ^{14}C-glucose (Persson, 1968).*

Fraction	Relative specific activity
Treatment A	
Cold 0·25M H_2SO_4 extract	1·41
Hot 0·5M NaOH extract	1·00
Residue	0·81
Humic Acid	0·55
Treatment B	
$Na_2P_2O_7$ extract	1·29
Residue	0·89
Treatment C	
Ethanol + benzene extract	2·31
0·275M H_2SO_4 extract	1·51
1·75M H_2SO_4 extract	1·58
12·5M H_2SO_4 extract	0·56
Humic Acid	0·18
Residue	0·37

glucose also gives rise to oligosaccharides such as raffinose, cellobiose and maltose (Macura and Kunc, 1961). The distribution of radioactivity amongst the neutral sugars on hydrolysis of the soils shows that hexose and deoxyhexose synthesis usually predominates (Cheshire *et al.*, 1973) (Table 5.8). A number of organic acids of low molecular weight, including gluconic and 2-ketogluconic acid are also synthesized (Louw and Webley, 1959; Webley and Duff, 1965).

Other monosaccharides

Comparatively little work has been done with monosaccharides other than glucose but their decomposition and transformation in soil is likely to be very similar to that of glucose. For example the rate of decomposition of 1% xylose in soil at 20°C has been found to be almost identical: after 224 d at 20°C, 77·1% of the xylose carbon had been evolved as CO_2 compared with 78·8% for glucose, and the distribution of activity in the monosaccharides of the hydrolysates was very similar (Table 5.9). The addition of glucose to mixtures of

TABLE 5.8 *Amounts and specific activities of sugars in soil during incubation with* [14]*C-glucose (Cheshire et al., 1971).*

	Specific activity (μ Ci g^{-1})			
	0 d	56 d	224 d	448 d
Glucose	147	25·2	8·5	8·7
Galactose	1·4	20·6	9·7	10·0
Mannose	3·5	21·6	10·3	10·8
Arabinose	1·8	3·9	2·2	3·0
Xylose	0·4	3·5	2·2	2·9
Rhamnose	1·7	18·6	19·4	15·9
Fucose	2·1	17·0	9·7	10·0
	Amount (mg g^{-1} soil)			
Glucose	13·4	5·5	4·1	4·1
Galactose	1·6	1·6	1·4	1·2
Mannose	1·4	1·5	1·1	0·9
Arabinose	1·2	1·0	0·9	1·2
Xylose	1·4	1·4	0·9	1·0
Rhamnose ⎱ Fucose ⎰	0·7	0·8	0·5	0·5

TABLE 5.9 *Percentage distribution of radioactivity in six sugars after 112 d incubation of soil with* [14]*C glucose or* [14]*C xylose (Cheshire et al., 1971).*

Sugar	Distribution of radioactivity (%)	
	^{14}C-glucose substrate	^{14}C-xylose substrate
Glucose	51·0	59·0
Galactose	16·6	19·0
Mannose	19·4	12·6
Arabinose	3·8	2·3
Xylose	4·0	3·8
Rhamnose	5·2	3·3

soil with galactose or lactose has been found to have a sparing action on these sugars, thought to be caused by inhibition of enzymes by metabolic products from the glucose (Macura and Kubatova, 1973).

Sucrose

Many soil microorganisms decompose sucrose to its constituent monosaccharides glucose and fructose, but some, such as *Bacillus subtilis*, synthesize levans, which are fructose polymers. Yet others give rise to dextrans (Quastel, 1965).

DECOMPOSITION AND TRANSFORMATION OF CARBOHYDRATE IN WHOLE PLANT MATERIAL

One method of measuring the rate of disappearance of carbohydrate substances from plant materials during incubation with soil has been by their selective extraction at intervals. For example, between 46 and 71% of the hot-water-soluble components of the leaves of four trees—beech, birch, maple, and poplar—was lost during incubation of soil over 10 months (Coldwell and Delong, 1950) and about 66% of the water-soluble components of immature Sann hemp were lost within 15 d incubation (Table 5.10). Mature plants lost less of each

TABLE 5.10 *Decomposition of Sann Hemp during incubation in soil (Ghildyal and Gupta, 1959).*

Incubation period (d)	Loss (%)		
	Water-soluble substances	Pentosans	Cellulose
	3-week-old plants		
15	66	60	50
120	85	—	90
	7-week-old plants		
15	51	50	41
120	79	64	75

component over the same period, indicating either a change in quality of the substances or the accompanying plant components. In a study more specifically related to carbohydrates, the decomposition in Arousa leaves was measured by hydrolysis and analysis of sugars in extracts of a soil–leaf mixture (Jain and Bhattacharya, 1960). After 3 months incubation more carbohydrate was lost from the alcohol-soluble fraction than from the hemicellulose fraction and water-soluble extracts (Table 5.11).

Decomposition studies such as those described above, which depend on solubility to define the plant components under study, are liable to error because of changes in solubility brought about by microbial attack. For example using ^{14}C-labelled fractions of plants it has been shown (Sørensen, 1963) that activity initially present in hemicellulose or cellulose fractions of straw is, after incubation, found distributed in all the soil organic matter fractions, with various solubilities corresponding to the various carbohydrate fractions of plants. The disappearance of carbohydrate from incubation mixtures at least allows an estimate to be made of the minimum decomposition that has occurred. Minimum decomposition rates for the carbohydrates in litter–soil mixtures were studied by Sowden and Ivarson (1962a) using coniferous and deciduous litter. Up to 6 months, the loss of sugars, 9 and 14% respectively, paralleled the loss of carbon, but after that time the loss of sugars became proportionally greater. In the coniferous litter the proportion of the various sugars remained fairly constant during the first 6 months, but thereafter there was a relative decrease in the glucose, xylose and uronic acid contents. Galactose was the most persistent of the sugars. In the deciduous litter, of which 80% was maple leaves, glucose, xylose and uronic acid decreased over

TABLE 5.11 *Decomposition of the carbohydrate of Arousa leaves during incubation in soil (Jain and Bhattacharya, 1960).*

Leaf fraction	Loss (%)
Alcohol-soluble	97·7
Cold water extract	33·3
Hot water extract (pectin)	28·1
5% NaOH extract (hemicellulose)	55·0

TABLE 5.12 *Effect of incubation on the amounts of sugars present in a deciduous litter–soil mixture (Sowden and Ivarson, 1962a).*

	0 d	(mg g^{-1}) 165 d	1200 d
Galactose	30	34	24
Glucose	258	135	92
Mannose	16	19	17
Arabinose	27	32	25
Ribose	2	2	2
Xylose	77	57	22
Rhamnose	15	16	14
Uronic acid	82	53	24

the whole incubation period (Table 5.12). On the assumption that the glucose content in excess of the values for mannose and galactose is a measure of that in cellulose, about 75 and 83% of the cellulose in the coniferous and deciduous litters, respectively, were decomposed after 3 years.

The loss of uronic acid contrasted with previous results for incubation of plant materials, reported by Tepper (1957) and Springer (1959), where the proportion of uronic acid appeared to increase with time. It is very likely that it was not uronic acid that was being measured in these earlier experiments, but easily decarboxylated humic substances.

Soil pH conditions may favour the growth of particular organisms and this may be reflected in the different rates of decomposition of added substrate. For example, more cellulose and hemicellulose were decomposed to CO_2 at pH 7 than at pH 4 (Table 5.13), indicating that bacteria were responsible. The similarity in the rates of decomposition of pectin observed for the two pH values suggested that it is prone to attack by fungi which, unlike most bacteria, thrive under such acid conditions.

It should now be obvious that chemical analysis of residues from natural substrates can give only limited information, but the actual fate of plant components in soil may be resolved by the use of radioactively labelled substrates. Transformations of individual sugars in whole plant material, on incubation with soil, have been

TABLE 5.13 *Effect of soil* pH *on the rate of decomposition of carbohydrates (Benoit and Starkey, 1968).*

Carbohydrate	Soil pH	% decomposed to CO_2 in 3 weeks
Cellulose	7·0	20
	4·0	8
Hemicellulose	7·0	63
	4·0	49
Pectin	7·0	67
	4·0	66

followed using ^{14}C-labelled cereal rye straw. An attempt was made to distinguish the glucose in cellulose by differential hydrolysis and, as incubation under similar conditions with ^{14}C-labelled glucose had shown little if any synthesis of xylose, residual activity in this sugar was regarded as being from undegraded xylan in the plant material. On this basis, rates of decomposition for cellulose and xylan were obtained (Table 5.2) which showed that over 20% of both polysaccharides still remained after 16 months incubation of soil mixed with 0·2% cereal rye (Cheshire *et al.*, 1973). When whole fresh young ryegrass plants were incubated in soil for 1 year similar rates of decomposition of the predominant sugars were observed. The percentage of glucose, arabinose and xylose remaining were 18, 22 and 26% respectively (Cheshire *et al.*, 1979b).

The picture obtained by chemical analysis appears to be confirmed by microscopic observations on the decomposition of ^{14}C-labelled leaves (Grossbard, 1969). Autoradiography indicated a piecemeal decomposition with the complete disappearance of small sections of leaf.

The residual plant materials probably retain much of their original physical structure and, because soluble components are most easily decomposed, it is not surprising that the distribution of activity amongst the sugars in the humin fraction of a soil has been found to be closer to that in the straw substrate than to that in the soluble fulvic polysaccharide (Guckert, 1972). Table 5.14 summarizes results of investigations of the persistence of plant tissue components in soil

TABLE 5.14 *Loss of CO_2 from various plant polysaccharides upon incubation.*

	Incubation period	Substrate evolved as CO_2 (%)	Reference
Holocellulose	6 weeks	62	Martin *et al.* (1974)
Cellulose	1·5 years	13·9	Sørensen (1974)
		13·4	
	3·3 years	23·9	
Cellulose	10 d	25–40	Sørensen (1975)
	30 d	48–66	
	90 d	55–73	
Cellulose	118 d	61	Sørensen (1963)
Hemicellulose	118 d	65	

using ^{14}C-labelled whole plant material, and measuring the loss of [14]CO_2.

It must be stressed that these values only indicate the minimum decomposition that has occurred. It will be seen from Table 5.2 that the decomposition of hemicellulose is 70% after 448 d as measured by loss of CO_2, whereas the loss of the major component of the hemicellulose, the xylan, is about 98%.

Paul and van Veen (1978) have calculated first order decomposition rate constants for numerous organic substrates, making the assumption that microorganisms retain a proportion of the metabolized substrate carbon within their tissue. For many substrates, more accurate values must surely be obtained by applying this correction. They recognize the difficulties of measuring residual substrate when it is similar to newly synthesized microbial products as is the case with many carbohydrate materials. Using [14]CO_2 evolution, first-order rate constants for cellulose were only slightly lower than for hemicellulose over the first 10–14 d incubation in soil, but a closer examination of the hemicellulose decomposition reveals that although 67·9% of the initial ^{14}C activity remains in the soil only 19% of the xylose, the chief constituent, has survived (Cheshire *et al.*, 1974b). A corresponding value for the chief constituent of cellulose, glucose, is not available for the same period, although after 56 d, 44% is still present in cellulose-like form. This example illustrates how materials which decompose at different rates may appear to have similar stabilities in soil because of the recycling of substrate carbon by microorganisms.

DECOMPOSITION AND TRANSFORMATION OF THE CARBOHYDRATE OF MICROORGANISMS

In incubations of ^{14}C-glucose in soil there is a rapid growth of labelled microflora. The subsequent transformations of the radioactive substances may be taken to indicate the fate of microbial tissue in soil. Usually it is more labile than the native soil polysaccharide. Many polysaccharides produced by microorganisms are rapidly degraded in soil, for example fructosans (Martin and Richards, 1963), but others are as resistant as those in lignified plant tissue. The stability of the polysaccharides from *Chromobacterium violaceum* and *Azotobacter indicus* probably depends on a polysaccharide structure which is particularly resistent to enzymatic attack as well as on interaction with other soil components through the carboxyl groups of the constituent uronic acids (Martin *et al.*, 1965).

A radioactively labelled lipoprotein polysaccharide complex from *Pseudomonas fluorescens* containing glucose, rhamnose, fucose, glucosamine and N-acetylglucosamine lost all the fucose on incubation with soil for 100 d (Mayaudon and Simonart, 1965) whereas the other sugars were still strongly labelled. Light coloured fungal hyphae are more prone to decomposition in soil than those with a dark colour (Webley and Jones, 1971) and this is thought to be because of protection by melanin. Some other fungi appear to resist lysis because of an unusual carbohydrate composition involving fucose (Pengra *et al.*, 1969).

Chemical degradation

The non-enzymic decomposition of polysaccharides by acid or alkaline conditions is much slower than enzymic decomposition. Nevertheless, because of the long residence time individual polymers have in the soil it may be relatively important. In acid conditions, furanoside polymers would be liable to hydrolysis (Finch *et al.*, 1971). In alkaline conditions sugar residues at the ends of polymers are successively removed as saccharinic acids.

THE ORIGIN OF SOIL POLYSACCHARIDE

The first step in considering the origin of soil polysaccharide was to compare its nature and composition with those of the polysaccharides contributed to soil by various organisms. The chief difficulty arose from ignorance about the polysaccharides in the organisms. Methods of analysis were not always sufficiently sensitive to show the presence of minor sugar components which may nevertheless accumulate in soil, whilst the more obvious major components are more rapidly oxidized. At one time, for instance, it was considered that the presence of fucose indicated a microbial origin (Duff, 1952a) because this sugar had not been found in plant tissue (Bacon and Cheshire, 1971). Similar arguments are still being used about the mannose in soil (Folsom *et al.*, 1974) although it is now established that mannose is a major component of the cell-wall polysaccharides of plants (Albersheim, 1965). The soil polysaccharide also contains very small proportions of 2-*O*-methyl rhamnose, 2-*O*-methylxylose, 3-*O*-methylxylose and 4-*O*-methylgalactose which have all been found in plants (Aspinall and McKay, 1958; Bacon and Cheshire, 1971). These may have gradually accumulated in soil from plant sources. Although Forsyth (1954) considered it inconceivable that methyl sugars were derived from vegetation, an examination of many soil organisms was necessary before one was found which produced a sugar with an R_F value and a colour reaction similar to one of the methyl sugars of soil (Forsyth and Webley, 1949 a,b; Forsyth, 1954). Latterly 3-*O*-methyl-*D*-xylose has been found in two soil organisms (Weckesser *et al.*, 1971). Such methyl sugars are probably widespread in both plants and microorganisms. It is most likely that soil glucosamine and galactosamine originate in microorganisms and animals, where these sugars have a major structural role. Glucosamine is present in the chitin of fungi whereas the galactosamine probably derives from bacterial cell-wall polysaccharides. Nevertheless these amino sugars are often minor components in plants and no definite conclusion about their origin in soil can be reached from such considerations. The occurrence of sugars in soils, plants and microorganisms is shown in Table 5.15.

The only polysaccharide in soil which has yet been identified is cellulose. This probably originates from plants but it could also be synthesized by microorganisms (Hepper, 1975; Deinema and Zevenhuisen, 1971). It has been pointed out that soil polysaccharide is somewhat similar in composition to the hemicellulose of plants (Bacon, 1967). The similarity to a holocellulose is shown in Table 5.16. The high proportion of galacturonic acid in soils, a major component of the pectic fraction of plant polysaccharide, indicates

TABLE. 5.15 *Occurrence of sugars in plants, soils and microorganisms.*

	Plants	Soils	Microorganisms
Glucose	+	+	+
Galactose	+	+	+
Mannose	+	+	+
Arabinose	+	+	+
Xylose	+	+	+
Ribose	+	+	+
Rhamnose	+	+	+
Fucose	+	+	+
Fructose	+	+	+
Glucuronic acid	+	+	+
Galacturonic acid	+	+	+
4-O-methyl glucuronic acid	+		+
Glucosamine	+	+	+
Galactosamine	+	+	+
Apiose	+		
2-O-methylrhamnose	+	+	
3-O-methylrhamnose	+		
2-O-methylfucose	+		
3-O-methylfucose	+		
2-O-methylxylose	+	+	
3-O-methylxylose	+	+	+
2-O-methylgalactose	+		
3-O-methylgalactose	+		
4-O-methylgalactose	+	+	
Muramic acid			+
Fucosamine			+
Neuraminic acid			+
Sulphated polysaccharides		+	+

TABLE 5.16 *Relative proportions of sugars in soils and holocellulose (Bacon, 1967).*

	Soil under grass	Fallow soil	Holocellulose from *Tilia* leaves
Galactose	37	26	14
Glucose	100	100	100
Mannose	32	21	4·2
Arabinose	26	29	15
Xylose	32	24	22
Fucose	5·2	5·9	2·1
Rhamnose	16	12	3·5

the presence of the latter. However, 4-O-methyl glucuronic acid, which is a common component of plant hemicelluloses, has not been observed in soil hydrolysates.

A detailed study of the sugar composition of peat, and of the plants from which it is derived has been made at two sites (Lucas, 1970) and illustrates the difficulties of interpreting the composition in terms of origin. In one, the surface litter layer reflected the composition of the plants in being rich in glucose and xylose. With increasing depth, usually meaning increasing age, there was an increase in galactose, mannose and particularly rhamnose and these sugars were considered to have been derived from microorganisms. At the second site the surface litter had a more mixed composition but with increasing depth glucose and xylose became predominant. At a depth of 3.6 m there was two or three times as much xylose as any other sugar. The composition of various fractions of the peats varied considerably. The soluble fraction of the lower horizons of one peat appeared to be particularly rich in ribose whilst in the other the proportion of mannose was very high. The fulvic acid fraction of the peats both had a high proportion of uronic acid and glucose in the form of cellulose formed a predominant part of the humin fraction. Lucas concluded that although some plant polysaccharides survive in the peat a substantial fraction of the carbohydrates is microbial in origin, as shown by the relatively high abundancies of the mannose, rhamnose, fucose and ribose.

Analysis of the bond linkages present in the fraction of soil polysaccharide which is soluble in alkali (Table 4.29, Cheshire et al., 1979a) shows that in the hexose components $(1\rightarrow3)$ and $(1\rightarrow4)$

linkages predominate and are present in very similar amounts. Both
$(1\rightarrow6)$ and $(1\rightarrow2)$ linkages are also present, and 21% of hexose
residues form branching points (Table 4.30). $(1\rightarrow3)$ is the most
common linkage in fungal cell-wall glucans. Other hexose linkages
frequently occurring in microorganisms are $(1\rightarrow4)$ and $(1\rightarrow6)$ and
there is often branching. Deoxyhexose sugar residues in the soil
polysaccharide showed an even greater degree of branching, 24%.
Most of the pentose residues in the soil polysaccharide are present in
$(1\rightarrow4)$ linkages and a much lower proportion are at branching points
(11%). This would correspond to the $(1\rightarrow4)$ xylan in plant matter, but
xylose in $(1\rightarrow4)$ linkage is also a constituent of some yeasts. Thus
much of the polysaccharide resembles microbial products, but
products of a plant origin are not excluded.

On the subject of composition, only a very small proportion of the
organisms isolated from soil are capable of synthesizing polysaccharides with an overall complexity as great as soil polysaccharides
(Cheshire et al., 1976; Webley et al., 1965). Bacteria and fungi usually
have three or four different types of sugar in their polysaccharide
(Stacy and Barker, 1960; Salton, 1964; Gorin and Spencer, 1968) and
only algae commonly have the six or seven found in soil polysaccharide. The number of algae in soil is small, however, $(10^5 \, g^{-1})$, and
although individually they are large in relation to bacteria, the
relative proportions by weight in the surface layers of soil is only
about 1:100 bacteria.

Another approach to the problem of the origin has been to consider
evidence for a particular source of soil carbohydrate from examination of the composition of soils with known histories. It has been
calculated that carbohydrate in the residues of plants added to soil in
1 year is equivalent to about 50% of the carbohydrate content of the
soil and in addition large quantities of carbohydrates are added to
soil from plant roots during the life of the plant. These carbohydrates
have been thought to be exudates but some studies have suggested
that they arise from autolysis of root cortex cells (Martin, 1977).
Where there is an unusually large addition of plant material, or an
accumulation, it would be expected that this might be reflected in the
composition of the soil polysaccharide, assuming plant polysaccharides to be different from those of microorganisms. Differences
between soils have been observed by Oades (1972) who found higher
proportions of glucose, arabinose and xylose in Australian soils rich

in plant remains. On the other hand little difference could be detected in the relative proportions of the sugars in hydrolysates of Rothamsted soils kept fallow for 10 years, when these were compared with adjacent cropped soils (Cheshire and Anderson, 1975).

Microbial polysaccharides may not be the only alternative to those of plants or animals in soils. Various workers have observed the synthesis of oligosaccharides from monosaccharides in sterilized soil (Kiss and Peterfi, 1959; Hoffman, 1963) and it has been suggested that enzymes in soil may synthesize polysaccharide using free sugars at random as these become available (Haworth et al. 1946; Martin, 1971). This would create diverse polysaccharides atypical of any organism and could explain the apparently rather unique composition.

The most rewarding studies on the origin of soil polysaccharide have come from the use of labelled substrates in biosynthetic studies. Simple ^{14}C-labelled compounds such as glucose have been used in incubation experiments and these clearly show that the microorganisms are capable of synthesizing the common soil sugars (Keefer and Mortensen, 1963) for, after a few weeks, label is present in all the different sugars in the alkali soluble dialysed polysaccharide. But when the distribution of the label amongst the various sugars is measured quantitatively (Cheshire et al., 1969; Oades and Wagner, 1970; 1971) the synthesized polysaccharide is shown to be deficient in the pentoses xylose and arabinose when compared with native soil polysaccharide (Table 5.17). These observations are supported by studies on the composition of individual microbial products which showed that hexoses are the predominant constituents of polysaccharides produced by microorganisms of the rhizosphere of grasses (Webley et al., 1965) and bacterial and filamentous fungal isolates of soils (Cheshire et al., 1976). It therefore appears that overall composition of the microbial polysaccharides is different from that of soil polysaccharide and can only account for a part of it. Calculation shows that the live biomass can account for no more than 2·5% of the soil carbohydrate. It might be argued that only with the passage of time will the composition of the microbially synthesized polysaccharide come to resemble the native soil polysaccharide more closely. When incubation with ^{14}C-glucose are continued over a long predominance of labelling in the hexoses is still maintained,

TABLE 5.17 *Comparison of the relative proportions of sugars in soil polysaccharide with those synthesized from ^{14}C-glucose (Cheshire, 1977).*

	Relative proportion	
	Native	Synthetic
Glucose	100	100
Galactose	39	34
Mannose	34	27
Arabinose	29	10
Xylose	34	8
Rhamnose	15	16
Fucose	5	4

(Table 5.8), the most stable of the synthesized sugars being the deoxyhexoses rhamnose and fucose. Even after 16 months the labelled microbially synthesized polysaccharide is being decomposed at a faster rate than the native soil polysaccharide.

The microbially synthesized polysaccharides also differ from the native soil polysaccharides in the way they fractionate. In the first example to be described, fractionation was by anion exchange chromatography (Chahal, 1968). A soil was incubated with labelled glucose at its natural pH of 4·75 and also after adjustment to pH 7·4 and 8·9. The isolated polysaccharide was applied to a column of DEAE cellulose and eluted with a phosphate–chloride buffer and a concentration gradient of NaOH. Four fractions were isolated, varying in activity, the two most active being eluted last (Table 5.18). The hexoses galactose, glucose and mannose were usually the most strongly labelled sugars in the fractions.

In a second example after a similar incubation experiment Oades

TABLE 5.18 *Relative specific activity of polysaccharide fractions from soils incubated with glucose (Chahal, 1968).*

pH of soil	Fraction 1	Fraction 2	Fraction 3	Fraction 4
4·75	0·95	0·50	2·02	1·93
7·4	0·25	0·06	1·39	0·66
8·9	0·09	0·12	1·59	1·46

TABLE 5.19 *Proportion of labelled sugars derived from glucose in the humin (Guckert et al., 1971).*

Incubation period (weeks)	% of ^{14}C sugars in the humin
1	33·2
2	37·4
6	38·6

(1974) found that sugars with the greatest activity occurred in a fulvic acid fraction of very low molecular weight. It is notable that xylose was one of the strongly labelled sugars in this fraction, although insufficient data is provided to calculate the overall proportion in the soil.

Thirdly, it was observed that approaching 40% of the residual radioactivity in sugars was present in the soil humin after incubating ^{14}C-glucose with soil for up to 6 weeks (Table 5.19) whereas the humin also contained 60–70% of the native soil sugars. Once again a greater proportion of the ^{14}C label was found to be present in the hexoses and deoxyhexoses than in the pentoses arabinose and xylose.

Most of the experiments using ^{14}C-labelled glucose have involved incubation within the temperature range of 20–30°C which is an unnaturally high temperature for temperate regions. At natural Scottish winter temperatures, using fresh moist soil, the distribution of radioactivity amongst the sugars is similar to that found at 20°C (Cheshire et al., 1978) but using predried remoistened soil there is a synthesis of xylose equivalent to that of the hexoses mannose and

TABLE 5.20 *Relative proportion of sugars derived from-glucose during incubation at natural winter temperatures with predried soil.*

	Relative proportion
Glucose	100
Galactose	14
Mannose	10
Arabinose	1
Xylose	11
Rhamnose	1

TABLE 5.21 *Effect of temperature on the numbers of various organisms in soil incubations using remoistened soil (Cheshire et al., 1978).*

	Initial count (number g^{-1})	Final count (number g^{-1})	
		14 d 5°C	7 d 25°C
Bacteria	$8 \cdot 67 \times 10^4$	40×10^4	6920×10^4
Fungi	$<0 \cdot 1 \times 10^4$	$8 \cdot 0 \times 10^4$	200×10^4
Yeasts	$<0 \cdot 1 \times 10^4$	135×10^4	76×10^4

galactose (Table 5.20). The xylose synthesis is thought to be caused by yeasts, because under these conditions yeasts thrive and bacteria appear to be inhibited (Table 5.21) and 85% of isolated yeasts synthesized xylose-containing polysaccharide when cultured on glucose or sucrose, compared with only 3% of the bacteria and none of the filamentous fungi. It would seem unlikely that these particular conditions of drying, cold and the ready availability of simple metabolites are ever coincidentally present to allow a predominant growth of yeasts to occur. Nevertheless yeasts are present in the soil and over a long period may significantly contribute xylans to the soil polysaccharide. The only microorganisms found in soil to date which can synthesize all the sugars common to the soil polysaccharide, are the algae. Incubation of surface soil in indirect sunlight to encourage algal growth, has, however, little effect on the pattern of labelling of the sugars during the incubation of glucose. Other soil conditions which have been varied but found to have little influence have been pH and degree of aeration.

The kind of substrate may determine the nature of the substances formed. In incubations with labelled glucose the relatively poor distribution of ^{14}C amongst the amino acids compared with the sugars indicates that the six carbon chain of glucose is preserved in the transformation to other hexoses and deoxyhexoses (Table 5.22); but neither acetate nor xylose nor starch (Cheshire *et al.*, 1969; 1971) nor dextran substrates (Oades and Wagner, 1972) gives rise to a different pattern of labelling.

Good evidence for the microbial synthesis of the amino sugars glucosamine and galactosamine in a soil during incubation has been

TABLE 5.22 *Distribution of radioactivity after incubating ^{14}C-glucose in soil (Cheshire et al., 1971).*

	Specific activity (μ Ci g^{-1}C)
Glutamic acid	10·8
Asparatic acid	2·2
Leucine	2·9
Glycine	4·7
Threonine	3·2
Valine	3·3
Proline	3·1
Alanine	4·2
Serine	1·2
Arginine	6·3
Lysine	4·7
Phenylalanine	7·0
γ-amino butyric acid	1·3
Tyrosine	0·8
Histidine	4·5
Glucose	65·0
Galactose	37·2
Mannose	29·2
Arabinose	6·2
Xylose	7·2
Rhamnose	16·2
Fucose	17·0

obtained by Marumoto *et al.* (1974). They showed that label was associated with these two amino sugars in the hydrolysate of a soil which had been incubated with ^{14}C-ryegrass at 30°C for 6 months but not in the hydrolysate of the substrate itself.

The facts presented so far, clearly show that microbial synthesis can account only for a part of the soil polysaccharide. Add to this the known considerable persistence of plant material described in the previous section and the conclusion is inescapable that the carbohydrate is a mixture of microbial and plant polysaccharides, most of those containing xylose and arabinose being of plant origin and those containing galactose, mannose, rhamnose, fucose, galactosamine and glucosamine of microbial origin. A large proportion of the glucose probably persists as plant cellulose but much of the remainder will be in microbial products.

It is interesting to consider how the conclusion fits in with the composition of fractions described in pp. 95–105). Although the chemical methods of fractionation provide scant evidence of any distinction between plant and microbial polysaccharides, physical separation on the basis of density does indicate a relationship between the amounts of xylose and arabinose and another between the amounts of galactose and mannose (Table 4.12).

CONCLUSION

Soil scientists tend to treat the carbohydrate in soil as if it were part of a discrete polysaccharide entity, with uniform composition. This is quite understandable when one considers the uniformity of composition of the extractable polysaccharide material, the difficulty of fractionation, together with the relatively simple constituent building blocks of which it is made. This artificial concept has been used throughout this work because of the need to report and discuss the collective behaviour and properties of soil polysaccharide but it should be remembered that the substance must be a very great mixture of polymers from both living and dead organisms.

Chapter 6

The Significance of Soil Carbohydrate

SOIL STRUCTURE AND SOIL POLYSACCHARIDE

Good soil structure is important for soil aeration, root development and ease of cultivation (Quastel and Webley, 1947; Low, 1973). It may also prevent soil erosion losses (Greenland, 1971). There is still much to understand about the various mechanisms involved in the aggregation of soil particles and the contribution made by soil organic matter, particularly polysaccharide. A number of reviews of this subject have been compiled (Martin *et al.*, 1955; Low, 1962; Griffiths, 1965; Guckert, 1968).

When soil structure first began to be studied it was soon appreciated that for many soils there was a relationship between the total organic matter content and the degree of aggregation. Baver (1940) obtained correlation coefficients of about 0·65–0·75 for aggregates less than 0·05 mm diameter and organic carbon for more than 70 soils. This relationship has become more obvious recently on certain sandy soils where, because of a change in farming practice from mixed animal and arable production to continuous arable cropping, the level of organic matter in the soil has decreased to less than 2%, and a breakdown of soil structure has occurred. A good crumb structure may be restored by a period under grass, which probably owes its effectiveness to changes associated with an increase in the organic matter content of the soil brought about by the large number of very fine roots the grass produces (Low, 1962).

The addition of plant material as a manure increases the organic matter content of soils and also the aggregation (Martin, 1945), but it has been found that the organic matter with greatest effect is that which is most readily decomposed. In sterilized soil there is little, if any, change in aggregation on incubation (Martin and Waksman, 1940; Peele, 1940; McCalla *et al.*, 1957: Martin *et al.*, 1959). With unsterilized soil containing fermentable organic matter, aggregation rapidly reaches a maximum and then slowly decreases. This pattern of change is characteristic of microbial activity and so Martin (1945) examined the effects of incubating particular organisms in previously sterilized soil with simple substances such as sucrose. He found that ten out of the 11 organisms tested had an effect, increasing the aggregation from 28% in the control to 67% in the most effective. The effects of two organisms, *Bacillus subtilis* and *Cladosporium*, were studied further. When the culture fluid and cells of the bacterium *B. subtilis* were added to soil, aggregation was increased. When the bacterial cells were separated from the culture fluid and the nondialysable matter isolated so that each part could be tested separately, it was found that the dialysed medium accounted for 80% of the effectiveness (Table 6.1). Polysaccharide was shown to be

TABLE 6.1 *Effect of* Bacillus subtilis *cells and culture medium on the aggregation of soil particles (Martin, 1945).*

	Aggregation (%)	
	Sandy loam	Declo loam
Water	26	26
Whole culture	73	78
Cells alone	55	36
Dialysed medium minus cells	66	67

present in the dialysed medium by precipitation with ethanol and the release of reducing sugars on hydrolysis. With the fungus *Cladosporium*, the isolated culture fluid was much less effective, accounting for 50% of the aggregating effect. Since these early experiments the extracellular polysaccharides produced by many different organisms have been shown to have soil particle binding properties (Geohegan and Brian, 1948a,b; Martin and Richards,

1963; Martin *et al.*, 1965; Griffiths and Jones, 1965; Martin, 1971). Baver (1968) has summarized some observations on the general effectiveness of different types of organisms (Table 6.2). The

TABLE 6.2 *Order of effectiveness of types of microorganisms on the stability of soil aggregates (Baver, 1968).*

Fungi	Actinomycetes	Bacteria		Yeasts	Reference
		Gum producing	Other		
1	—		2	—	Peele and Beale (1940)
1	—	1	2	—	Martin (1945)
1	2	3	5	4	McCalla (1946)
1	2	2	3	3	Swaby (1949)
1	—		2	—	Muller (1958)
1	2	1	3	—	Watson and Stojanovic (1965)

relationship between aggregation and soil organic matter content could thus be attributable to the polysaccharide component which forms a remarkably constant proportion of the organic matter.

Following up Martin's findings, an attempt was made by Rennie *et al.* (1954) and Chesters *et al.* (1957) to measure the microbial gum in soils by alkali extraction and isolation using Forsyth's technique or precipitation by acetone, and to relate the amounts found to the degree of aggregation. Figures obtained for four soils indicated a fairly low, statistically significant positive relationship (Table 6.3). Such soil polysaccharides when remixed with soil immediately form stable aggregates. Thus 0·02% of the polysaccharide added to a silt loam soil increased the number of aggregates greater than 0·1 mm diameter by about 50% (Rennie *et al.*, 1954). The greatest increase, to 70% compared with a base value of 45%, was caused by an addition of 0·5%. This compares with a native gum content of 1% for the soil. Whistler and Kirby (1956) tested soil polysaccharide in kaolin–sand mixtures and found it effective in increasing aggregation, particularly at 1% concentration.

These findings present a strong argument for polysaccharide being a major cause of aggregation, and it is relevant that Combeau (1965)

TABLE 6.3 *Correlation between aggregation and microbial gum, clay and Fe contents (Rennie et al., 1954; Chesters, et al., 1957).*

Aggregate diameter (mm)	Partial coefficients of correlation		
	Microbial gum	Clay	Fe
>0·5	0·436	0·118	0·241
>0·25	0·415	0·137	0·195
>0·1	0·260	0·152	0·176

found a greater correlation of structural stability with the non-humified organic matter of soil than with the total. Nevertheless the argument is not all one-sided. Changes in the total polysaccharide in a clay soil during the 4 months of the growing season appeared quite unrelated to changes in the aggregation of the soil (Webber, 1965) whereas there was a relationship between aggregation and the stage of growth. Also Mehta *et al.* (1960) questioned the causal relationship using the argument that the results may be interpreted to show that good aggregation may encourage the growth of polysaccharide-producing organisms. They also pointed out the inefficiency with which the soil polysaccharide 'gum' is extracted, and the high proportion of non-carbohydrate matter it contains. They suggested that a direct chemical treatment be applied to soil aggregates to test the involvement of polysaccharides. Soil aggregates of a Swiss braunerde, 2–4 mm diameter, were therefore subjected to various chemical treatments. Care was needed because of the destruction which could result from the expansion of entrapped air. Treatment of polysaccharide with periodate at room temperature usually produces oxidized polysaccharides which are unstable under alkaline conditions such as in borate buffer at pH 9·6. Synthetic aggregates made with soil polysaccharide or guar gum were completely destroyed by this treatment, but natural aggregates were not. Other differences were found with hot dilute HCl, NaCl and ClO_2 (Table 6.4). It was concluded that the natural aggregation was not caused by polysaccharide. It was argued that most of the native soil polysaccharide is destroyed by the periodate because very little

TABLE 6.4 *Effect of chemical treatments on the stability of soil aggregates (Mehta et al., 1960).*

Treatment	Natural	Synthetic
0·05 M NaIO$_4$	Stable	Unstable
1·0 M HCl 100°	Stable	Unstable
1·0 M NaCl	Unstable	Stable
ClO$_2$	Unstable	Stable

polysaccharide could be extracted from the periodate-treated soil and only a small amount of sugar was left in the residue on hydrolysis (Mehta *et al.*, 1960). The hydrolysis conditions using M HCl at 100°C were rather mild, however, and would not have completely hydrolysed all types of polysaccharide, for example not chitin. Synthetic aggregates made with chitosan were destroyed and it was considered that sufficient hydrolysis would have occurred to make such polysaccharides in natural aggregates soluble, but this cannot be presumed. It was suggested that the effectiveness of ClO$_2$ might be because it could attack humic acid which could act as a binding agent, but it was not possible to identify any one single cause of aggregation. Aggregates in various other Swiss soils showed a similar behaviour.

Greenland *et al.* (1962) extended the use of periodate oxidation to cover a range of natural and cultivated soils and found considerable differences. Aggregates from Australian red–brown earths, under pasture for 4 years or less, were completely destroyed, whereas with older pastures only a small reduction in stability occurred. On the other hand there was agreement with the findings of Mehta *et al.* for a Rendzina soil in which the aggregates were unaffected by periodate. Greenland *et al.* (1961) explained these differences by the hypothesis that the stable aggregation was caused by fungal hyphae which were resistant to periodate. This explanation was engendered by the visual observations of fungal hyphae reported by Bond (1960).

The increased stability of aggregates in soils which have been incubated with organic substrate, may be destroyed with periodate (Harris *et al.*, 1963; Watson and Stojanovic, 1965). These effects would correspond to that found for young pastures. Bond and Harris (1964) summarized these ideas. They considered that the initial development of structure largely depends on microbial gums which produce only a temporary effect. More permanent aggregation is the result of the development of a network of fungal mycelia. In Australian red–brown earths, 6 years under pasture was necessary for the accumulation of sufficient mycelia to give good structure which persists even when the mycelia are no longer viable. Fungal hyphae contain chitin, which is β-(1→4)-N-acetyl-D-glucosamine.

and also (1→3) glucan, and the latter, having no vicinal OH groups, should be stable to periodate oxidation. Whether these polysaccharides play any part in the more permanent aggregation associated with fungal development has yet to be determined.

Here it should be mentioned that Salomon (1962) failed to obtain a correlation between hexosamine content and aggregate size.

More recent studies emphasize that only a part of the structural stability of soil aggregates depends on polysaccharide (Stefanson, 1968; Hamblin and Greenland, 1977) and the proportion varies with soil treatment. Polysaccharides appear to be relatively more important in stabilizing aggregates in soils with low organic matter contents (Greenland et al., 1962; Clapp and Emerson, 1965a; Hamblin and Greenland, 1977). With successive years under pasture a far greater residual stability, unaffected by periodate or pyrophosphate treatment, develops (Fig. 10) (Stefanson, 1971). The concept that this does not involve polysaccharide relies almost entirely on the presumption that periodate is able to destroy the polysaccharide. It is unfortunate that no estimates of residual polysaccharide after periodate treatment have been reported since those given by Mehta and co-workers.

Paul and Kullman (1968) found evidence that a nitrogen-containing glucan was involved in crumb formation but they did not consider that specific substances were responsible, rather that active aggregating substances were characterized by a linear structure, and the ability to form colloids and chemical linkages.

It is not known precisely how polysaccharide causes aggregation. Aspiras et al. (1971) suggest that soil particles adhere to mucilage-covered hyphae. In more explicit terms the polysaccharide is thought to be adsorbed on clay mineral particles and form interparticle bonds (Greenland, 1971). Although each of the bonds may be relatively weak, there are large numbers of them and the overall result is a very stable association. Greenland (1965a,b) finds little evidence for the suggestion that there is hydrogen bonding (Emerson, 1959). A distinction should be made between the polysaccharide associated with living structures, such as fungal hyphae, and the more discrete polysaccharide arising from bacterial or plant sources, despite the

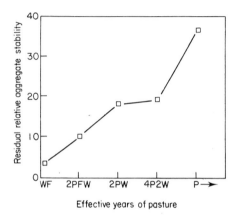

Fig. 10 Relationship between residual relative aggregate stability to periodate and pyrophosphate and the effective years of pasture in a rotation on Urrbrae red–brown earth; W wheat, P pasture, F fallow (Stefanson, 1971).

fact that both may have some types of bonding with soil particles in common. Greenland (1971) attributed the effectiveness of pasture in creating good aggregation to the extensive distribution of fine roots as substrate for the bacterial synthesis of polysaccharide of just the right size in just the right place to form effective interparticle bonds.

Greenland (1965a) suggested that significant adsorption of anionic polymers by clay minerals can occur through bridge linkages involving polyvalent cations. That periodate and the complexing reagent pyrophosphate have a common ability to destroy part of the aggregate stability, suggested to Stefanson (1971) that some of the polysaccharide–mineral bonds may involve polyuronide–polyvalent cation bridges. The idea that the aggregating activity of soil polysaccharide is related to its uronic acid component has frequently been tested, but the results have been inconclusive. Harris *et al.* (1966) and Dhoot *et al.* (1974) found that larger aggregates contained greater amounts of uronic acids, but Saloman (1962) could find no difference and Kullman and Keopke (1961) were unable to relate aggregate stability to the uronic acid content. When bacterial polysaccharides have been used to make synthetic aggregates, the most stable have been found to be those with the greatest viscosity and molecular weight (Geohegan and Brian, 1948) whereas the presence of carboxyl groups in the polysaccharide lowers the stability (Clapp *et al.*, 1962). Although carboxyl groups, being present in native soil polysaccharide, are likely to be involved in aggregation, it should be bourne in mind that the hydroxyl groups of polysaccharides may also react with metal ions to form stable complexes. The increased adsorption of uncharged polysaccharide on particular forms of clay supports the idea that such polysaccharide–metal bonds might be important in forming stable aggregates. In fact, removal of iron and aluminium by extraction with acetylacetone in benzene does not destroy aggregates, but they become very much less stable (Giovannini and Sequi, 1976a,b). From studies of aggregation caused by microorganisms both Harris *et al.* (1963) and Monnier (1965) observed greater stability as a result of anaerobic conditions. Monnier suggested that the products of the anaerobic fermentation solubilizes iron in an organic matter complex which subsequently precipitates and confers stability on the aggregating substances. The persistence of natural aggregates may be increased by some kind of protection by phenolic materials. It has been found that with the addition of tannic acid to aggregates made by incubating glucose in soil for 28 or 42 d there was no diminished aggregation over a period of 6 months whereas the untreated aggregates were almost destroyed (Griffiths and Burns, 1972). It is postulated that in natural conditions humic compounds could play a similar stabilizing role, and there is some experimental evidence to support this. When soil containing additional adsorbed humic acid was incubated with glucose, the stable aggregates formed persisted for long periods in contrast to those formed without the humic acid (Swift and Chaney, 1979). Guckert *et al.* (1975a) distinguished two stages in the formation of a stable structure, an aggregating stage involving polysaccharide, and a stabilizing stage, the latter involving more humified material. Complete dispersion of aggregates from grassland soil on treatment with periodate was only obtained after a preliminary extraction with neutral pyrophosphate solution (Clapp and Emerson, 1965a,b) which would dissolve some humic substances, and this supports the concept. Stefanson (1971) found that a sequence of periodate:borate and pyrophosphate treatments was more effective than either treatment applied alone but showed that the order did not matter. It is not easy to relate this finding to the two-stage idea presented above.

Investigations of soil structure are not helped by the difficulty of quantifying the measurement of stability. Many methods have depended on a physical disruption with water either by dripping water on to aggregates (Low, 1954) or shaking the aggregates in sieves under water (Yoder, 1936; Mehta et al., 1960; Greenland et al., 1962) or just simply suspending the soil in water (Emerson, 1967; Greenland et al., 1975). A method which is a little more flexible in relation to chemical treatment of the aggregates is the measurement of changes in the permeability caused by the dispersion brought about by dilute solutions of sodium chloride (Dettman and Emerson, 1959; Greenland et al., 1961; Hamblin and Greenland, 1977).

THE RELATIONSHIP BETWEEN SOIL POLYSACCHARIDE AND WATER RETENTION IN SOILS

The extracellular polysaccharides of bacteria are hygroscopic. It has been suggested that this would allow cells to retain water or to rehydrate slowly after desiccation (Hepper, 1975) and also help to retain water in the soil (Haworth et al., 1946; Geohegan and Brian, 1946).

THE CONTRIBUTION OF SOIL POLYSACCHARIDE TO THE EXCHANGE CAPACITY AND THE COMPLEXING OF ELEMENTS

Uronic carboxyl groups of soil polysaccharide provide exchange sites for cations and, from titration studies, the exchange capacity is found to be in the region of 50–100 meq $100g^{-1}$. This value is low in comparison with other soil organic fractions such as the humic acid

(300 meq $100g^{-1}$) and forms a very small percentage of the whole.

In acidic conditions the amino groups, if free, will contribute to the anion exchange capacity of soils and this, though smaller than the cation exchange capacity of the polysaccharide, is a more significant contribution to the total anion exchange capacity of the soil.

Little attention has been given to metal complexes of polysaccharide in soil because they would be expected to be weak in comparison with those of other complexing substances in soil. Most of the polysaccharide complexes described in pp. 118–9 were created by the addition of metal ions or by chemical changes brought about by the soil extraction process. There is little direct evidence for the existence of metal–polysaccharide complexes in soil. Certain soil bacteria have been found to produce 2-ketogluconic acid in glucose-enriched soil (Duff and Webley, 1959), which may be important in chelating ions such as calcium in soil minerals, so causing the release of phosphate (Duff et al., 1963). The release of calcium may, however, be caused by the increased acidity of the medium (Moghimi and Tate, 1978).

The availablity of boron in soils to plants is partly related to the organic matter content of the soil, particularly in soils rich in organic matter (Berger and Truog, 1945; Berger, 1949, 1962; Gupta, 1968; Harada and Tamai, 1968; John et al., 1977). This is explained as being the effect of boron adsorption by the organic matter from which it appears readily available. It seems very likely that the soil polysaccharide is involved in this adsorption because stable complexes are formed between borates and the vicinal hydroxyl groups of sugar (Parks and White, 1952).

It is very probable that uronic acids alter soil minerals. Hernandez and Roberts (1975) drew attention to the similarity of the product of the reaction of galacturonic acid with biotite to smectite in the A_2 horizon of podzols.

VALUE OF SOIL CARBOHYDRATE AS A SOURCE OF FOOD FOR SOIL ORGANISMS

Of all the organic matter fractions in soil, the polysaccharide fraction is likely to be the most readily available food for organisms. Other, more humified, compounds are thought to have more varied structures which could account for their biological stability (Kleimhempel, 1971). Polysaccharides, containing a large number of repeating units, should in theory give a good yield of metabolizable substances released by the action of extracellular degradative enzymes. There is an abundance of polysaccharide, but in practice the availability to organisms is restricted as discussed in Chapter 5.

SOIL CARBOHYDRATE AND THE ORIGIN OF HUMIC SUBSTANCES

One of the roles attributed to carbohydrates in soil has been as a precursor of humic material. Because the way in which humic compounds originate is still unknown, it is of value to consider these early theories involving a carbohydrate origin (Waksman, 1926), even though most of them are now disfavoured. For example, in the last century it was thought that cellulose is slowly oxidized to ulmic acid and ulmin, dark brown polymers which on further oxidation are transformed to humic acid and humin, and that these steps could be reproduced in the laboratory by the action of acids on carbohydrates. But eventually it was recognized that there were differences between the natural and synthetic materials. Another sequence of events proposed, cellulose→oxycellulose→humal acid→humic acid (Marcusson, 1926), included humal acid which has since been characterized as hemicellulose (Waksman, 1936). Another theory was that humic acid resulted from the polymerization of furfural derivatives (Beckley, 1921). Although polymers made synthetically by heating glucose or cellulose with 6N HCl resemble humic acid in colour, they show little evidence of the characteristic free radicals on examination by EPR spectroscopy (Atherton et al., 1967).

Maillard showed that dark brown polymers resembling humic substances form when sugars are heated with amino acids. The chief objection to this as a soil mechanism was the absence of large quantities of free sugars or amino acids in soil. Now that more is known of the slow rate of turnover of organic matter in soil this is less restricting, and Stevenson (1960) has speculated that fulvic acid may be the product of a reaction between carbohydrate and amino compounds.

The initial step is the formation of an aldosylamine.

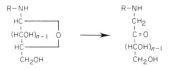

This undergoes an Amadori rearrangement to form a keto derivative.

Such products undergo dehydration and fragmentation before polymerizing to form brown products (Stevenson, 1973). Hodge (1955) presents evidence for the enol being the reactive form.

Modern ideas on humic matter recognize that part is contributed by lignin and part by the polymerization of phenolic compounds synthesized by microorganisms. These are responsible for some of the aromatic components. Humic compounds are also partly aliphatic, and some, such as fulvic acid, predominantly so. Polymers formed from maleic acid anhydride have been considered a useful model (Anderson *et al.*, 1976) but the nature of the polymer is still unknown. It is thought most unlikely that acid anhydrides could form the starting material for the synthesis of these types of polymer under natural conditions in the soil. The attraction of the proposition that substances derived from carbohydrate may fulfill this role is in the large quantities of them available.

Recently claims have been made that oxidation of humic acid from peat by peracetic acid leads to the release of amino acids and sugars, the latter being equivalent to 65% of the humic acids (Meneghel *et al.*, 1971; 1972). A short hydrolysis with water was needed before the reducing sugars could be detected. It was also claimed that heating a peat humic acid with HI liberated a lipid and a carbohydrate composed of seven sugars and a uronic acid. The authors considered that their humic acid was constructed of chains of high molecular weight fatty acids, esterified and etherified by condensed sugar molecules (Bloch *et al.*, 1970). The experimental findings may be partly explained by the association of carbohydrate with humic acid; Haworth (1971) envisaged humic acid as having a complex aromatic core to which is attached, chemically or physically, polysaccharides, proteins, simple phenols and metals. The effect of the enzymes hemicellulase or pronase on a fulvic acid was to release both sugars and amino acids (Majumdar and Rao, 1978) and this finding was interpreted to mean that some sugar units are linked to the fulvic acid core through amino acids and some amino acids are linked through sugars.

Scheffer and Kickuth (1961) isolated cellulose from humic acid, but from the ease with which polysaccharide may be separated from humic acid Felbeck (1965) surmised that carbohydrate, apart from hexosamines (Felbeck, 1971), was not an integral part of humic acid.

It is concluded that there is no firm evidence that carbohydrate contributes in any specific way to the origin of humic substances.

References

Acharya, C. N. (1937). Determination of the furfuraldehyde yield of soils and of plant materials admixed with soil. *Biochemical Journal* **31**, 1800.

Acton, C. J. Paul, E. A. and Rennie, D. A. (1963a). Measurement of the polysaccharide content of soils. *Canadian Journal of Soil Science* **43**, 141–150.

Acton, C. J., Rennie, D. A. and Paul, E. A. (1963b). The relationship of polysaccharides to soil aggregation. *Canadian Journal of Soil Science* **43**, 201–209.

Adams, A. P., Bartholemew, W. V. and Clark, F. E. (1954). Measurement of nucleic acid components in soil. *Soil Science Society of America Proceedings* **18**, 40–46.

Adams, G. A. and Castagne, A. E. (1948). Some factors affecting the determination of furfural. *Canadian Journal of Research* **B26**, 314–324.

Albersheim, P. (1965). Biogenesis of the cell wall. *In* "Plant Biochemistry" (Eds J. Bonner and J. E. Varner) Ch. 13. Academic Press, New York.

Allen, A. L., Stevenson, F. J. and Kurtz, L. T. (1973). Chemical distribution of residual fertilizer nitrogen in soil as revealed by nitrogen-15 studies. *Journal of Environmental Quality* **2**, 120–124.

Alvsaker, E. (1948). A modified Waksman procedure and its application to soil samples from Western Norway. *Universitetet i Bergen skrifter Nr. 23.*

Alvsaker, E. and Michelsen, K. (1957). Carbohydrates in a cold water extract of a pine forest. *Acta Chemica Scandinavica* **11**, 1794–1795.

Anderson, G. (1956). The identification and estimation of soil inositol phosphates. *Journal of the Science of Food and Agriculture* **6**, 437–444.

Anderson, G. (1970). The isolation of nucleoside diphosphates from alkaline extracts of soil. *Journal of Soil Science* **21**, 96–104.

Anderson, G. and Hance, R. J. (1963). Investigation of an organic phosphorous component of fulvic acid. *Plant and Soil* **19**, 296–303.

Anderson, G. and Malcolm, R. E. (1974). The nature of alkali-soluble soil organic phosphates. *Journal of Soil Science* **25**, 282–297.

Anderson, H. A and Russell, J. D. (1976). Possible relationship between soil fulvic acid and polymaleic acid. *Nature, London* **260**, 597.

Anderson, H. A., Fraser, A. R., Hepburn, A. and Russell, J. D. (1977). Chemical and infrared spectroscopic studies of fulvic acid fractions from a podzol. *Journal of Soil Science* **28**, 623–633.

Aspinall, G. O. and McKay, J. E. (1958). The hemicelluloses of European larch (*Larix decidua*) I. The constitution of a xylan. *Journal of the Chemical Society*, 1059–1066.

Aspiras, R. B., Allen, O. N. and Harris, R. F. (1971). Aggregate stabilization by filamentous microorganisms. *Soil Science* **112**, 282–284.

Atherton, N. M., Cranwell, P. A., Floyd, A. J. and Haworth, R. D. (1967). Humic acid – I E.S.R. spectra of humic acids. *Tetrahedron* **23**, 1653–1667.

Bachelier, G. (1966). Les sucres dans le sol et leur dosage global. *Cahiers O.R.S.T.O.M. serie Pedologie* **4**, 9–22.

Bacon, J. S. D. (1967). The chemical environment of bacteria in soil. *In* "The Ecology of Soil Bacteria" (Ed. T. R. G. Gray), 25–43. Liverpool University Press.

Bacon, J. S. D. and Cheshire, M. V. (1971). Apiose and mono-O-methyl sugars as minor constituents of the leaves of deciduous trees and various other species. *Biochemical Journal* **124**, 555–562.

Bacon, J. S. D., Gordon, A. H., Morris, E. J. and Farmer, V. C. (1975). Acetyl groups in cell-wall preparations from higher plants. *Biochemical Journal* **149**, 485–487.

Barker, S. A., Finch, P., Hayes, M. H. B., Simmonds, R. G. and Stacey, M. (1965). Isolation and preliminary characterisation of soil polysaccharides. *Nature, London* **205**, 68–69.

Barker, S. A., Hayes, M. H. B., Simmonds, R. G. and Stacey, M. (1967). Studies on soil polysaccharides I. *Carbohydrate Research* **5**, 13–24.

Baver, L. D. (1940). "Soil physics." John Wiley, New York.

Baver, L. D. (1968). The effect of organic matter on soil structure. *Pontificiae Academiae Scientiarum Scripta Varia* **32**, 383–413.

Beckley, V. A. (1921). The formation of humus. *Journal of Agricultural Science* **11**, 69–77.

Belcher, R., Nutten, A. J. and Sambrook, C. M. (1954). The determination of glucosamine. *Analyst* **79**, 201–8.

Benoit, R. E. and Starkey, R. L. (1968). Inhibition of decomposition of cellulose and some other carbohydrates by tannin. *Soil Science* **105**, 291–296.

Berger, K. C. (1949). Boron in soils and crops. *In* "Advances in Agronomy" (Ed. A. G. Norman), Vol. 1, 321–351. Academic Press, New York.

Berger, K. C. (1962). Micronutrient deficiencies in the United States. *Journal of Agricultural and Food Chemistry* **10**, 178–181.

Berger, K. C. and Truog, E. (1945). Boron availability in relation to soil reaction and organic matter content. *Soil Science Society of America Proceedings* **10**, 113–116.

Bernier, B. (1956). A study of some aspects of the significance of carbohydrates in soil processes. D Phil Thesis, University of Oxford.

Bernier, B. (1958). Characterization of polysaccharides isolated from forest soils. *Biochemical Journal* **70**, 590–598.

Black, W. A. P., Cornhill, W. J. and Woodward, F. N. (1955). A preliminary investigation on the chemical composition of sphagnum moss and peat. *Journal of Applied Chemistry* **5**, 484–492.

Bloch, J-M., Hubler, J-C., Meneghel, R., Petit-Sarlotte, C. and Wilhelm, J-C. (1970). Sur la nature glyco-lipidque d'acides humiques extraits de tourbes. *Compte Rendu Hebdomadaire des Séances de l'Académie des Sciences, Paris* **270**, 1030–1032.

Bond, R. D. (1960). C.S.I.R.O. Division of Soils Div. Rep. 10/60.

Bond, R. D. and Harris, J. R. (1964). The influence of the microflora on physical properties of soils. *Australian Journal of Soil Research* **2**, 111–122.

Bondietti, E., Martin, J. P. and Haider, K. (1972). Stabilization of amino sugar units in humic-type polymers. *Soil Science Society of America Proceedings* **36**, 597–602.

Bouhours, J-F. and Cheshire, M. V. (1969). The occurrence of 2-O-methylxylose and 3-O-methylxylose in peat. *Soil Biology and Biochemistry* **1**, 185–190.

Bower, C. A. (1945). Separation and identification of phytin and its derivatives from soils. *Soil Science* **59**, 277–285.

Bremner, J. M. (1949). Studies on Soil Organic Matter. Part I. The chemical nature of soil organic nitrogen. *Journal of Agricultural Science* **39**, 183–193.

Bremner, J. M. (1951). A review of recent work on soil organic matter. *Journal of Soil Science* **2**, 67–82.

Bremner, J. M. (1958). Amino sugars in soils. *Journal of the Science of Food and Agriculture* **9**, 528–532.

Bremner, J. M. and Shaw, K. (1954). Studies on the estimation and decomposition of amino sugars in soil. *Journal of Agricultural Science* **44**, 152–159.

Brink, R. H., Dubach, P. and Lynch, D. L. (1960). Measurement of carbohydrates in soil hydrolysates with anthrone. *Soil Science* **89**, 157–166.

Burges, A. (1967). The decomposition of organic matter in the soil. *Soil Biology* **16**, 479–492.

Burt, J. R. (1964). Automated analysis of sugar phosphates. *Analytical Biochemistry* **9**, 293–302.

Campbell, C. A., Paul, E. A., Rennie, D. A. and McCallum, K. J. (1967). Factors affecting the accuracy of the carbon-dating method in soil humus studies. *Soil Science* **104**, 81–85.

Carballas, T., Carballas, M. and Jacquin, F. (1978). Biodégradation et humification de la matière organique des sols humiferes atlantiques. *Anales de edafología y agrobiología* **37**, 205–212.

Casagrande, J. and Park, K. (1978). Muramic acid levels in bog soils from the Okefenokee Swamp. *Soil Science* **125**, 181–183.

Chahal, K. S. (1968). "Isotopes and radiation in soil organic matter studies." International Atomic Energy Agency (Vienna), 207–218.

Chahal, K. S. and Wagner, G. H. (1965). Decomposition of organic matter in Sanborn field soils amended with C^{14} glucose. *Soil Science* **100**, 96–103.

Chahal, K. S., Mortensen, J. L. and Himes, F. L. (1966). Decomposition products of carbon-14 labelled rye tissue in a peat profile. *Soil Science Society of America Proceedings* **30**, 217–220.

Chalmot, G. de (1894). Pentosans in plants. *American Chemical Journal* **16**, 229.

Chen, Y., Sowden, F. J. and Schnitzer, M. (1977). Nitrogen in Mediterranean soils. *Agrochimica* **21**, 7–14.

Cheshire, M. V. (1977). Origins and stability of soil polysaccharide. *Journal of Soil Science* **28**, 1–10.

Cheshire, M. V. and Anderson, G. (1975). Soil polysaccharides and carbohydrate phosphates. *Soil Science* **119**, 356–362.

Cheshire, M. V. and Mundie, C. M. (1966). The hydrolytic extraction of carbohydrates from soil by sulphuric acid. *Journal of Soil Science* **17**, 372–381.

Cheshire, M. V. and Thompson, S. J. (1972). Configuration of soil arabinose. Proceedings of the Biochemical Society, 1–2 June, 1972. *Biochemical Journal* **129**, 19P.

Cheshire, M. V., Mundie, C. M. and Shepherd, H. (1969). Transformation of ^{14}C glucose and starch in soil. *Soil Biology and Biochemistry* **1**, 117–130.

Cheshire, M. V., Mundie, C. M. and Shepherd, H. (1971). The origin of the pentose fraction of soil polysaccharide. *Journal of Soil Science* **22**, 222–236.

Cheshire, M. V., Mundie, C. M. and Shepherd, H. (1973). The origin of soil polysaccharide: transformation of sugars during the decomposition in soil of plant material labelled with [14]C. *Journal of Soil Science* **24**, 54–68.

Cheshire, M. V., Greaves, M. P. and Mundie, C. M. (1974a). Decomposition of soil polysaccharide. *Journal of Soil Science* **25**, 483–498.

Cheshire, M. V., Mundie, C. M. and Shepherd, H. (1974b). Transformation of sugars when rye hemicellulose labelled with ^{14}C decomposes in soil. *Journal of Soil Science* **25**, 90–98.

Cheshire, M. V., Greaves, M. P. and Mundie, C. M. (1975). The microbial degradation of soil polysaccharide. Rapport du Premier Colloque International Biodégradation et Humification, Nancy, France (1974), 148–156.

Cheshire, M. V., Greaves, M. P. and Mundie, C. M. (1976). The effect of temperature on the microbial transformation of (^{14}C) glucose during incubation in soil. *Journal of Soil Science* **27**, 75–88.

Cheshire, M. V., Berrow, M. L., Goodman, B. A. and Mundie, C. M. (1977). Metal distribution and nature of some Cu, Mn and V complexes in humic and fulvic acid fractions of soil organic matter. *Geochimica et Cosmochimica Acta* **41**, 1131–1138.

Cheshire, M. V., Sparling, G. P., Mundie, C. M., Shepherd, H. and Murayama, S. (1978). Effect of temperature and soil drying on the transformation of (^{14}C) glucose in soil. *Journal of Soil Science* **29**, 360–366.

Cheshire, M. V., Bracewell, J. M., Mundie, C. M., Robertson, G. W.,

Russell, J. D. and Fraser, A. R. (1979a). Structural studies on soil polysaccharide. *Journal of Soil Science* **30**.

Cheshire, M. V., Mundie, C. M. and Shepherd, H. (1979b). Transformation of carbohydrate constituents of grass during decomposition in soil. *Journal of the Science of Food and Agriculture* **30**, 330.

Chesters, G., Attoe, O. J. and Allen, O. N. (1957). Soil aggregation in relation to various soil constituents. *Soil Science Society of America Proceedings* **21**, 272–277.

Clapp, C. E. (1957). High molecular weight water-soluble muck; isolation and determination of constituent sugars of a borate complex-forming polysaccharide employing electrophoretic techniques. PhD Thesis, Cornell University.

Clapp, C. E. and Emerson, W. W. (1965a). The effect of periodate oxidation on the strength of soil crumbs. I. Qualitative studies. *Soil Science Society of America Proceedings* **29**, 127–130.

Clapp, C. E. and Emerson, W. W. (1965b). The effect of periodate oxidation on the strength of soil crumbs. II. Quantitative studies. *Soil Science Society of America Proceedings* **29**, 130–134.

Clapp, C. E., Davis, R. J. and Waugaman, S. H. (1962). The effect of rhizobial polysaccharides on aggregate stability. *Soil Science Society of America Proceedings* **26**, 466–469.

Clapp, C. E., Olness, A. E. and Hoffman, D. J. (1968). Adsorption studies of a dextran on montmorillonite. Trans. Ninth International Congress of Soil Science, Adelaide **1**, 627–634.

Clark, F. E. and Tan, K. H. (1969). Identification of a polysaccharide ester linkage in humic acid. *Soil Biology and Biochemistry* **1**, 75–81.

Coffin, D. E. and DeLong, W. A. (1960). Extraction and characterization of organic matter of a podzol B horizon. Trans. Seventh International Congress of Soil Science **2.13**, 91–97.

Coldwell, B. B. and DeLong, W. A. (1950). Studies of the composition of deciduous forest tree leaves before and after partial decomposition. *Scientific Agriculture* **30**, 456–465.

Combeau, A. (1965). Variations saisonnières de la stabilité structurale du sol en region temperée (comparison avec la zone tropicale). *Cahiers O.R.S.T.O.M. Serie Pedologie* **3**, 123–135.

Cortez, J. (1977). Biodégradation, in vitra, de deux lipopolysaccharides bactériens ^{14}C dans un sol rouge méditerranéen. *Geoderma* **18**, 177–192.

Cosgrove, D. J. (1970). Origin of inositol polyphosphates in soil. International symposium on Hydrogeochemistry and Biogeochemistry, Tokyo, Japan.

Cosgrove, D. J. and Tate, M. E. (1963). Occurrence of *neo*-inositol hexaphosphate in soil. *Nature, London* **200**, 568.

Cross, C. F. and Bevan, E. J. (1903). "Cellulose, an outline of the structural elements of plants." Longmans, London.

Daji, J. A. (1932). The determination of cellulose in soil. *Biochemical Journal* **26**, 1275–1280.

Davies, R. I. (1971). Relation of polyphenols to decomposition of organic matter and to pedogenic processes. *Soil Science* **111,** 80–85.

Decau, J. (1968). Les polysaccharides du sol: origine, évolution et rôle. *Annales agronomiques* **19,** 65–82.

De Datta, S. K., Franklin, R. E. and Himes, F. L. (1967). Partial characterization of soil polysaccharide-strontium 90-yttrium 90 complexes. *Soil Science* **103,** 47–55.

Deinema, M. H. and Zevenhuisen, L. P. T. M. (1971). Formation of cellulose fibrils by Gram-negative bacteria and their role in bacterial flocculation. *Archiv für Mikrobiologie* **78,** 42–57.

Deriaz, R. E. (1961). Routine analysis of carbohydrates and lignin in herbage. *Journal of the Science of Food and Agriculture* **12,** 152–160.

Dettman, M. G. and Emerson, W. W. (1959). A modified permeability test for measuring the cohesion of soil crumbs. *Journal of Soil Science* **10,** 215–226.

Dhoot, J. S., Singh, N. T. and Brar, S. S. (1974). Note on the influence of organic matter on polyuronides of soil aggregates. *Indian Journal of Agricultural Sciences* **44,** 243–244.

Dische, Z. and Shettles, L. B. (1948). Specific color reaction of methyl pentoses and a spectrophotometric micromethod for their determination. *Journal of Biological Chemistry* **175,** 595–603.

Dormaar, J. F. (1967). Polysaccharides in chernozemic soils of Southern Alberta. *Soil Science* **103,** 417–423.

Dormaar, J. F. and Lynch, D. L. (1962). Amendments to the determination of "uronic acids" in soils with carbazole. *Soil Science Society of America Proceedings* **26,** 251–254.

Doutre, D. A., Hay, G. W., Hood, A. and VanLoon, G. W. (1978). Spectrophotometric methods to determine carbohydrates in soil. *Soil Biology and Biochemistry* **10,** 457–462.

Drobnik, J. (1960a). Primary oxidation of organic matter in the soil. I. The form of respiration curves with glucose as the substrate. *Plant and Soil* **12,** 199–211.

Drobnik, J. (1960b). Primary oxidation of organic matter in the soil. II. Influence of various kinds of preincubations. *Plant and Soil* **12,** 212–222.

Dubach, P. and Lynch, D. L. (1959). Comparison of the determination of "uronic acids" in soil extracts with carbazole and by decarboxylation. *Soil Science* **87,** 273–275.

Dubach, P., Zweifel, G., Bach, R. and Deuel, H. (1955). Untersuchungen an der Fulvosaure-Fraktion einiger schweizerischer Boden. *Zurich-Schweiz Universitatsstr.* 2. **69,** 97–107.

Dubach, P., Mehta, N. C., Jakab, T., Martin, F. and Roulet, N. (1964). Chemical investigations on soil humic substances. *Geochimica et Cosmochimica Acta* **28,** 1567–1578.

Duff, R. B. (1952a). The occurrence of L-fucose in soil, peat, and in a polysaccharide synthesized by soil bacteria. *Chemistry and Industry,* 1104.

Duff, R. B. (1952b). The occurrence of methylated carbohydrates and rhamnose as components of soil polysaccharides. *Journal of the Science of Food and Agriculture* **3**, 140–144.

Duff, R. B. (1961). Occurrence of 2-*O*-methylrhamnose and 4-*O*-methylgalactose in soil and peat. *Journal of the Science of Food and Agriculture* **12**, 826–831.

Duff, R. B. and Webley, D. M. (1959). 2-Ketogluconic acid as a natural chelator produced by soil bacteria. *Chemistry and Industry*, 1376–1377.

Duff, R. B. Webley, D. M. and Scott, R. O. (1963). Solubilization of minerals and related materials by 2-ketogluconic acid-producing bacteria. *Soil Science* **95**, 105–114.

Easterwood, V. M. and Huff, B. J. L. (1969). Carbohydrate analysis by gas chromatography of acetylated aldonontriles. *Svensk papperstidning årsbok* **72**, 768–772.

Egawa, T. and Sekiya, K. (1956). Studies on humus and aggregate formation. *Soil Science and Plant Nutrition* **2**, 75–88.

Emerson, W. W. (1959). The structure of soil crumbs. *Journal of Soil Science* **10**, 235–244.

Emerson, W. W. (1963). The effect of polymers on the swelling of montmorillonite. *Journal of Soil Science* **14**, 52–63.

Emerson, W. W. (1967). A classification of soil aggregates based on their coherence in water. *Australian Journal of Soil Research* **5**, 47–57.

Feilitzen, H. von and Tollens, B. (1898). Uber den Gehalt des Torfes an Pentosan und anderen Kohlenhydraten. *Journal für Landwirtschaft* **46**, 17–23.

Felbeck, G. T. (1965). Structural chemistry of soil humic substances. *Advances in Agronomy* **17**, 327–368.

Felbeck, G. T. (1971). Structural hypotheses of soil humic acids. *Soil Science* **111**, 42–48.

Finch, P., Hayes, M. H. B. and Stacey, M. (1966). Studies on soil polysaccharides and on their interaction with clay preparations. Meetings of Commissions II and IV of the International Society of Soil Science, Aberdeen II, 19–32.

Finch, P., Hayes, M. H. B. and Stacey, M. (1968). Studies on the polysaccharide constituents of an acid extract of a Fenland muck soil. Trans. Ninth International Congress of Soil Science, Adelaide **3**, 193–201.

Finch, P., Hayes, M. H. B. and Stacey, M. (1971). The Biochemistry of soil Polysaccharides. *In* "Soil Biochemistry" (Eds A. D. McLaren and J. Skujins), Vol. 2, 257–319. Marcel Dekker, New York.

Florajanczyk, S. (1965). Separation of so called fulvic acids on a strongly basic anion exchanger. *Roczniki gleboznawcze* **15**, 409–440.

Folsom, B. L., Wagner, G. H. and Scrivner, C. L. (1974). Comparison of soil carbohydrate in several prairie and forest soils by gas–liquid chromatography. *Soil Science Society of America Proceedings* **38**, 305–309.

Ford, G. W., Greenland, D. J. and Oades, J. M. (1969). Separation of the light fraction from soils by ultrasonic dispersion in halogenated hydrocarbons containing a surfactant. *Journal of Soil Science* **20,** 291–296.

Forsyth, W. G. C. (1947). Studies on the more soluble complexes of soil organic matter. *Biochemical Journal* **41,** 176–181.

Forsyth, W. G. C. (1948). Carbohydrate Metabolism in the soil. *Chemistry and Industry,* 515–519.

Forsyth, W. G. C. (1950). Studies on the more soluble complexes of soil organic matter. *Biochemical Journal* **46,** 141–146.

Forsyth, W. G. C. (1954). The synthesis of polysaccharides by bacteria isolated from a tropical soil. Fifth International Congress of Soil Science **3,** 119–122.

Forsyth, W. G. C. and Webley, D. M. (1949a). Polysaccharides synthesized by aerobic mesophilic spore-forming bacteria. *Biochemical Journal* **44,** 455–459.

Forsyth, W. G. C. and Webley, D. M. (1949b). The synthesis of polysaccharides by bacteria isolated from soil. *Journal of General Microbiology* **3,** 395–399.

Gallali, T., Guckert, A. and Jacquin, F. (1972). Étude des polysaccharides et de leur rôle au sein de la matière organique humifiée des sols. *Bulletin de l'École Nationale Superieure Agronomique de Nancy* **14,** 207–219.

Gaponenkov, T. K. and Shatsman, L. I. (1964). Aluminium-uronic complexes of soils. *Pochvovedenie* **12,** 84–88.

Garyayev, P. P., Vladychenskiy, A. S., Deyanova, S. A., Kaloshin, P. M. and Poglazov, B. F. (1973). Separation and purification of some organic compounds from a sod-podzolic soil and volcanic soil material. *Soviet Soil Science,* 232–239. Translated from *Pochvovedenie* 1973 **4,** 134–140.

Geohegan, M. J. and Brian, R. C. (1946). Influence of bacterial polysaccharides on aggregate formation in soils. *Nature, London* **158,** 837.

Geohegan, M. J. and Brian, R. C. (1948a). Aggregate formation in soil. 1. Influence of some bacterial polysaccharide on the binding of soil particles. *Biochemical Journal* **43,** 5–13.

Geohegan, M. J. and Brian. R. C. (1948b). Aggregate formation in soil. 2. Influence of various carbohydrates and proteins on aggregation of soil particles. *Biochemical Journal* **43,** 14.

Ghildyal, B. P. and Gupta, U. C. (1959). A study of the biochemical and microbiological changes during the decomposition of *Crotolaria juncea* (Sann Hemp) at different stages of growth in soil. *Plant and Soil* **11,** 312–330.

Giovannini, G. and Sequi, P. (1976a). Iron and aluminium as cementing substances of soil aggregates I. Acetylacetone in benzene as an extractant of fractions of soil iron and aluminium. *Journal of Soil Science* **27,** 140–147.

Giovannini, G. and Sequi, P. (1976b). Iron and aluminium as cementing substances of soil aggregates II. Changes in stability of soil aggregates following extraction of iron and aluminium by acetylacetone in a non-polar solvent. *Journal of Soil Science* **27,** 148–153.

Given, P. H. and Dickensen, C. H. (1975). Biochemistry and Microbiology of Peats. *In* "Soil Biochemistry" (Eds E. A. Paul and A. D. McLaren), Vol. 3, Ch. 4. Marcel Dekker, New York.

Gorin, P. A. J. and Spencer, J. F. T. (1968). Structural chemistry of fungal polysaccharides. *Advances in Carbohydrate Chemistry* **23**, 367–417.

Graveland, D. N. and Lynch, D. L. (1961). Distribution of uronides and polysaccharides in the profiles of a soil catena. *Soil Science* **91**, 162–165.

Greenland, D. J. (1956). The adsorption of sugars by montmorillonite. I. X-ray studies. *Journal of Soil Sciences* **7**, 319–333.

Greenland, D. J. (1965a). Interaction between clays and organic compounds in soils. Part I. Mechanisms of interaction between clays and defined organic compounds. *Soils and Fertilizers* **28**, 415–425.

Greenland, D. J. (1965b). Interaction between clays and organic compounds in soils. Part II. Adsorption of soil organic compounds and its effect on soil properties. *Soils and Fertilizers* **28**, 521–532.

Greenland, D. J. (1971). Changes in the nitrogen status and physical condition of soils under pastures, with special reference to the maintenance of the fertility of Australian soils used for growing wheat. *Soils and Fertilizers* **34**, 237–251.

Greenland, D. J. and Ford, G. W. (1964). Separation of partially humified organic materials from soils by ultrasonic dispersion. Trans. Eighth International Congress of Soil Science **2, 15,** 137–147.

Greenland, D. J. and Oades, J. M. (1975). Saccharides. *In* "Soil Components" (Ed. J. E. Geiseking), Vol. 1, 213–257. Springer, New York.

Greenland, D. J., Lindstrom, G. R. and Quirk, J. P. (1962). Organic materials which stabilize natural soil aggregates. *Soil Science Society of America Proceedings* **26**, 366–371.

Greenland, D. J., Rimmer, D. and Payne, D. (1975). Determination of the structural stability class of English and Welsh soils, using a water coherence test. *Journal of Soil Science* **26**, 294–303.

Griffiths, D. A. and Dobbs, C. G. (1963). Relationships between Mycostasis and free monosaccharides in soils. *Nature, London* **199**, 408.

Griffiths, E. (1965). Micro-organisms and soil structure. *Biological Revues* **40**, 129–142.

Griffiths, E. and Burns, R. G. (1972). Interaction between phenolic substances and microbial polysaccharides in soil aggregation. *Plant and Soil* **36**, 599–612.

Griffiths, E. and Jones, D. (1965). Microbiological aspects of soil structure. I. Relationships between organic amendments, microbial colonization and changes in aggregate stability. *Plant and Soil* **23**, 17–33.

Grossbard, E. (1969). A visual record of the decomposition of ^{14}C-labelled fragments of grasses and rye added to soil. *Journal of Soil Science* **20**, 38–51.

Grov, A. (1963). Carbohydrates in cold water extracts of a pine forest soil. *Acta Chemica Scandinavica* **17**, 2301–2306.

Guckert, A. (1968). Influence de la matière organique du sol sur la stabilité structurale. *Bulletin de l'Ecole Nationale superieure Agronomique de Nancy* **9,** 58–68.

Guckert, A. (1972). Note sur l'èxtraction et la caractérisation de polysaccharides ^{14}C formés dans le sol au cour de la biodégradation de végétaux marqués. *Bulletin de l'Ecole Nationale superiere Agronomique de Nancy* **14,** 69–73.

Guckert, A. (1975). Origine et devenir des polysaccharide du sol. Rapport du premier Colloque International Biodegradation et Humification, Nancy, France, 1974, 116–127.

Guckert, A., Nussbaumer, E. and Jacquin, F. (1970). Étude comparée de l'action du glucose ^{14}C et de la paille ^{14}C sur la stabilité structurale d'un sol limoneux acide. *Bulletin de l'Ecole Nationale superieure Agronomique de Nancy* **12,** 26–49.

Guckert, A., Curé, B. and Jacquin, F. (1971). Comparative evolution of the polysaccharides of the humin after incubation of glucose ^{14}C and straw ^{14}C. Trans. International Symposium "Humus et Planta V", Prague, 155–160.

Guckert, A., Chone, T. and Jacquin, F. (1975a). Microflore et stabilité structurale des sols. *Revue d'Écologie et de Biologie du sol* **12,** 211–223.

Guckert, A., Valla, M. and Jacquin, F. (1975b). Adsorption of humic acids and soil polysaccharides on montmorillonite. *Soviet Soil Science* 89–95. Translated from *Pochvovedenye* 1975 **2,** 41–47.

Guckert, A., Tok, H. H. and Jacquin, F. (1977). Biodégradation de polysaccharides bacteriens adsorbés sur une montmorillonite. "Soil organic matter studies" Vol. 1, 403–411. International Atomic Energy Agency, Braunschweig.

Gupta, U. C. (1967). Carbohydrates. In "Soil Biochemistry" (Eds A. D. McLaren and G. H. Peterson), Vol. 1, 91–118. Marcel Dekker, New York.

Gupta, U. C. (1968). Relationship of total and hot-water soluble boron and fixation of added boron, to properties of podzol soils. *Soil Science and Society of America Proceedings* **32,** 45–48.

Gupta, U. C. and Sowden, F. J. (1963). Occurrence of free sugars in soil organic matter. *Soil Science* **96,** 217–218.

Gupta, U. C. and Sowden, F. J. (1964). Isolation and characterization of cellulose from soil organic matter. *Soil Science* **97,** 328–333.

Gupta, U. C. and Sowden, F. J. (1965). Studies on methods for the determination of sugars and uronic acids in soils. *Canadian Journal of Soil Science* **45,** 237–240.

Gupta, U. C., Sowden, F. J. and Stobbe, P. C. (1963). The characterization of carbohydrate constituents from different soil profiles. *Soil Science Society of America Proceedings* **27,** 380–382.

Halstead, R. L. (1954). Soil aggregation as influenced by microbial gums, type of crop, lime and fertilizers. PhD Thesis, University of Wisconsin.

Hamada, R., Yoshizaki, K., Sakagami, K. and Kurobe, T. (1976). Soil polysaccharides in buried humic horizons of ashitaka loam formation. *Environmental Biogeochemistry* **1,** 191–202.

Hamblin, A. P. and Greenland, D. J. (1977). Effect of organic constituents and complexed metal ions on aggregate stability of some East Anglian soils. *Journal of Soil Science* **28,** 410–416.

Handley, W. R. C. (1954). Mull and Mor formation in relation to forest soils. Forestry Commission Bulletin No. 23. H.M.S.O.

Harada, T. and Tamai, M. (1968). Some factors affecting behaviour of boron in soil. I. Some soil properties affecting boron adsorption of soil. *Soil Science and Plant Nutrition* **14,** 215–224.

Hardisson, C. and Robert-Gero, M. (1966). Synthese de substances parahumiques par azobacter chroococum. II. Etude chimique comparative avec des extraits humiques de sols. *Annals de l'Institute Pasteur* **111,** 486–496.

Harris, R. F., Allen, O. N., Chesters, G. and Attoe, O. J. (1963). Evaluation of microbial activity in soil aggregate stabilization and degradation by the use of artificial aggregates. *Soil Science Society of America Proceedings* **27,** 542–546.

Harris, R. F., Chesters, G., Allen, O. N. and Attoe, O. J. (1964). Mechanisms involved in soil aggregate stabilization by fungi and bacteria. *Soil Science Society of America Proceedings* **28,** 529–532.

Harris, R. F., Chesters, G. and Allen, O. N. (1966). Dynamics of soil aggregation. *Advances in Agronomy* **18,** 107–169.

Havrankova, J. (1967). Share of carbohydrates in the humus substances composition. *Biologia Plantarum* **9,** 102–108.

Haworth, R. D. (1971). The chemical nature of humic acid. *Soil Science* **111,** 71–79.

Haworth, W. N., Pinkard, R. W. and Stacey, M. (1946). Function of bacterial polysaccharide in soil. *Nature, London* **158,** 836–837.

Hayashi, T. and Nagai, T. (1962). On the components of soil humic acids (part 9). The carbohydrate composition of different components. *Soil Science and Plant Nutrition* **8,** 22–27.

Hayes, M. H. B. and Mortensen, J. L. (1963). Role of biological oxidation and organic matter solubilization in the subsidence of rifle peat. *Soil Science Society of America Proceedings* **27,** 666–668.

Hayes, M. H. B. and Swift, R. S. (1978). The chemistry of soil organic colloids. *In* "The Chemistry of Soil Constituents" (Eds D. J. Greenland and M. H. B. Hayes) 180–320. John Wiley, New York.

Hayes, M. H. B., Stacey, M. and Standley, J. (1968). Studies on the humification of plant tissue. Trans. Ninth International Congress of Soil Science, Adelaide **3,** 247–255.

Hayes, M. H. B., Stacey, M. and Swift, R. S. (1975). Techniques for fractionating soil polysaccharides. Supplementary volume Trans. Tenth International Congress of Soil Science, Moscow **12,** 75–81.

Hepper, C. M. (1975). Extracellular polysaccharides of soil bacteria. *In* "Soil Microbiology, a critical review" (Ed. N. Walker), 93–110.

Butterworth, London; Boston, USA.

Herman, W. A., McGill, W. B. and Dormaar, J. F. (1977). Effects of initial chemical composition on decomposition of roots of three grass species. *Canadian Journal of Soil Science* **57**, 205–15.

Hernandez, M. A. V. and Robert, M. (1975). The profound alteration of micas by galacturonic acid. The smectites of podzols. *Comptes Rendus Hebdomadaires des Seances de l'Academie des Sciences*, D **281**, 523–526.

Hodge, J. E. (1955). The amadori rearrangement. *Advances in Carbohydrate Chemistry* **10**, 169–205.

Hoffman, G. (1963). Synthetic effects of soil-enzymes. *Recent progress in Microbiology*, 230–234.

Hoffman, W. W. (1937). Photo-electric determination of glucose in blood and wine. *Journal of Biological Chemistry* **120**, 51–5.

Huntjens, J. L. M. and Albers, R. A. J. M. (1978). A model experiment to study the influence of living plants on the accumulation of soil organic waste in pastures. *Plant and Soil* **50**, 411–418.

Inoko, A. and Murayama, S. (1973). Chromatographic purification and fractionation of soil fulvic acid using synthetic adsorbent. *Journal of the Science of Soil and Manure* **44**, 11–17.

Ivarson, K. C. (1977). Changes in decomposition rate, microbial population and carbohydrate content of an acid peat bog after liming and reclamation. *Canadian Journal of Soil Science* **57**, 129–137.

Ivarson, K. C. and Sowden, F. J. (1962). Methods for the analysis of carbohydrate material in soil. I. Colorimetric determination of uronic acids, hexoses and pentoses. *Soil Science* **94**, 245–250.

Jacquin, F., Guckert, A. and Gallali, T. (1974). Evolution des polysaccharides et des aminopolysaccharides au cours de l'humification. Trans. Tenth International Congress Soil Science, Moscow **2**, 137–144.

Jain, H. K. and Bhattacharya. A. K. (1960). Studies on the chemical constituents of some local weeds in the aspect of agriculture. Part I. Humification of Arousa (*Justicia adhotoda*) leaves under constant moisture level and variations in the amounts of carbohydrates. *Zeitschrift für Pflanzenernährung, Düngung und Bodenkunde* **91**, 233–240.

Jenkinson, D. S. (1968). Chemical tests for potentially available nitrogen in soil. *Journal of the Science of Food and Agriculture* **19**, 160–168.

Jenkinson, D. S. (1970). Rothamsted Report for 1970. Part 1, 74–75.

Jenkinson, D. S. (1971). Studies on the decomposition of C^{14}-labelled organic matter in soil. *Soil Science* **111**, 64–69.

Joffe, J. S. (1955). Green manuring viewed by a pedologist. *Advances in Agronomy* **7**, 141–187.

John, M. K., Chuah, H. H. and Van Laerhoven, C. J. (1977). Boron response and toxicity as affected by soil properties and rates of boron. *Soil Science* **124**, 34–39.

Johnston, H. H. (1961). Common sugars found in soils. *Soil Science Society of America Proceedings* **25**, 415.

Juo, P. S. and Stotzky, G. (1967). Interference by nitrate and nitrite in the determination of carbohydrates by anthrone. *Analytical Biochemistry* **21**, 149–151.

Juvvik, P. (1965). Sodium chlorite delignification of soil in the investigation of soil polysaccharides. *Acta Chemica Scandinavica* **19**, 645–652.

Kadner, R., Fischer, W. and Schlungbaum, G. (1962). Zur Kenntnis der chemischen Zusammensetzung von Torfen aus Vorkommen in der Deutschen Demokratischen Republik. III. Qualitative papierchromatographischer Nachweis von Aminosauren und Kohlenhydraten in Torfen. *Freiberger Forschungshefte A* **263**, 159–173.

Kanke, B. and Yamane, I. (1974a). Form of sugars soluble in water and ethanol in lowland rice soils. *Journal of the Science of Soil and Manure* **45**, 263–267.

Kanke, B. and Yamane, I. (1974b). Variation in water soluble sugars in dried soil after submersion. *Journal of the Science of Soil and Manure* **45**, 539–540.

Kanke, B. and Yamane, I. (1974c). Effect of sonication and pretreatment with ethanol on water soluble sugars in lowland rice soils. *Journal of the Science of Soil and Manure* **45**, 577–581.

Keefer, R. F. and Mortensen, J. L. (1963). Biosynthesis of Soil Polysaccharides. I. Glucose and alfalfa tissue substrates. *Soil Science Society of America Proceedings* **27**, 156–160.

Khan, S. U. (1969). Some carbohydrate fractions of a gray wooded soil as influenced by cropping systems and fertilizers. *Canadian Journal of Soil Science* 49, 219–224.

Kibblewhite, M. G. (1977). PhD Thesis, University of Aberdeen.

Kilburn, D. (1971). Unpublished data. Quoted by Finch, P., Hayes, M. H. B. and Stacey, M. (1967).

Kiss, S. and Peterfi, S. (1959). Synthetic action of maltase and invertase of the soil. *Studia Universitatis Babes-Bolyai*, Ser. 2, 179–181.

Kiss, S., Dragan-Bularda, M. and Radulescu, D. (1975). Biological significance of enzymes accumulated in soil. *Advances in Agronomy* **27**, 25–87.

Kleinhempel, D. (1971). Theoretical aspects of the persistence of organic matter in soils. *Pedobiologia. Bd.* **11**, 425–429.

Kobo, K. and Wada, H. (1969). Water-soluble organic substances in soils (part 1). *Journal of the Science of Soil and Manure* **40**, 288–292.

Kohl, R. A. and Taylor, S. A. (1961). Hydrogen bonding between the carbonyl group and Wyoming bentonite. *Soil Science* **91**, 223–227.

Kullmann, A. and Koepke, K. (1961). Uber Untersuchungen des Uronsauregehaltes und der Wasserstabilitat von Bodenaggregaten. *Zeitschrift für Pflanzenernährung, Düngung und Bodenkunde* **93**, 97–108.

L'Annuziata, M. F., Olivares, J. G. I. and Olivares, L. A. (1977). Microbial epimerization of myo-inositol to chiro-inositol in soil. *Soil Science Society of America Journal* **41**, 733–736.

Lattes, A., Gui, G., Magny, J. and Carles, J. (1963). Assessment of the structure of soil organic matter by means of a new method of extraction.

Compte Rendu Hebdomadaire des Séances de l'Académie d'Agriculture de France **49**, 1003–1008.

Leavitt, S. (1912). Studies on soil humus. *Journal of Industrial and Engineering Chemistry*, 601–604.

Lewis, J. A. and Starkey, R. L. (1968). Vegetable tannins, their decompositions and effects on decomposition of some organic compounds. *Soil Science* **106**, 241–247.

Linehan, D. J. (1977). A comparison of the polycarboxylic acids extracted by water from an agricultural top soil with those extracted by alkali. *Journal of Soil Science* **28**, 369–378.

Lodha, B. C. (1974). Decomposition of digested litter. *In* "Biology of Plant Litter Decomposition" (Eds C. H. Dickinson and G. J. F. Pugh), Vol. 1, Ch. 7. Academic Press, London; New York.

Louw, H. A. and Webley, D. M. (1959). A study of soil bacteria dissolving certain mineral phosphate fertilizers and related compounds. *Journal of Applied Bacteriology* **22**, 227–233.

Low, A. J. (1954). The study of soil structure in the field and the laboratory. *Journal of Soil Science* **5**, 57–74.

Low, A. J. (1962). The effects of organic matter decomposition on soil physical conditions, especially structure. Welsh Soils Discussion Group Report No. 3. Soil Organic Matter, 16–25.

Low, A. J. (1973). Soil structure and crop yield. *Journal of Soil Science* **24**, 249–259.

Lowe, L. E. (1965). Sulphur fractions of selected Alberta soil profiles of the chemical and podzolic orders. *Canadian Journal of Soil Science* **45**, 297–303.

Lowe, L. E. (1968). Soluble polysaccharide fractions in selected Alberta soils. *Canadian Journal of Soil Science* **48**, 215–217.

Lowe, L. E. (1969). Distribution and properties of organic fractions in selected Alberta soils. *Canadian Journal of Soil Science* **49**, 129–141.

Lowe, L. E. (1972). Aspects of chemical variability in forest humus layers under a mature western hemlock – western red cedar stand. *Canadian Journal of Forest Research* **2**, 487–489.

Lowe, L. E. (1975). Fractionation of acid-soluble components of soil organic matter using polyvinyl pyrrolidone. *Canadian Journal of Soil Science* **55**, 119–126.

Lowe, L. E. (1978a). Monosaccharide distribution in selected coniferous forest humus layers in British Columbia and its relationship to vegetation and degree of decomposition. *Canadian Journal of Soil Science* **58**, 19–25.

Lowe, L. E. (1978b). Carbohydrates in soil. *In* "Soil Organic Matter" (Eds M. Schnitzer and S. U. Khan), 65–93. Elsevier, New York.

Lowe, L. E. and Turnbull, M. E. 1968. Removal of interference in the determination of uronic acids in soil hydrolysates. *Soil Science* **106**, 312–313.

Lucas, A. J. (1970). Chemistry of carbohydrates in some organic sediments of the Florida everglades. PhD Thesis, Pennsylvania State University.

Lueken, H., Hutcheon, W. L. and Paul, E. A. (1962). The influence of nitrogen on the decomposition of crop residues in the soil. *Canadian Journal of Soil Science* **42**, 276–288.

Lynch, D. L. and Cotnoir, L. J. (1956). The influence of clay minerals on the breakdown of certain organic substrates. *Soil Science Society of America Proceedings* **20**, 367–370.

Lynch, D. L. and Graveland, D. N. (1962). Some organic matter-clay mineral relationships in four Alberta soil profiles. *Canadian Journal of Soil Science* **42**, 68–76.

Lynch, D. L., Wright, L. M. and Cotnoir, L. J. (1956). The adsorption of carbohydrates and related compounds on clay minerals. *Soil Science Society of America Proceedings* **20**, 6–9.

Lynch, D. L., Hearns, E. E. and Cotnoir, L. J. (1957a). The determination of polyuronides in soils with carbazole. *Soil Science Society of America Proceedings* **20**, 367–370.

Lynch, D. L., Wright, L. M. and Olney, H. O. (1957b). Qualitative and quantitative chromatographic analyses of the carbohydrate constituents of the acid-insoluble fraction of soil organic matter. *Soil Science* **84**, 405–411.

Lynch, D. L., Olney, H. O. and Wright, L. M. (1958). Some sugars and related carbohydrates found in Delaware soils. *Journal of the Science of Food and Agriculture* **9**, 56–60.

McCalla, T. M. (1946). Influence of some microbial groups on stabilizing soil structure against falling water drops. *Soil Science Society of America Proceedings* **11**, 257–263.

McCalla, T. M., Haskins, F. A. and Fralik, E. F. (1957). Influence of various factors on aggregation of Peorian loess by microorganisms. *Soil Science* **84**, 155–161.

McGrath, D. (1971). The determination of uronic acid (and sugar) in soil. *Geoderma* **5**, 261–269.

McGrath, D. (1973). Sugars and uronic acids in Irish soils. *Geoderma* **10**, 227–235.

MacKenzie, A. F. and Dawson, J. E. (1962). A study of organic soil horizons using electrophoretic techniques. *Journal of Soil Science* **13**, 160–166.

McKercher, R. B. and Anderson, G. (1968a). Content of inositol penta- and hexaphosphates in some Canadian soils. *Journal of Soil Science* **19**, 47–55.

McKercher, R. B. and Anderson, G. (1968b). Characterisation of the inositol penta- and hexaphosphate fractions of a number of Canadian and Scottish soils. *Journal of Soil Science* **19**, 302–310.

Macura, J. and Kubatova, Z. (1973). Control of carbohydrate utilization by soil microflora. *Soil Biology and Biochemistry* **5**, 193–204.

Macura, J. and Kunc, F. (1961). Continuous flow method in soil microbiology. II. Observations on glucose metabolism. *Folia Microbiologica* **6**, 398–407.

Macura, J., Szolnoki, J., Kunc, F., Vancura, V. and Babicky, A. (1965).

Decomposition of glucose continuously added to soil. *Folia Microbiologica* **10**, 44–54.

Majumdar, S. K. and Rao, C. V. N. (1978). Physico-chemical studies on enzyme-degraded fulvic acid. *Journal of Soil Science* **29**, 489–497.

Majumdar, S. K., Banerjee, S. K. and Rao, C. V. N. (1974). Studies on soil humic substances. Part 1. Fractionation of fulvic acid and characteriation of sugar components. *Journal of the Indian Chemical Society* **51**, 686–690.

Maksimova, A. Ye (1973). The chemical composition of humic acids in forest litter. *Soviet Soil Science*, 187–195. Translated from *Pochvovedenie* 1973 **3**, 97–105.

Marcusson, J. (1926). Lignin and oxycellulosetheorie. *Zeitschrift für Angewandte Chemie* **39**, 898–900.

Martin, F. (1976). The determination of polysaccharides in fulvic acids by *pyrolysis gas chromatography*. *In* "Analytical Pyrolysis", Proc. Third International Symposium on Analytical Pyrolysis, Amsterdam, 1976. (Eds C. E. R. Jones and C. A. Cramers), 179–187. Elsevier, Amsterdam.

Martin, J. K. (1977). Factors influencing the loss of organic carbon from wheat roots. *Soil Biology and Biochemistry* **9**, 1–7.

Martin, J. P. (1945). Microorganisms and soil aggregation: I. Origin and nature of some of the aggregating substances. *Soil Science* **59**, 163–174.

Martin, J. P. (1971). Decomposition and binding action of polysaccharides in soil. *Soil Biology and Biochemistry* **3**, 33–41.

Martin, J. P. and Richards, S. J. (1963). Decomposition and binding action of a polysaccharide from *Chromobacterium violaceum* in soil. *Journal of Bacteriology* **85**, 1288–1294.

Martin, J. P. and Waksman, S. A. (1940). Influence of microorganisms on soil aggregation and erosion. *Soil Science* **50**, 29–47.

Martin, J. P., Martin, W. P., Page, J. B., Raney, W. A. and DeMent, J. D. (1955). Soil aggregation. *Advances in Agronomy* **7**, 1–37.

Martin, J. P., Ervin, J. O. and Shepherd, R. A. (1959). Decomposition and aggregating effect of fungus cell material in soil. *Soil Science Society of America Proceedings* **23**, 217–220.

Martin, J. P., Ervin, J. O. and Shepherd, R. A. (1965). Decomposition and binding action of polysaccharides from *Azobacter indicus* (Beijerinckia) and other bacteria in soil. *Soil Science Society of America Proceedings* **29**, 397–399.

Martin, J. P., Ervin, J. O. and Shepherd, R. A. (1966). Decomposition of the iron, aluminium, zinc and copper salts or complexes of some microbial and plant polysaccharides in soil. *Soil Science Society of America Proceedings* **30**, 196–200.

Martin, J. P., Ervin, J. O. and Richards, S. J. (1972). Decomposition and binding action in soil of some mannose-containing microbial polysaccharides and their Fe, Al, Zn and Cu complexes. *Soil Science* **113**, 322–327.

Martin, J. P., Haider, K., Farmer, W. J. and Fustec-Mathon, E. (1974). Decomposition and distribution of residual activity of some ^{14}C-microbial polysaccharides and cells, glucose, cellulose and wheat straw in soil. *Soil Biology and Biochemistry* **6**, 221–230.

Martin, J. P., Parsa, A. A. and Haider, K. (1978). Influence of intimate association with humic polymers on biodegradation of (^{14}C) labelled organic substances in soil. *Soil Biology and Biochemistry* **10**, 483–486.

Marumoto, T., Furukawa, K., Yoshida, T., Kai, H., Yamada, Y. and Harada, T. (1974). Contribution of microbial cells and their cell walls to the accumulation of soil organic matter decomposable on drying. 1. Changes in the contents of individual amino acids and amino sugars in the organic nitrogen fraction as a result of the decomposition of applied ryegrass residues. *Journal of the Science of Soil and Manure* **45**, 23–28.

Mayaudon, J. (1966). Humification des complexes microbiens dans le sol I. Le lipoproteinepolysaccharide carbone-14 du pseudomonas fluorescens. The use of isotopes in soil organic matter. Studies FAO/IAEA Technical meeting, 259–269. Brunswick-Volkenrode, 1963. Pergamon, London.

Mayaudon, J. (1968). Étude radiorespirometrique compareé de la mineralization dans le sol du glucose marqué en (1), (2), (3–4), (6) et (u). *Annales de L'Institut Pasteur* **4**, 710–730.

Mayaudon, J. (1971). Use of radiorespirometry in soil microbiology and biochemistry. *In* "Soil Biochemistry" (Eds A. D. McLaren and J. Skujins) Vol. 2, 202–256. Marcel Dekker, New York.

Mayaudon, J. and Simonart, P. (1958). Study of the decomposition of organic matter in soil by means of radioactive carbon. *Plant and Soil* **9**, 376.

Mayaudon, J. and Simonart, P. (1959a). Étude de la decomposition de la matière organique dans le sol au moyen de carbone radioactif: III. Décomposition des substances solubles dialysables des proteins et des hemicelluloses. *Plant and Soil* **11**, 170–5.

Mayaudon, J. and Simonart, P. (1959b). Étude de la decomposition de la matière organique dans le sol au moyen de carbone radioactif: V. Decomposition de cellulose et de lignine. *Plant and Soil* **11**, 181–192.

Mayaudon, J. and Simonart, P. (1965). Humification in the soil of a C^{14}-labelled polysaccharide complex of microbial origin. *Mededelingen van de Landbouwhoogeschool en der opzoekingsstations van de Staat te Gent.* **30**, 941–955.

Mehta, N. C. and Deuel, H. (1960). Zur Pentosanbestimmung in Boden. *Zeitschrift für Pflanzenernährung, Düngung und Bodenkunde* **90**, 209–218.

Mehta, N. C., Streuli, H., Muller, M. and Deuel, H. (1960). Role of polysaccharides in soil aggregation. *Journal of the Science of Food and Agriculture* **11**, 40–47.

Mehta, N. C., Dubach, P. and Deuel, H. (1961). Carbohydrates in the soil. *Advances in Carbohydrate Chemistry* **16**, 335–355.

Meneghel, R., Petit-Sarlotte, C. and Bloch, J. M. (1971). Attempts to characterize the degradation products of a humic acid – Evidence for glucides and amino acids. *Analytical letters* **4**, 317–323.

Meneghel, R., Petit-Sarlotte, C. and Bloch, J. M. (1972). Sur la caractérisation et l'isolement des produits de dégradation d'un acide humique après oxydation peracétique. *Bulletin de la Société chimique de France* **7**, 2997–3001.

Millar, W. N. and Casida, L. E. (1970). Evidence for muramic acid in soil. *Canadian Journal of Microbiology* **16**, 299–303.

Mingelgrin, U. and Dawson, J. E. (1973). The isolation and properties of a neutral polysaccharide from a woody peat soil. *Soil Science* **116**, 36–43.

Moghimi, A., Tate, M. E. and Oades, J. M. (1978). Characterization of rhizosphere products especially 2-ketogluconic acid. *Soil Biology and Biochemistry* **10**, 283–287.

Moghimi, A. and Tate, M. E. (1978). Does 2-ketogluconate chelate calcium in the pH range 2·4 to 6·4? *Soil Biology and Biochemistry* **10**, 289–292.

Molloy, L. F., Bridger, B. A. and Cairns, A. (1977). Studies on a climosequence of soils in tussock grasslands 13. Structural carbohydrates in tussock leaves, roots, and litter and in the soil light and heavy fractions. *New Zealand Journal of Science* **20**, 443–451.

Monnier, G. (1965). Action des matières organiques sur la stabilité des sols. Thèse, Faculty Science, Paris.

Monnier, G., Turc, L. and Jeanson-Lussinang, C. (1962). Une méthode de fractionnement densimétrique par centrifugation des matières organiques du sol. *Annales agronomiques* **13**, 55.

Montgomery, R. (1970). *In* "The Carbohydrates" (Eds W. Pigman and D. Horton), Vol. 2B.

Mortensen, J. L. (1960). Physico-chemical properties of a soil polysaccharide. Trans. Seventh International Congress of Soil Science **2**, 14, 98–104.

Mortensen, J. L. and Schwendinger, R. B. (1963). Electrophoretic and spectroscopic characterization of high molecular weight components of soil organic matter. *Geochimica et Cosmochimica* **27**, 201–208.

Muller, G. (1958). Beziehungen zwischen Biologie und Struktur des Bodens. Tagungsberichte Nr 13 Inst. für Acker und Pflanzeban Munchenberg, 167–192.

Muller, M., Mehta, N. C. and Deuel, H. (1960). Chromatographische Fraktionierung von Bodenpolysacchariden an Cellulose-Anionenaustauschern. *Zeitschrift für Pflanzenernährung, Düngung und Bodenkunde* **90**, 139–145.

Mundie, C. M. (1976). The identification and determination of glucuronic and galacturonic acids in Scottish soils and soil fractions using ion-exchange and gas-liquid chromatography. *Journal of Soil Science* **27**, 331–336.

Mundie, C. M., Cheshire, M. V., Anderson, H. A. and Inkson, R. H. E. (1976). Automated determination of monosaccharides using p-hydroxybenzoic acid hydrazide. *Analytical Biochemistry* **71**, 604–607.

Murayama, S. (1977a). An automated anion-exchange chromatographic procedure for the estimation of saccharides in acid hydrolysates of soil. *Soil Science and Plant Nutrition* **23**, 247–252.

G

Murayama, S. (1977b). Saccharides in some Japanese paddy soils. *Soil Science and Plant Nutrition* **23**, 479–489.

Murayama, S. and Inoko, A. (1975). Untersuchungen uber den Einfluss einer langjahrigen Kalkdungung auf die Humuszusammensetzung und den Kohlenhydratgehalt im reisboden. *Soil Science of Plant Nutrition* **21**, 239–251.

Muramaya, S., Cheshire, M. V., Mundie, C. M., Sparling, G. P. and Shepherd, H. (1979). Comparison of the contribution to soil organic matter fractions, particularly carbohydrates, made by plant residues and microbial products. *Journal of the Science of Food and Agriculture*.

Mutatker, V. K. and Wagner, G. H. (1967). Humification of Carbon-14 labelled glucose in soils of Sanborn Field. *Soil Science Society of America Proceedings* **31**, 66–70.

Nagar, B. R. (1962). Free monosaccharides in soil organic matter. *Nature, London* **194**, 896–897.

Neukom, H., Deuel, H., Heri, W. J. and Kundig, W. (1960). Chromatographische Fraktionierung von Polysacchariden an Cellulose-Anionenaustauschern. *Helvetica chimica acta* **43**, 64–71.

Neunylov, B. A. and Khavkina, N. V. (1968). Study of the rate of decomposition and conversion processes of organic matter tagged with C^{14} in the soil. *Soviet Soil Science*, 234–235. Translated from *Pochvovedenie* 1968 **2**, 103–108.

Nissenbaum, A. and Schallinger, K. M. (1974). The distribution of the stable carbon isotope ($^{13}C/^{12}C$) in fractions of soil organic matter. *Geoderma* **11**, 137–145.

Oades, J. M. (1967a). Gas–liquid chromatography of alditol acetates and its application to the analysis of sugars in complex hydrolysates. *Journal of Chromatography* **28**, 246–252.

Oades, J. M. (1967b). Carbohydrates in some Australian soils. *Australian Journal of Soil Research* **5**, 103–115.

Oades, J. M. (1972). Studies on soil polysaccharides III. Composition of polysaccharides in some Australian soils. *Australian Journal of Soil Research* **10**, 113–126.

Oades, J. M. (1974). Synthesis of polysaccharides in soil by micro-organisms. Trans. Tenth International Congress of Soil Science, Moscow **3**, 93–100.

Oades, J. M. and Swincer, G. D. (1968). Effect of time of sampling and cropping sequences of the carbohydrates in red brown earths. Trans. Ninth International Congress of Soil Science, Adelaide **3**, 183–192.

Oades, J. M. and Wagner, G. H. (1970). Incorporation of ^{14}C into sugars in a soil incubated with ^{14}C glucose. *Geoderma* **4**, 417–423.

Oades, J. M. and Wagner, G. H. (1971). Biosynthesis of sugars in soils incubated with ^{14}C glucose and ^{14}C dextran. *Soil Science Society of America Proceedings* **35**, 914–917.

Oades, J. M., Kirkman, M. A. and Wagner, G. H. (1970). The use of gas–liquid chromatography for the determination of sugars extracted from soils by sulfuric acid. *Soil Science Society of America Proceedings* **34**, 230–235.

Oden, V. S. and Lindberg, S. (1926). Einige Trofanalysen im Lichte neuzeitlicher Theorien der Kohlebildung. *Brennstoff – Chemie.* **7**, 165–180.

Ogston, A. G. (1958). Sedimentation in the ultracentrifuge. *Biochemical Journal* **70**, 598–599.

Olness, A. and Clapp, C. E. (1973). Occurrence of collapsed and expanded crystals in montmorillonite-dextran complexes. *Clays and Clay Minerals* **21**, 289–293.

Olness, A. and Clapp, C. E. (1975). Influence of polysaccharide structure on dextran adsorption by montmorillonite. *Soil Biology and Biochemistry* **7**, 113–118.

Orlov, D. S. and Sadovnikova, L. K. (1975). Content and distribution of carbohydrates in the major soil groups of the U.S.S.R. *Soviet Soil Science* 440–449. Translated from *Pochvovedenie* 1975 **8**, 81–90.

Orlov, D. S., Sadovnikova, L. K. and Sadovnikova, Y. N. (1975). Carbohydrates in soils. *Agrokhimiya* **3**, 139–152.

Otsuka, H. (1975). The state of humus in the soil profile and the sugar, uronic acid and amino acid contents and amino acid composition of fulvic acids. 5. Studies on a volcanic soil at Ohnobaru, Tarumizu City, Kagoshima Prefecture. *Journal of the Science of Soil and Manure* **46**, 180–184.

Parfitt, R. L. (1972). Adsorption of charged sugars by montmorillonite. *Soil Science* **113**, 417–421.

Parfitt, R. L. and Greenland, D. J. (1970). Adsorption of polysaccharides by montmorillonite. *Soil Science Society of America Proceedings* **34**, 862–865.

Parks, W. L. and White, J. L. (1952). Boron retention by clay and humus systems saturated with various cations. *Soil Science Society of America Proceedings* **16**, 298–300.

Parsons, J. W. and Tinsley, J. (1960). Extraction of soil organic matter with anhydrous formic acid. *Soil Science Society of America Proceedings* **24**, 198–201.

Parsons, J. W. and Tinsley, J. (1961). Chemical studies of polysaccharide material in soils and composts based on extraction with anhydrous formic acid. *Soil Science* **92**, 46–53.

Paul, E. and Kullmann, A. (1968). Die Bildung von Bodenkrumeln durch Beimischen von Luzernwurzeln bzw. deren Preßsaft. 2. Mitteilung: Qualitative und quantitative Untersuchungen uber Kohlenhydrate in Zusammenhang mit der Krumelbildung. *Albrecht-Thaer-Archiv.* **12**, 41–51.

Paul, E. A. and Veen, J. A. van (1978). The use of tracers to determine the dynamic nature of organic matter. Trans. Eleventh International Congress of Soil Science, Edmonton, Canada **3**, 61–102.

Peele, T. C. (1940). Microbial activity in relation to soil aggregation. *Journal of the American Society for Agronomy* **32**, 204–212.

Peele, T. C. and Beale, O. W. (1940). Influence of microbial activity upon aggregations and erodibility of lateritic soils. *Soil Science Society of America Proceedings* **5**, 33–35.

Pengra, R. M., Cole, M. A. and Alexander, M. (1969). Cells walls and lysis of *Mortierella parvispora* hyphae. *Journal of Bacteriology* **97**, 1056–1061.

Persson, J. (1968). Biological testing of chemical humus analysis. *Lantbrukshögskolans annaler*. **34**, 81–217.

Piper, T. J. and Posner, A. M. (1968). On the amino acids found in humic acid. *Soil Science* **106**, 188–192.

Quastel, J. H. (1946). Soil Metabolism. The Royal Institute of Chemistry of Great Britain and Ireland, London.

Quastel, J. H. (1965). Soil Metabolism. *Revue of Plant Physiology* **16**, 217–240.

Quastel, J. H. and Webley, D. M. (1947). The effects of the addition to soil of alginic acid and of other forms of organic matter on soil aeration. *Journal of Agricultural Science* **37**, 257–266.

Reese, E. T. (1968). Microbial transformation of soil polysaccharides. Organic Matter and Soil Fertility. *Pontificiae Academiae Scientiarum Scripta Varia* **32**, 535–577.

Rennie, D. A., Truog, E. and Allen, O. N. (1954). Soil aggregation as influenced by microbial gums, level of fertility and kind of crop. *Soil Science Society of America Proceedings* **18**, 399–403.

Robert, M. (1964). Étude biologique des sols au cours de l'épreuve d'incubation. *Annales de l'Institut Pasteur* **5**, 801–806.

Ross, D. J. and Molloy, L. F. (1977). Studies on a climosequence of soils in tussock grasslands 14. Water-soluble carbohydrates in tussock plant materials and in a light fraction from soil. *New Zealand Journal of Science* **20**, 453–459.

Roulet, N., Dubach, P., Mehta, N. C., Muller-Vonmoos, M. and Deuel, H. (1963a). Verteilung der organischen Substanz und der Kohlenhydrate bei der Gewinnung "wurzelfreien" Bodenmaterials durch Sclammsiebung. *Zeitschrift für Pflänzenernahrung, Düngung und Bodenkunde* **101**, 210–215.

Roulet, N., Mehta, N. C., Dubach, P. and Deuel, H. (1963b). Abtrennung von Kohlehydraten und Stickstoffverbindungen aus Huminstoffen durch Gelfiltration und Ionenaustausch-Chromatographie. *Zeitschrift für Pflanzenernährung, Düngung und Bodenkunde* **103**, 1–9.

Rovira, A. D. and Greacen, E. L. 1957. The effect of aggregate disruption on the activity of micro-organisms in the soil. *Australian Journal of Agricultural Research* **8**, 659–673.

Russell, J. D. and Anderson, H. A. (1974). Comment on 'Sceptroscopie infra-rouge de quelques acides humiques' by J. R. Bailly. *Plant and Soil* **41**, 695–696.

Sadovnikova, L. K. and Orlov, D. S. (1975). Carbohydrates in the soils of deciduous forests. *Moscow University Soil Science Bulletin* **30**, 44–45.

Sadovnikova, L. K. and Orlov, D. S. (1976). Group composition of humus of zonal soils and the role of carbohydrates in the formation of various fractions. *Moscow University Soil Science Bulletin* **31**, 53–65.

Saini, G. R. (1966). Sequestering of iron and aluminium by soil polysaccharides. *Current Science* **10**, 259–260.

Saini, G. R. and MacLean, A. A. (1966). Adsorption – flocculation reaction of soil polysaccharides with kaolinite. *Soil Science Society of America Proceedings* **30**, 697–699.

Saini, G. R. and Salonius, P. O. (1969). Relation between Staudinger indices and precipitabilities of polysaccharides extracted from forest soils. *Soil Science Society of America Proceedings* **33**, 693–695.

Sallans, H. R., Snell, J. M., Mackinney, H. W. and McKibbin, R. R. (1937). Water soluble acid substances in the raw humus of podzol soils. *Canadian Journal of Research* **15**, 315–320.

Salomon, M. (1962). Soil aggregation – Organic matter relationships in redtop-potato rotations. *Soil Science Society of America Proceedings* **26**, 51–54.

Salton, M. R. J. (1964). "The Bacterial Cell Wall." Elsevier, Amsterdam.

Sawyer, C. D. and Pawluk, S. (1963). Characteristics of organic matter in degrading chernozemic surface soils. *Canadian Journal of Soil Science* **43**, 275–286.

Scheffer, F. and Kickuth, R. (1961). Chemische Abbauversuche an einer naturlichen Huminsaure. *Zeitschrift für Pflanzenernährung, Düngung und Bodenkunde* **94**, 180–188.

Schlichting, E. (1953). The fulvic acid fraction. *Zeitschrift für Pflanzenernährung, Düngung und Bodenkunde* **106**, 97–107.

Schreiner, O. and Lathrop, E. C. (1911). The distribution of organic constituents in soils. *Journal of the Franklin Institute* **172**, 145–151.

Sequi, P., Guidi, G. and Petruzzelli, G. (1975). Distribution of amino acid and carbohydrate components in fulvic acid fractionated on polyamide. *Canadian Journal of Soil Science* **55**, 439–445.

Shorey, E. C. (1913). Some organic soil constituents. U.S. Department of Agriculture, Bureau of Soils – Bulletin No. 88.

Shorey, E. C. and Lathrop, E. C. (1910). Pentosans in soils. *Journal of the American Chemical Society* **32**, 1680–1683.

Shorey, E. C. and Martin, J. B. (1930). The presence of uronic acids in soils. *Journal of the American Chemical Society* **52**, 4907–4915.

Singh, S. and Bhandari, G. S. (1963). Investigations on acid stable reducing organic fractions in some soils of Rajasthan. *Journal of the Indian Society of Soil Science* **11**, 293–298.

Singh, S. and Singhal, R. M. (1974). Measurement of carbohydrates in the acid hydrolysates of some Outer Himalayan soils. *Journal of the Indian Society of Soil Science* **22**, 290–293.

Sinha, M. K. (1972). Organic matter transformations in soil. II. Nature of carbohydrates in soils incubated with C^{14}-labelled oat roots under aerobic and anaerobic conditions. *Plant and Soil* **36**, 295–299.

Skujins, J. and Pukite, A. (1970). Extraction and determination of *N*-

acetylglucosamine from soil. *Soil Biology and Biochemistry* **2**, 141–143.

Smith, D. G., Bryson, C., Thompson, E. M. and Young, E. G. (1958). Chemical composition of the peat bogs of the maritime provinces. *Canadian Journal of Soil Science* **38**, 120–127.

Smith, F. and Montgomery, R. (1959). "The chemistry of plant gums and mucilages." Reinhold, New York.

Smithies, W. R. (1952). Chemical composition of a sample of mycelium of *Penicillium griseofulvum* Dierckx. *Biochemical Journal* **51**, 259–264.

Sørensen, H. (1963). Studies on the decomposition of C^{14}-labelled barley straw in soil. *Soil Science* **95**, 45–51.

Sørensen, H. (1965). Fixation of metabolic products in the soil during decomposition of carbohydrates. *Nature, London* **208**, 97–98.

Sørensen, L. H. (1972). Role of amino acid metabolites in the formation of soil organic matter. *Soil Biology and Biochemistry* **4**, 245–255.

Sørensen, L. H. (1974). Rate of decomposition of organic matter in soil as influenced by repeated air-drying – rewetting and repeated additions of organic material. *Soil Biology and Biochemistry* **6**, 287–292.

Sørensen, L. H. (1975). The influence of clay on the rate of decay of amino acid metabolites synthesized in soils during decomposition of cellulose. *Soil Biology and Biochemistry* **7**, 171–177.

Sowden, F. J. (1959). Investigations on the amounts of hexosamines found in various soils and methods for their determination. *Soil Science* **88**, 138–143.

Sowden, F. J. (1977). Distribution of nitrogen in representative Canadian soils. *Canadian Journal of Soil Science* **57**, 445–456.

Sowden, F. J. and Ivarson, K. C. (1962a). Decomposition of forest litters III. Changes in the carbohydrate constituents. *Plant and Soil* **16**, 389–400.

Sowden, F. J. and Ivarson, K. C. (1962b). Methods for the analysis of carbohydrate material in soil: 2. Use of cellulose column and paper chromatography for determination of the constituent sugars. *Soil Science* **94**, 340–344.

Sowden, F. J., Griffiths, S. M. and Schnitzer, M. (1976). The distribution of nitrogen in some highly organic tropical volcanic soils. *Soil Biology and Biochemistry* **8**, 55–60.

Spakhov, Y. M. and Spakhova, A. S. (1970). Composition of free water-soluble compounds in the rhizosphere of some tree species. *Soviet Soil Science*, 703–710. Translated from *Pochvovdenie* **11**, 46–53.

Spiro, R. G. (1973). Glycoproteins. *Advances in Protein Chemistry* **27**, 349–467.

Springer, U. (1950). Stoffabbau und Humusaufbau bei der Zersetzung Landwirtschaftlich wichtiger organischer stoffe. *Landwirtschaftliche Forschung* **2**, 50–56.

Stacey, M. and Barker, S. A. (1960). "Polysaccharides of Microorganisms." Oxford.

Stefanson, R. C. (1968). Factors determining seasonal changes in the stabilities of soil aggregates. Trans. Ninth International Congress Soil Science, Adelaide **2**, 395–402.

Stefanson, R. C. (1971). Effect of periodate and pyrophosphate on the seasonal changes in aggregate stabilization. *Australian Journal Soil Research* **9**, 33–41.

Stevenson, F. J. (1954). Ion exchange chromatography of the amino acids in soil hydrolysates. *Soil Science Society of America Proceedings* **17**, 31–34.

Stevenson, F. J. (1956). Isolation and identification of some amino compounds in soil. *Soil Science Society of America Proceedings* **20**, 201–204.

Stevenson, F. J. (1957a). Investigations of aminopolysaccharides in soils: 1. Colorimetric determination of hexosamines in soil hydrolysates. *Soil Science* **83**, 113–122.

Stevenson, F. J. (1957b). Investigations of aminopolysaccharides in soil: 2. Distribution of hexosamines in some soil profiles. *Soil Science* **84**, 99–106.

Stevenson, F. J. (1957c). Distribution of the forms of nitrogen in some soil profiles. *Soil Science Society of America Proceedings* **21**, 283–287.

Stevenson, F. J. (1960). Chemical nature of the nitrogen in the fulvic fraction of soil organic matter. *Soil Science Society of America Proceedings* **24**, 472–477.

Stevenson, F. J. (1973). Nonbiological transformations of amino acids in soils and sediments. *Advances in Organic Chemistry*, 701–714.

Stotzky, G. and Norman, A. G. (1961). Factors limiting microbial activities in soil. I. Level of substrate, nitrogen and phosphorus. *Archiv für Mikrobiologie* **40**, 341–369.

Strelkow, S. S. (1958). Die Kohlenhydrate des Torfes und ihre Verwertung. *Torfyanaya promȳshlennost'*. H7, 7–10.

Streuli, V. H., Mehta, N. C., Muller, M. and Deuel, H. 1958. Polysaccharides in the soil. *Mitteilungen aus dem Gebiete der Lebensmitteluntersuchung und Hygiene* **49**, 396–397.

Swaby, R. J. (1949). The relationship between microorganisms and soil aggregation. *Journal of General Microbiology* **3**, 236–254.

Swift, R. S. and Chaney, K. (1979). The role of soil organic colloids in the formation and stabilization of soil aggregates. *Journal of the Science of Food and Agriculture* **30**, 329–330.

Swincer, G. D., Oades, J. M. and Greenland, D. J. (1968a). Studies on soil polysaccharides: I. The isolation of polysaccharide from soil. *Australian Journal of Soil Research* **6**, 211–224.

Swincer, G. D., Oades, J. M. and Greenland, D. J. (1968b). Studies on soil polysaccharides: II. The composition and properties of polysaccharides in soils under pasture and under a fallow-wheat rotation. *Australian Journal of Soil Research* **6**, 225–235.

Swincer, G. D., Oades, J. M. and Greenland, D. J. (1969). The extraction, characterization and significance of soil polysaccharides. *Advances in Agronomy* **21**, 195–235.

Tan, E. L. and Loutit, M. W. (1976). Concentration of molybdenum by extra-cellular material produced by rhizosphere bacteria. *Soil Biology and Biochemistry* **8**, 461–464.

Tan, K. H. and Clark, F. E. (1968/69). Polysaccharide constituents in fulvic and humic acids extracted from soil. *Geoderma* **2**, 245–255.

Tepper, E. Z. (1957). Dynamics of uronic acids during aerobic decomposition of clover and timothy roots. *Pochvovedenie* **6**, 95–99, *Soils and Fertilizers* **20**, 337.

Theander, O. (1952). Studies on Sphagnum peat. I. Preliminary studies on the carbohydrate constituents. *Svensk kivinsk. tidskrift.* **64**, 197–199.

Theander. O. (1954). Studies on Sphagnum peat. III. A quantitative study on the carbohydrate constituents of Sphagnum mosses and Sphagnum peat. *Acta Chemica Scandinavica* **8**, 989–1000.

Thomas, R. L. (1964). Fractionation and characterization of the polysaccharides extracted from the Brookston soil. *Dissertation Abstracts* **24**.

Thomas, R. L. and Lynch, D. L. (1961). A method for the quantitative estimation of pentoses in soil. *Soil Science* **91**, 312–316.

Thomas, R. L. and Mortensen, J. L. (1964). The electrophoretic characterization of soil polysaccharide fractions. *Agronomy Abstracts* **26**.

Thomas, R. L., Mortensen, J. L. and Himes, F. L. (1967). Fractionation and characterization of a soil polysaccharide extract. *Soil Science Society of America Proceedings* **31**, 568–570.

Toogood, J. A. and Lynch, D. L. (1959). Effect of cropping systems and fertilizers on mean weight-diameter of aggregates of Breton plot soils. *Canadian Journal of Soil Science* **39**, 151–156.

Tracey, M. V. (1950). A colorimetric method for the determination of pentoses in the presence of hexoses and uronic acids. *Biochemical Journal* **47**, 433–436.

Tschaikowa, W. D. and Rakowski, W. E. (1959). Untersuchungen der Kohlenhydratkomplexe der Hochmoore nach der Methode der Papierchromatographie. *Arbeiten des Instituts für Torf der Akademie der Wisenschaften der BSSR. Abt. Physiko-math. und techn. Wissenschaften. Minsk* **7**, 11–18.

Turchenek, L. W. and Oades, J. M. (1974/5). Studies on soil organic materials. P. 29 Biennial Report. Waite Agricultural Research Institute.

Visser, S. A. (1968). Stability of different types of carbohydrates in various tropical soils. *Annales de l'Institute Pasteur* **115**, 766–786.

Wagner, G. H. and Tang, D-E. W. (1976). Soil polysaccharides synthesized during decomposition of glucose and dextran and determined by ^{14}C-labelling. *Soil Science* **121**, 222–226.

Waksman, S. A. (1926). The origin and nature of the soil organic matter or soil "humus". I. Introduction and historical. *Soil Science* **22**, 123–162.

Waksman, S. A. (1936). "Humus." Bailliere, Tindall and Cox, London.

Waksman, S. A. and Diehm, R. A. (1931a). On the decomposition of hemicelluloses by microorganisms. I. Nature, occurrence, preparation and decomposition of hemicelluloses. *Soil Science* **32**, 73–95.

Waksman, S. A. and Diehm, R. A. (1931b). On the decomposition of hemicelluloses by microorganisms. II. Decomposition of hemicelluloses by fungi and actinomyces. *Soil Science* **32**, 97–117.

Waksman, S. A. and Diehm, R. A. (1931c). On the decomposition of hemicelluloses by microorganisms. III. Decomposition of various hemicelluloses by aerobic and anaerobic bacteria. *Soil Science* **32**, 119–139.

Waksman, S. A. and Stevens, K. R. (1930). A critical study of the methods for determining the nature and abundance of soil organic matter. *Soil Science* **30**, 97–116.

Watson, J. H. and Stojanovic, B. J. (1965). Synthesis and bonding of soil aggregates as affected by microflora and its metabolic products. *Soil Science* **100**, 57–62.

Webber, L. R. (1965). Soil polysaccharides and aggregation in crop sequences. *Soil Science Society of America Proceedings* **29**, 39–42.

Webley, D. M. and Duff, R. B. (1965). The incidence in soils and other habitats of micro-organisms producing 2-ketogluconic acid. *Plant and Soil* **22**, 307–313.

Webley, D. M. and Jones, D. (1971). Biological transformation of microbial residues in soil. *In* "Soil Biochemistry" (Eds A. D. McLaren and J. Skujins), Vol. 2. Marcel Dekker, New York.

Webley, D. M., Duff, R. B., Bacon, J. S. D. and Farmer, V. C. (1965). A study of polysaccharide-producing organisms occurring in the root region of certain pasture grasses. *Journal of Soil Science* **16**, 149–157.

Weckesser, J., Rosenfelder, G., Mayer, H. and Luderitz, O. (1971). The identification of 3-*O*-methyl-D-xylose and 3-*O*-methyl-L-xylose as constituents of the lipopolysaccharides of *Myxococcus fulvus* and *Rhodopseudomonas viridis*, respectively. *European Journal of Biochemistry* **24**, 112–115.

Wen, C. H. and Chen, L. L. (1962). The determination of pentoses in soils with aniline. *Acta Pedologica Sinica* **10**, 220–226.

Went, J. C. and De Jong, F. (1966). Decomposition of cellulose in soils. *Antonie van Leeuwenhoek* **32**, 39–56.

Whistler, R. L. and Kirby, K. W. (1956). Composition and behaviour of soil polysaccharides. *Journal of the American Chemical Society* **78**, 1755–1759.

Whitehead, D. C., Buchan, H. and Hartley, R. D. (1975). Components of soil organic matter under grass and arable cropping. *Soil Biology and Biochemistry* **7**, 65–71.

Wrenshall, C. L. and Dyer, W. J. (1941). Organic phosphorus in soils: II. The nature of the organic phosphorus compounds. A. Nucleic acid derivatives. B. Phytin. *Soil Science* **51**, 235–248.

Wright, J. R., Schnitzer, M. and Levick, R. (1958). Some characteristics of the organic matter extracted by dilute inorganic acids from a podzolic B horizon. *Canadian Journal of Soil Science* **38**, 14–22.

Yoder, R. E. (1936). A direct method of aggregation analysis of soils and a study of the physical nature of erosion losses. *Journal of the American*

Society of Agronomy **28,** 337–51.

Yoshida, R. K. (1940). Studies on organic phosphorus compounds in soil; isolation of inositol. *Soil Science* **50,** 81–89.

Yoshizaki, K., Sakagami, K., Hamada, R., Kurobe, T. and Matsui, T. (1970). A characterization of polysaccharides by U.V. method and nature of humus obtained from buried humic horizons of ashitaka loam formation. Reprinted from Miscellaneous Reports of the Research Institute for Natural Resources No. 74, 32–43.

Yukhnin, A. A., Zaslavskiy, YE. M. and Ammosova, YA. M. (1973). Determination of carbohydrates in soils and soil components. *Biologicheskie Nauki* (*Moscow*) **16,** 131–134.

APPENDIX TABLE I *Carbohydrate content of soil determined by colorimetric methods of analysis.*

Soil description	C (%)	Carbohydrate C	Pentose C as % of total C	Uronic acid C	Hexosamine N as % of total N	Reference
Silt loam	1·87	6·24	—	—	—	Brink et al. (1960)
Sandy loam	2·78	3·99	—	—	—	Brink et al. (1960)
			Acrisols			
Black ash soil	2·06	—	1·50	—	—	Wen and Chen (1962)
			Andosols			
Pontcassé 0–15 cm	—	—		—	7·7	Sowden et al. (1976)
La Plaine 0–15	—	—		—	5·9	
15–25	—	—		—	6·3	
Boetica 0–20	—	—		—	7·1	
20–40	—	—		—	8·0	
Montreal 0–8	—	—		—	7·6	
30–45	—	—		—	6·6	
Sandy loam	—	—		—	7·5	Bremner and Shaw (1954)
			Cambisols			
Sol brun acide A₁	2·1	12·0	—	—	7·0	Gallali et al. (1972)
B	0·12	12·1	—	—	8·0	Jacquin et al. (1974)
Soil developed in Eocene	0·95	14·32	—	1·14	8·24	Whitehead et al. (1975)
deposits overlying chalk	1·73	14·85	—	1·46	7·86	
	4·77	16·10	—	1·36	8·30	
Brown forest soil	—	—	—	4·0	—	Mundie (1976)
Braunerde	3·2	—	4·2	—	—	Roulet et al. (1963a)
Nine pasture soils (eutric dystrochrepts)	4·6	10·8	—	1·04	—	McGrath (1973)
Brown humus soil	9·2	12·1	1·4	2660 p.p.m.	7·4	Ivarson and Sowden (1962)
Upper humus layer						Sowden (1959)
Brown earth	3·2	—	2·4	—	—	Mehta and Deuel (1960)
Kedslie	—	—	—	3·8	—	Mundie (1976)

APPENDIX TABLE I (*continued*)

Soil description	C (%)	Carbohydrate C	Pentose C as % of total C	Uronic acid C	Hexosamine N as % of total N	Reference
			Cambisols			
Carpow	—	—	—	2·3	—	Gallali et al. (1972)
			Chernozems			
Chernozem A₁	3·6	13·8	—	—	8·7	Jacquin et al. (1974)
Orthic black Ah	4·21	8·1	2·7	2·5	—	Lowe (1969)
Lacustrine chernozem brown zone Ah	2·30	2·3	—	—	—	Dormaar (1967)
Eolian chernozem brown zone Ah	1·49	1·8	—	—	—	Dormaar (1967)
Black cultivated soil 0–15·2 cm	6·3	12·0	1·8	1400 p.p.m. of the soil	7·4	Ivarson and Sowden (1962), Sowden (1959)
Degraded black	—	—	—	2515 p.p.m. of the soil	—	Dormaar and Lynch (1962)
Chernozem 0–20 A₁	4·09	15·5ᵃ	—	—	—	Orlov and Sadovnikova (1975)
2–50 A₁	6·41	11·6ᵃ	—	—	—	
0–38 A₁	1·24	22·6ᵃ	—	—	—	
0–50 A₁	1·79	11·8ᵃ	—	—	—	
0–26 Ap	2·67	14·7ᵃ	—	—	—	
0–24 Ap	1·86	19·1ᵃ	—	—	—	
0–28 Ap	2·38	12·5ᵃ	—	—	—	
0–23 Ap	3·51	6·5ᵃ	—	—	—	
			Gleysols			
Montserrat 0–15cm	—	—	—	—	5·6	Sowden et al. (1976)
30–60cm	—	—	—	—	4·0	
Gleysol 0–10 A₁	1·03	18·3ᵃ	—	—	—	Orlov and Sadovnikova (1975)
0–3 A₁	1·71	32·5ᵃ	—	—	—	
0–20 A₁	3·76	29·3ᵃ	—	—	—	
Sod meadow 0–10 Ap	3·33	16·6ᵃ	—	—	—	
Gleysol 0–10 Ap	6·09	10·8ᵃ	—	—	—	
10–20 Ap	3·71	15·1ᵃ	—	—	—	
Gleysol 0–10 Ap	4·66	16·2ᵃ	—	—	—	
Mountain meadow clay loam 6–20 A₁	5·02	16·8ᵃ	—	—	—	
Eight meadow soils (typic haplaquepts)	5·4	14·5	—	—	—	McGrath (1973)

Soil						Reference
Grey forest soils A_1	2·7–4·2	16·3–25·0[a]	—	—	—	Sadovnikova and Orlov (1975)
Greyzem 1–20 A_1	1·91	20·9[a]	—	—	—	
			Ferralsols			
Red soil	0·90	—	5·23	—	—	Wen and Chen (1962)
Brick red soil	3·52	—	3·53	—	—	Orlov and Sadovnikova (1975)
Yellow earth 0–15 A_1	1·90	25·9[a]	—	—	—	
Red earth 0–14 A_1	4·58	25·5[a]	—	—	—	
Red clay 2–18 A_1	2·16	23·9[a]	—	—	—	
			Fluvisols			
Paddy soil	1·00	—	5·31	—	—	Wen and Chen (1962)
Red paddy soil	0·71	—	7·98	—	—	
			Histosols			
Fen	—	—	—	—	7·1	Bremner and Shaw (1954)
Low moor peat	—	—	—	—	7·4	
Mountain peat	—	—	—	—	10·2	
Phragmities peat 130–180cm	52·6	—	6·0	—	—	Cheshire and Mundie (1966)
Peat AT 7–20	21·8	20·2[a]	—	—	—	Orlov and Sadovnikova (1975)
			Kastanozems			
Orthic brown Ah	2·75	6·6	1·6	1·6	—	Lowe (1969)
Dark brown soil, virgin 0–15·2cm	5·7	14·4	2·7	1600 p.p.m. of the soil	9·0	Ivarson and Sowden (1962) / Sowden (1958)
Kastanozem 0–22 A_1	1·39	21·9[a]	—	—	—	Orlov and Sadovnikova (1975)
			Luvisols			
Dark grey wooded 6·5–5cm L	32·0	6·8	—	1·1	—	Sawyer and Pawluck (1963)
5–0cm FH	19·3	5·4	—	3·7	—	
0–4cm Ah	10·3	11·0	—	7·6	—	
4–8cm Ah	5·8	8·6	—	1·9	—	
8–12cm Ah	3·8	6·6	—	1·6	—	
12–16cm Ahe	2·2	5·0	—	2·4	—	
16–20cm Ahe	2·3	5·4	—	3·3	—	

Soil description	C (%)	Carbohydrate C	Pentose C as % of total C	Uronic acid C	Hexosamine N as % of total N	Reference
			Luvisols			
20–24cm Aeh	1·2	4·2	—	2·1	—	Sadovnikova and Orlov (1975)
24–28cm Aeh	1·0	5·4	—	2·2	—	
28–32cm Ae	1·0	4·0	—	2·1	—	
36–40cm Ae	0·8	5·0	—	2·0	—	
Light grey forest soils A₁	3·3–3·8	12·3–21·0[a]	—	—	—	
Dark grey forest soils A₁	6·1–7·5	11·6–13·0[a]	—	—	—	
Brown forest soils A₁	2·3–12·3	7·1–25·5[a]	—	—	—	
Dark brown forest soil A₁	4·3	17·4[a]	—	—	—	
Soil brun lessivé A₁	2·7	13·0	—	—	6·0	Gallali et al. (1972) Jacquin et al. (1974)
B	0·8	20·5	—	—	7·1	
Grey wooded L–H	30·1	—	—	5·02	—	Gupta and Sowden (1965)
Ah	2·4	—	—	4·58	—	
Orthic grey wooded Ah	8·86	12·2	3·0	2·0	—	Lowe (1969)
Grey wooded	1·42	7·32	3·56	2·46	4·16	Khan (1969)
Pasture soils	1·14	7·16	2·95	2·11	3·73	
Five pasture soils (typic hapludalfs) and (typic ochraqualfs)	6·2	14·3	—	1·32	—	McGrath (1973)
Grey wooded	—	—	—	1100 p.p.m. of the soil	—	Dormaar and Lynch (1962)
1–19cm A₁	3·56	10·6[a]	—	—	—	Orlov and Sadovnikova (1975)
Broadbalk clay loam	—	—	—	—	7·0	Bremner and Shaw (1954)
			Phaeozems			
Brunizem 0–16in A₁	—	—	—	—	9·5	Stevenson (1957b)
16–21in A₃	—	—	—	—	9·3	
21–25in B₁	—	—	—	—	9·4	
25–36in B₂	—	—	—	—	10·9	
36–47in B₃	—	—	—	—	10·7	

Sample						Reference
Phaeozems						
47–53in C_1	—	—	—	—	9·5	Brink et al. (1960)
53–68in C_{2-1}	—	—	—	—	9·1	
68–82 C_{2-2}	—	—	—	—	6·5	
82–92in D	—	7·59	—	—	6·8	
Prairie soil	3·03	—	—	—	—	
Planosols						
Planosol 3–6in A_1	—	—	—	—	11·3	Stevenson (1957b)
6–12in	—	—	—	—	12·4	
12–15in A_2	—	—	—	—	13·3	
15–22in	—	—	—	—	17·4	
22–25in B_2	—	—	—	—	20·2	
25–33in	—	—	—	—	21·1	
33–36in B_3	—	—	—	—	24·3	
36–45in	—	—	—	—	19·2	
45–48in C	—	—	—	—	20·8	
Grey solod Ah	6·3	8·1	2·6	1·8	—	Lowe (1969)
Black solod Ah	5·02	9·3	3·0	2·6	—	
Brown solod Ah	2·79	9·3	2·4	2·0	—	
Black solodic	—	—	—	4290 p.p.m. of the soil	—	Dormaar and Lynch (1962)
Podzols						
Podzol 0–5 in A_1	—	—	—	—	8·6	Stevenson (1957b)
6–12in B_1	—	—	—	—	6·5	
Northern podzol F	46·1	—	—	4·43	—	Gupta and Sowden (1965)
H	34·3	—	—	5·83	—	
Orthic podzol Bh	8·0	—	—	5·38	—	
Sandy granitic loam 5–10cm	3·6	8·0	3·6	3·7	—	Mundie (1976) / Cheshire and Mundie (1966)
Podzol	12·9	—	1·0	—	—	Roulet et al. (1963a)
Podzol B	5·79	3·51	—	—	—	Brink et al (1960)
Podzol A_2	9·6	12·6	1·7	3560 p.p.m. of the soil	10	Ivarson and Sowden (1962) / Sowden (1959)

APPENDIX TABLE I (continued)

Soil description	C (%)	Carbohydrate C	Pentose C as % of total C	Uronic acid C	Hexosamine N as % of total N	Reference
			Podzols			
Podzol	1·8	—	1·5	—	—	Mehta and Deuel (1960)
Podzol	6·0	—	2·1	—	—	
Podzol	1·9	—	1·1	—	—	
Podzol 3–15 A$_1$A$_2$	0·41	6·9[a]	—	—	—	Orlov and Sadovnikova (1975)
2–9 A$_1$	1·97	6·9[a]	—	—	—	
0–3 A$_1$	3·93	15·2[a]	—	—	—	
0–10 A$_2$	2·72	20·6[a]	—	—	—	
Forest soil L	53·7	18·8	3·84	—	1·4	Lowe (1972)
F	52·2	15·1	3·27	—	2·8	
H	53·2	10·3	2·37	—	6·8	
			Regosols			
Coniferous forest soil F	51·1	—	6·3	—	—	Cheshire and Mundie (1966)
			Rendzinas			
Rendzina A$_1$	6·2	—	8·9	—	7·0	Gallali et al. (1972) / Jacquin et al. (1974)
Humus carbonate soil	—	14·6	—	3·0	—	Roulet et al. (1963a)
Rendzina	—	5·8	—	5·4	—	
			Solonetz			
Solonetz 0–5in A$_1$	—	—	—	—	8·0	Stevenson (1957b)
5–7in A$_2$	—	—	—	—	6·0	
7–9in B$_{21}$	—	—	—	—	6·4	
9–12in B$_{22}$	—	—	—	—	5·6	
12–22in B$_2$sa	—	—	—	—	5·5	
22–36in Csa$_1$	—	—	—	—	4·6	
36–50in Csa$_2$	—	—	—	—	5·9	
50–60in Cca	—	—	—	—	6·1	
Orthic 0–12in A$_1$	5·7	6·6	—	—	—	Graveland and Lynch (1961)
Solod 0–11in A$_1$	4·0	7·2	—	—	—	
Solodized solonetz 0–14in A$_1$	6·3	6·4	—	—	—	
Solodized solonetz Ah	4·1	—	—	5·61	—	Gupta and Sowden (1965)

			Solonetz			
Grey solonetz Ah	21·3	10·7	3·0	1·6	—	Lowe (1969)
Black solonetz Ah	4·49	11·4	3·1	1·9	—	
Brown solonetz Ah	3·77	8·0	3·6	2·2	—	
Black solodized solonetz	—	—	—	1110 p.p.m. of the soil	—	Dormaar and Lynch (1962)
Solonetz 0–10 A$_1$	4·68	17·0[a]	—	—	—	Orlov and Sadovnikova (1975)
			Vertisols			
Pelosol A$_1$	4·3	10·2	—	—	6·8	Gallali et al. (1972)
Vertisol A$_1$	3·7	10·9	—	—	10·2	Jacquin et al. (1974)
			Xerosols			
0–10 Ap	0·25	14·4[a]	—	—	—	Orlov and Sadovnikova (1975)
0–7 A$_1$	0·51	18·8[a]	—	—	—	
0–20 A$_1$	1·45	25·1[a]	—	—	—	

[a]Values by phenol-sulphuric acid.

APPENDIX TABLE II. Sugar composition of soils and peats.

Soil	Horizon	Relative proportion (%)								Total sugar (%)	Reference[a]
		Gal.	Glc.	Man.	Ara.	Rib.	Xyl.	Rha.	Fuc.		
Marshall series phaeozem, USA	Ap	17·4	27·6	17·5	15·0	1·8	12·9	5·5	2·3	0·727	A
	B21	18·9	25·0	21·9	15·2	2·0	8·6	5·3	3·0	0·529	—
	B22	20·3	21·8	22·3	16·2	1·8	9·1	5·2	3·4	0·350	—
Grundy series phaeozem, USA	Ap	16·7	29·2	16·8	15·5	2·0	10·7	6·4	2·6	0·740	—
	B1	15·9	26·7	22·0	15·7	1·6	9·6	5·6	2·9	0·701	—
	B21	16·9	24·5	22·5	15·3	1·9	9·8	6·1	3·1	0·624	—
Mexico series planosol, USA	A1	19·5	28·8	17·7	14·0	0	12·2	5·2	2·7	0·463	—
	B + A	13·8	29·3	28·4	17·5	0	6·5	4·5	0	0·267	—
	B21	14·8	25·0	25·7	18·8	1·9	7·5	4·1	2·2	0·409	—
Mexico series planosol, USA	A1	15·6	33·1	19·5	11·2	1·3	9·5	6·4	3·4	0·914	—
	A2	15·8	31·4	23·2	13·5	1·4	7·6	5·0	2·2	0·434	—
	B2	17·9	25·4	20·4	16·7	2·3	7·5	5·5	4·2	0·407	—
Menfro series luvisol, USA	Ap	18·8	30·7	16·1	12·4	1·1	11·4	6·4	3·2	0·417	—
	B1	16·3	27·2	25·8	11·6	2·2	10·1	4·0	2·7	0·148	—
	B21	18·3	23·7	24·7	11·9	3·1	9·4	5·8	3·1	0·111	—
Weldon series luvisol, USA	Ap	17·4	30·1	13·5	15·1	1·0	12·7	6·9	3·3	0·483	—
	A3	16·9	23·7	19·2	18·1	1·7	10·9	6·4	3·0	0·208	—
	B2	16·5	24·0	23·2	15·3	2·1	10·5	5·4	3·0	0·151	—
Marion series luvisol, USA	Ap	20·0	28·5	16·8	11·2	1·6	6·5	4·1	1·6	0·624	—
	A2	14·6	30·4	19·1	11·8	2·0	10·7	6·5	5·0	0·173	—
	B2	16·2	24·2	23·4	15·7	1·9	9·5	6·6	2·4	0·161	—
Gasconade series luvisol, USA	A1	18·7	22·9	15·6	11·7	3·9	17·2	5·9	4·1	1·065	—
	B	18·1	21·3	18·0	14·7	3·7	12·5	7·4	4·2	0·753	—
Bardley series luvisol, USA	A1	14·5	40·0	16·7	10·4	1·2	8·1	6·3	2·8	1·070	—
	B2	13·8	27·5	26·5	10·1	1·7	10·2	6·1	4·1	0·344	—
	B21	14·6	27·2	24·3	11·6	2·2	10·9	5·5	3·8	0·286	—

Series	Horizon										
Clarkesville series *luvisol*, USA	A₁	11.6	39.1	19.6	8.8	1.3	9.6	5.0	5.1	0.818	—
	B₁	11.8	37.6	20.9	9.7	1.4	9.0	5.5	4.3	0.367	—
	B₂₁	13.0	29.8	23.2	11.4	1.6	10.5	5.8	4.8	0.203	—
Coulstone series *luvisol*, USA	A₁	9.8	36.3	23.6	4.8	0	13.1	3.9	8.5	0.622	—
	A₂₂	14.1	41.5	18.0	5.5	1.7	7.9	3.9	6.9	0.319	—
	A₂₃	14.6	22.4	26.4	10.4	1.8	11.1	6.1	7.1	0.126	—
Coulstone series *luvisol*, USA	A₁	13.3	38.9	19.5	8.1	1.1	8.2	6.5	4.4	0.709	—
	A₂₂	15.9	36.4	18.7	9.2	1.2	9.2	4.9	4.5	0.401	—
	A₂₃	11.6	39.3	20.5	9.0	1.2	8.3	4.0	6.1	0.206	—
Coulstone series *luvisol*, USA	A₁	9.8	39.4	19.8	5.9	0.7	11.4	5.2	7.7	0.135	—
	A₂₂	8.3	47.2	19.2	5.4	1.7	8.7	3.3	6.3	0.483	—
	A₂₃	7.8	49.6	21.6	6.9	0	7.9	3.4	2.8	0.282	—
Orthic podzol *podzol*, Canada	L–H	14.9	54.0	14.9	5.0	—	3.7	4.3	3.1[b]	—	B
Northern podzol *podzol*, Canada	F	14.8	41.7	14.8	10.4	—	7.8	7.0	3.5[b]	—	—
Orthic grey wooded *luvisol*, Canada	L–H	15.2	43.9	11.6	12.2	—	8.5	4.9	3.6[b]	—	—
Brown forest *cambisol*, Canada	Ah	14.7	31.9	12.9	14.9	—	10.4	10.4	4.8[b]	—	—
Grenville series *cambisol*, Canada	Upper layer	15.1	40.7	12.4	9.7	1.2	8.6	7.7	4.6[c]	1.704	C
Lennoxville series *podzol*, Canada	A₂	13.6	44.1	10.2	11.6	1.1	9.7	5.7	4.2[c]	2.065	—
Scott series *kastanozem*, Canada	0–6 in	17.2	36.0	15.3	10.7	0.9	11.1	5.3	3.5[c]	1.404	—
Lacombe series *chernozem*, Canada	0–6 in	20.1	36.4	12.5	10.1	1.1	10.6	5.3	3.7[c]	1.262	—
Arago series *podzol*, Canada	0	15	54	15	5	—	4	4	3[b]	11.79	D
	B₂	16	35	16	9	—	9	9	6[b]	1.11	—
	0₁	15	42	15	10	—	8	7	3[b]	9.22	—
Inuvik series *podzol*, Canada	0₂	13	50	14	10	—	7	4	3[b]	6.14	—
	B₂	9	30	17	13	—	12	10	9[b]	0.162	—
	C₁	16	22	20	8	—	11	13	10[b]	0.225	—

APPENDIX TABLE II (*continued*)

Soil	Horizon	Relative proportion (%)								Total sugar (%)	Reference[a]
		Gal.	Glc.	Man.	Ara.	Rib.	Xyl.	Rha.	Fuc.		
Bratnober series *luvisol*, Canada	O	15	44	12	12	—	9	5	4b	8·59	—
	A₁	14	48	13	10	—	8	5	2b	0·517	—
	A₂	15	37	14	13	—	10	7	4b	0·124	—
	B₂	19	33	14	13	—	8	9	4b	0·256	—
	C	17	28	15	13	—	11	13	3b	0·0923	—
Trossachs series *solonetz*, Canada	A₁	16	28	18	17	—	9	8	4b	0·972	—
	B₂	14	24	18	18	—	10	8	8b	0·234	—
Oxbow series *chernozem*, Canada	A₁	14	36	16	15	—	8	8	3b	0·589	—
	B	15	34	14	12	—	8	12	5b	0·118	—
	C	20	26	20	12	—	10	8	4b	0·020	—
Granby series *gleysol*, Canada	Ap	15	31	15	15	—	11	8	5b	0·374	—
Urrbrae series *luvisol*, Australia	0–6cm	12·2	39·3	19·4	11·7	1·9	7·5	4·4	3·6	0·56	E
Urrbrae series *luvisol*, Australia	0–6cm	13·0	43·1	18·5	9·8	0·8	8·7	2·6	3·5	0·17	—
Lateritic podzolic *podzol*, Australia	—	6	32	6	20	2	27	4	2	—	F
Sandy podzol *podzol*, Australia	—	11	49	13	13	0	8	3	2	—	—
Solodized solonetz *solonetz*, Australia	—	12	35	9	19	0	19	5	1	—	—
Ando *andosol*, Australia	—	17	40	24	7	0	5	5	2	—	—
Solonized brown soil *kastanozem*, Australia	—	10	58	9	10	1	8	2	1	—	—

Soil	Depth										
Terra Rossa *luvisol*, Australia	—	15	40	17	10	1	11	6	1	—	—
Rendzina *rendzina*, Australia	—	15	50	11	10	0	9	5	2	—	—
Krasnozem *nitosol*, Australia	—	17	30	19	17	2	9	4	2	—	—
Black earth *vertisol*, Australia	—	16	39	17	9	2	9	4	4	—	—
Grey clay *vertisol*, Australia	—	17	37	14	11	3	10	4	2	—	—
Tropical red earth *Ferralsol*, Australia	—	14	29	21	12	5	8	5	5	—	—
Pine Forest Soil *podzol*, Norway	F	11	48	9	11	—	17	5	—	19·7	G
Frilsham series *cambisol*, England	0–15cm	14·4	41·3	10·8	12·7	—	14·7	5·1	1·1	—	H
Frilsham series *cambisol*, England	0–15cm	14·8	39·2	9·2	11·9	—	18·8	4·3	1·9	—	—
Frilsham: Windsor series *cambisol: gleysol*, England	0–15cm	14·3	39·2	10·4	11·5	—	16·9	5·8	2·0	—	—
Countesswells series *podzol*, Scotland	0–25cm	11·3	42·3	12·2	12·2	—	13·0	6·5	2·4	1·23	I
Nagano series *gleysol*, Japan	0–15cm	13	47	11	9	1	13	6	—	0·410	J
Saga series *gleysol*, Japan	0–10cm	12	47	11	9	1	14	6	—	0·499	—
Miyazaki series *gleysol*, Japan	0–17cm	13	43	13	10	1	12	7	—	0·558	—
Fen peat *histosol*, Australia	—	13	41	19	9	0	9	7	2	—	F
Iwanuma series *histosol*, Japan	0–10cm	13	40	13	10	1	15	8	—	1·421	J

APPENDIX TABLE II (continued)

Soil	Horizon	Gal.	Glc.	Man.	Ara.	Rib.	Xyl.	Rha.	Fuc.	Total sugar (%)	Reference[a]
Iwanuma peat *histosol*, Japan	20–46cm	10	36	4	14	1	32	3	—	5·298	—
Cairn O Mount peat *histosol*, Scotland	—	7	80	6	1	—	3	3	—	22·47	L
Loxahatchee peat *histosol*, USA	0 m	3·8	79·3	1·9	0·7	0·5	9·1	3·6	1·1	—	M
	0·3	9·2	59·5	6·0	0·7	3·3	8·7	4·7	7·9	—	—
	0·6	10·5	54·5	7·7	1·0	1·1	8·2	12·9	3·9	—	—
	1·5	8·5	42·4	7·3	2·2	3·6	9·6	21·9	4·4	—	—
Little shark river peat *histosol*, USA	0 m	23·4	35·2	11·7	8·6	3·1	15·6	0·9	1·4	—	—
	1·2	12·3	46·6	3·4	1·8	2·8	24·7	3·6	4·8	—	—
	2·1	5·0	53·4	3·7	1·9	2·2	23·5	7·0	3·2	—	—
	3·6	9·7	13·1	10·0	2·7	1·8	44·5	10·2	8·1	—	—
Kiyukwi series *fluvisol*, Japan	0–15cm	14	41	12	11	2	14	7	—	0·285	J
Insch series *cambisol*, Scotland	—	11	39	12	14	—	13	8	3	1·52	K
Mean of all values	—	14·7	35·4	17·1	11·9	1·5	10·4	5·9	3·4	—	—
Mean of upmost horizons	—	14·4	38·7	14·7	11·5	1·3	11·1	5·6	3·1	—	—

[a]References
A   Folsom *et al.* (1974)   I   Cheshire *et al.* (1974a)
B   Gupta and Sowden (1965)   J   Murayama (in press)
C   Sowden and Ivarson (1962a)   K   Cheshire *et al.* (1974b)
D   Gupta and Sowden (1963)   L   Cheshire (unpublished)
E   Swincer *et al.* (1968b)   M   Lucas (1970)
F   Oades (1972)   b + rib
G   Juvvik (1965)   + high R_G
H   Whitehead *et al.* (1975)

APPENDIX TABLE III. *Amounts of minor sugar components with high R_{Rha} values in soils (Gupta et al., 1963)*.

Soil	R_{Rha} Probable R_G value	Sugar (%)							Total (%)
		0·9 0·27	1·1 0·37	1·2 0·40	1·4 0·42	1·5 0·45	1·6 0·48	1·8 0·54	
Orthic podzol O Arago, *podzol*		0	0·08	0·10	0·12	0·18	0·08	0·17	0·73
Podzol with permafrost O1 Inuvik, *podzol*		0	0·09	0·12	0·18	0·18	0·06	0·25	0·88
Orthic grey wooded O Bratnober, *luvisol*		0	0·15	0·10	0·06	0·15	0·09	0·13	0·68
Solodized solonetz A1 Trossachs, *solonetz*		0·09	0·011	0·015	0·023	0·034	0·016	0·022	0·130
Orthic dark grey Ap gleysolic Granby, *gleysol*		0	0	0·014	0·014	0·020	0·009	0	0·057

Index

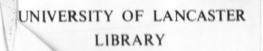

THE ISRAELI AIR FORCE STORY

The story of the Israeli Air Force has become one of the great fighting legends of the world. Israel's little band of pilots are as vital to their country's survival as the Few were to Britain's: and the Battle of Israel has continued, not just for a month, but for twenty years. Reservists, leaving their homes and jobs for a few hours, fly out on operational sorties alongside the regular pilots. While the Arabs have more machines than men capable of flying them, Israel always has more volunteers than planes. The Israeli Air Force is indeed a *corps d'élite*.

In this enthralling book Squadron Leader Robert Jackson tells the whole story for the first time; a story which begins with Israel's birth in a crucible of violence, when planes had to be smuggled in from any part of the world where they could be found; includes the Suez invasion and the over-whelming triumph of the Six Day War; and traces the development of the perilous situation which exists today.

We are also given periodic glimpses 'over the hill' at what has been going on in the Arab air forces.

Squadron Leader Jackson allows full reign to the drama of his material, but he supports it with a wealth of scrupulously researched technical detail. The result is a real contribution to the history of our times and to the history of warfare. And it is a book for everyone, of whatever political affiliations, who can appreciate a true tale of ingenuity and high courage.

Robert Jackson is a well-known specialist in aviation and military studies. An active pilot and a Squadron Leader in the Royal Air Force Volunteer Reserve, he is presently involved in training air cadets. He is married with two small children.

He has contributed articles on air and military subjects to newspapers and magazines all over the world. His most recent books were *Strike from the Sea—a history of British Naval air power* and *The Red Falcons* (the Soviet Air Force in action since 1918).

THE ISRAELI
AIR FORCE STORY

*The struggle for Middle East
aircraft supremacy since 1948*

by

ROBERT JACKSON

TOM STACEY LTD

Tom Stacey Ltd
28 Maiden Lane, London W.C.2, England
First published 1970
Copyright © Robert Jackson

SBN 85468 014 4

Printed in Great Britain by
Northumberland Press Limited
Gateshead

CONTENTS

Prologue

The morning trembled with sound. Familiar, re-assuring sound to the inhabitants of Tel Aviv and the towns and villages scattered along the Plain of Sharon; the crackling thunder of the Atar turbojets that powered the Mirages and Mystères of the Air Force's combat squadrons. The sound had become part of their lives, as had the sight of the needle-nosed fighters streaking across the sky in vigilant pairs.

Today, somehow, it was different. There was a sense of grim urgency in the thunder that rolled across Israel's coastal plain. This was not the usual morning patrol, but something far bigger. From Hatzor, Ekron, Lod and Ramat David—the airfield complex clustered like a shield around the heart of Israel—the jets were taking off in a steady stream, scorching the concrete runways with the spitting flame of their afterburners.

The aircraft kept low, forming-up at an altitude of three hundred feet, hugging the shadow of the Judaean Hills. For millennia the hills had brooded, unchanging, over the destiny of this land, their barren faces, eroded by the sun and the scorching Khamsin, turned towards the fertile plain below, as though thirsting for a breath of the cooler Mediterranean breezes that bathed it. For four thousand years and more, the angular rock walls had echoed to the clash of sword and spear; noises of war joined more recently by the crackle of gunfire, and now by the roar of massed jet engines.

On the other side of those hills lay danger. Beyond the morning mist that still lay over the valley of the Jordan, at Ajlun on the slopes of the 4,000 foot Jebel Umed Daraj, the scanner of an ultra-modern Jordanian Marconi 547 surveillance radar conducted a never-ending search of the sky over Israel. Only by keeping down behind the hills, beyond the penetration of the radar's probing beams, could the Israeli aircraft hope to escape detection.

Below the line of the horizon, the first wave of aircraft, Mirages, sped low over the sweeping curve of the Israeli coast just south of Yafo. There were twenty-four of them in battle formation; six flights of four aircraft, flying in loose 'finger fours'.* This was the classic combat unit developed and perfected by the Luftwaffe during the Spanish Civil War and subsequently used by every air force in the world.

The aircraft swept out over the Mediterranean at a steady 550 knots, 150 feet above the waves, rocking slightly as they skimmed through patches of turbulence. They were heavily laden with fuel; in addition to the 730 gallons carried in its internal tanks, each aircraft carried two auxiliary fuel pods slung beneath its wings.

Nestling beside these pods, one under each wing, was the war-load: a slim, eight-foot bomb of a type never before used under operational conditions. The weapon, originally projected by Engins Matra of France and subsequently developed in Israel, was

* 'Finger Four' formation: so called because it resembles out-stretched fingers of right hand, palm downover, with each fingertip representing an aircraft.

designed to be dropped from a low-level strike aircraft flying at between 300 and 500 knots, at a
height of 300 feet above ground level. The bomb
was fitted with a battery of four small retro-rockets,
set to fire automatically exactly three-tenths of a
second after release, slowing the weapon and causing
it to nose down towards its objective, the concrete
surface of a runway, at an angle of 80 degrees.
Another six-tenths of a second and a small drogue
parachute would deploy, stabilizing the bomb in its
descent. Then, just under two seconds after launch,
would come the last stage in the automatic sequence
of events: the firing of four solid-fuel booster
rockets, fitted in the tail section of the device. These
would burn away the drogue parachute and
accelerate the bomb to a speed of over 500 feet per
second. Under the drive of the rockets, the bomb's
365-pound high explosive warhead would break
through the runway surface and explode, causing a
shallow crater and distortion of the concrete over a
radius of several yards.

Two of these weapons, dropped on a runway
intersection, would be sufficient to prevent the runways being used by jet combat aircraft for several
hours. The main advantage of the weapon, which
weighed about 1,000 pounds, over more conventional
bombs was that it could be delivered with great
accuracy even by a pilot of average skill, thanks to an
advanced sighting system.

The pilots who were carrying the bomb into
action this brilliant morning were more than
average. They were drawn from the ranks of the
most experienced aircrews in an air force which
boasted the highest standard of training and
the world's finest combat aircraft: the Mirage

IIICJ, a machine capable of doubling as an interceptor, strike or reconnaissance aircraft, depending
on the role operational necessity demanded of it.
From the famous stable of France's Marcel
Dassault, it was a thoroughbred in the truest sense
of the word.

Thirty-five minutes after take-off, as though held
together by invisible threads, the 24 Mirages began
a gentle, beautifully-executed turn to port, coming
round on to a new heading of 175 degrees. Ahead
of them now, 100 miles away, lay the coast of
Egypt. Twenty-four gloved hands reached out and
flicked off safety catches, arming each Mirage's twin
30 mm DEFA cannon; the last check before combat.
As the minutes sped by, the Mirages' electronic
equipment reeled off a constantly-changing stream
of data: geographical co-ordinates, heading to steer,
time to target.

Eight minutes since the turning-point. Ahead of
them now, emerging from the southern horizon
under a thin film of blue haze, the pilots made
out a rapidly-expanding line of coast. Scanning the
sky above it, they saw nothing. No filmy contrails,
no glitter of sunlight on wings; none of the tell-tale
signs that would have revealed a patrolling enemy.

Sixty seconds later the Mirages skimmed over the
coast, speeding across the railway line running
westwards from Alexandria. Away to starboard lay
El Alamein, symbol of victory in a bygone war. The
pilots now turned to port again, skirting the eastern
fringe of the Qattara Depression and opening the
throttles of their aircraft as they did so. In spite of
their load, the Mirages accelerated rapidly with the
help of their afterburners, the pilots taking them
up to 1,000 feet and then levelling. Visible, misty

shock waves danced briefly over their wings as their sharp noses speared through the compressibility that borders the speed of sound; in the cockpits, the needles of their mach-meters hung poised on Mach One for an instant, then flickered past it. Tawny sand and dried scrub merged into a continuous blur beneath them as they streaked on at nearly 800 knots, trailing a wake of thunder behind them in a deafening swathe that sliced through the morning silence of the desert.

At any minute now, the speeding jets would be picked up by the Egyptian radar complex around Cairo. But Cairo was not their target; they were heading for the four principal airfields in the Canal Zone: Abu Sueir, Deversoir, Fayid and Kabrit. Cairo West, Almaza, El Mansura and Inchas would be hit at exactly the same time by a second wave of aircraft, now thundering over the desert six minutes behind the first.

It was exactly 08.43, Cairo time. The date: Monday, June 5th, 1967. And the fastest, most devastating air strike in history was two minutes away.

CHAPTER ONE

A Wing and a Prayer

Pilot Ray Kurtz had lost count of the times he had nursed a crippled bomber safely back to its base, its engines smoking, oxygen system inoperative and instruments out of action. As a former B-17 Flying Fortress commander with the United States Air Force, he had often found himself in that situation after bombing targets in German-occupied Europe. But this time it was different; the war with Germany had been over for three years, and instead of returning from a mission with an aircraft that was barely capable of staggering along through the air, Kurtz was setting out on one. Ahead of him lay a nerve-racking flight over 600 miles of open sea and, at the end of it, the prospect of having to run the gauntlet of enemy fighters and anti-aircraft fire.

Kurtz's B-17 was one of three machines which had been purchased from US war surplus stocks with the help of a Jewish organization in America. The aircraft had been flown out via Panama, crossing the Atlantic to refuel in Portugal before flying on to a secret base in western Czechoslovakia. They had barely made the trans-Atlantic flight; their engines were worn out and they were plagued with a host of other mechanical troubles.

The three B-17s had arrived in Czechoslovakia during the last week of June, 1948, and the task of keeping them in some kind of flying condition had fallen to Ray Kurtz and a small band of determined airmen. A few weeks earlier, they had crossed the Atlantic with a batch of transport aircraft destined for Israel; after that, they had remained in Czechoslovakia to supervise ferrying operations.

Suddenly, on July 7th, Kurtz had received a top-level order from the Israeli Government. The Jewish-American pilot was given a mere five days to get the Flying Fortresses airworthy for a special mission—the bombing of Cairo, Gaza and El Arish, as a reprisal for Egyptian air raids on Israel. For five days and four nights, Kurtz and his men slaved to get the bombers in flying condition. On the morning of the sixth day miraculously the Fortresses were ready to go, their engines serviceable, carrying a maximum load of fuel and 500-pound bombs obtained from a Czech arms dump. Then, after all the feverish activity, came the crushing anti-climax; minutes before they were due to take off, Kurtz got an order cancelling the mission.

They waited for two more days. Early in the morning of July 15th, they finally received the order to go. Kurtz and his crews hurried into their aircraft, eager to get off the ground to forestall any counter-orders.

Trouble hit Kurtz's aircraft almost immediately. During the run-up before take-off, two amplifiers burnt out, the propeller-governor of number one engine froze at 2,400 r.p.m., the manifold pressure of number two engine refused to rise above 18 inches and the artificial horizon packed in completely. Kurtz would have been fully justified in aborting the mission; he did not. At 10.00 local time the last of the three aircraft lurched off the ground and began a laborious climb to 10,000 feet. The B-17s droned southwards in loose formation, the pilots juggling with the throttles to maintain the most economical cruise and to nurse engines that were already beginning to overheat. Kurtz was running into further mechanical difficulties; several more instruments had become inoperative and the port inner engine was running erratically, its oil pressure fluctuating wildly.

Then they flew into bad weather and severe turbulence over Austria, skirting the eastern fringe of the Alps; the three aircraft split up to reduce the risk of collision. With some of the Fortresses' most vital instruments out of action, flying through the murk was a nightmare. For 30 minutes Kurtz fought to hold the aircraft steady as it was buffeted by vicious currents of air, flying straight and level more or less by instinct. By the time they emerged into a clear sky, the pilot was almost exhausted and the other members of the crew were feeling ill and wretched; some of them had been violently airsick.

The Fortresses rendezvoused over the coast of Yugoslavia. Two hours later, over Albania, the crews got a nasty surprise when bursts of light anti-aircraft fire erupted around them, but the danger soon fell behind and there were no further incidents. The Fortresses cruised on steadily, the pilots nursing their engines and anxiously watching their instruments for any sign of trouble. Over Crete, the aircraft split up once more and headed for their separate targets on the other side of the Mediterranean.

Cairo was Ray Kurtz's target. Half-way across the Mediterranean, he put his aircraft into a steady climb, crossing the coast of North Africa at 25,000 feet. With every passing minute, Kurtz was finding it harder to keep the aircraft on a straight course; he was feeling light-headed, and some of the other eight crew members were experiencing the same symptoms. Suddenly, the navigator collapsed face-down over his chart table, and at last Kurtz realized what the trouble was. In Czechoslovakia, the aircraft had been fitted with cylinders containing welder's oxygen, which is totally unsuitable for breathing at high altitude. Kurtz ordered all the crew members to turn their oxygen on to 'emergency', and after a while the situation eased a little.

One of the crewmen worst affected by the lack of oxygen was Johnnie Adir, the bomb-aimer. Kurtz was worried about him; if he passed out completely, the mission would probably fail. Adir had rigged up a makeshift bomb-sight salvaged from a wrecked German aircraft, and no one else had any idea how to use it. Now, as they droned high above the sand and scrub of the Qattara Depression, Kurtz told Adir to rest as much as possible. Another hour or so and they would be over the target, and the bombardier would need to be fully alert.

In spite of the seriousness of the situation, Kurtz allowed himself a thin smile. He had no difficulty in following a straight heading now, thanks to the Egyptians. The Fortress's radio equipment was tuned in to a homing beacon beaming out a steady signal from Fayid in the Canal Zone; all he had to do was follow it and it would take him right over the top of Cairo. In any case, there was no possibility of missing the Egyptian capital; the sprawling city of Cairo was brightly lit beneath. As they approached, Johnnie Adir, somewhat recovered, clambered into his position in the bomber's nose. His target was Cairo's main airport, whose runway lights twinkled ahead. Following Adir's instructions, Kurtz began his run-in, still holding the aircraft at

25,000 feet. Behind him crew members were still fainting from lack of oxygen.

The aircraft gave a sudden lurch as Adir pressed the release and the small bomb-load fell away. Immediately, Kurtz pushed the Fortress's nose down and turned away northwards, heading out to sea. The Egyptians had been taken completely by surprise; not a single gun had opened up. As the aircraft came down into the denser layers of the atmosphere, the crew recovered from their ordeal. The Fortress crossed the coast near Port Said and turned in towards Israel. At 22.45, local time, Kurtz brought the bomber in to land at Ekron airfield near Tel Aviv. The other two Fortresses touched down minutes later; they had also raided their targets successfully. The crews were half dead with fatigue, but they were jubilant. Although they knew that the damage inflicted by the couple of 500-pounders dropped by each aircraft must have been negligible, it didn't matter. What did matter was that for the first time, Israeli aircraft had struck at the heart of Egypt.

In February 1947, over a year earlier, the British Foreign Secretary had announced that Britain intended to renounce her Mandatory Administration and withdraw her forces from Palestine. Later, on November 29th of the same year, the United Nations had passed a resolution calling for the creation of two independent states in Palestine: one Jewish, one Arab. This resolution prompted an intensification of Arab terrorism against the country's Jewish community.

At this stage, none of the Jewish leaders in Palestine believed that the Palestinian Arabs would be in a position to mount a full-scale offensive against the Jewish community, even with the help of neighbouring Arab nations; nor indeed, with the UN resolution supported by both the United States and the Soviet Union, would they be prepared to risk such a move. The Jewish leaders were convinced that fighting on a large scale would break out as soon as the British withdrew and the State of Israel was proclaimed, but they believed that it would involve only guerilla warfare. Consequently, the organization was geared to deal with this type of conflict.

Haganah—the name means simply 'Defence'—was a clandestine organization that dated back to 1920, when it had been organized by the Zionist leader Vladimir Jabotinsky following an Arab attack on the Jewish community in Jerusalem. During the years of riot and rebellion between 1936 and 1939, when Palestine was torn by a wave of Arab nationalism, Haganah's role had been mainly one of passive defence; its policy was to fight only when sections of the Jewish community were attacked. Later, with the increasing disintegration of law and order, the Haganah High Command set up a Special Actions Branch with the task of organizing punitive and reprisal raids against the Arabs on their own territory.

During the Second World War, Haganah co-operated actively with the British against the Germans and thereby acquired substantial quantities of arms and ammunition from British Army sources and from the battlegrounds of the Middle East. An Arms Acquisition Organization (Rekhesh) was set up under Haganah control, and by 1945, thanks to its efforts, Haganah possessed a considerable stockpile of small arms and ammunition. This was augmented in 1945-46 with purchases of war-surplus material from various European countries. This was stored secretly near Milan with the connivance of the Italian authorities, and most of it was smuggled into Palestine during 1947, carefully hidden in crates packed with machinery parts. By the end of the year, 200 Bren guns, 1,500 rifles, 400 sub-machine guns, 500 revolvers and 1,500,000 rounds of ammunition had reached Palestine from Italy in six separate consignments.

It was encouraging, but one man in Palestine realized only too well that it was not enough. That man was David Ben-Gurion, Chairman of the Jewish Agency and head of its Security Department. He, and perhaps he alone, foresaw the possibility of a full-scale war engulfing Jews and Arabs in Palestine, and he campaigned ceaselessly to turn Haganah into a fully-organized fighting force, an army equipped with artillery and tanks, and supported by an air arm. Already, during a visit to the United States in 1945, Ben-Gurion had made great efforts to gain the support of influential Jewish leaders in setting up a military industry in Palestine. A group of 20 American Zionist businessmen, under the direction of Haim Slavin, had subsequently raised several million dollars and purchased a large quantity of heavy plant; some of this arrived in Palestine before the British Mandate expired and was later used for concerns such as Bedek Aircraft Ltd, which eventually became Israel Aircraft Industries.

The Haganah High Command, however, was reluctant to share Ben-Gurion's point of view. It agreed that the organization should be brought to a greater state of readiness, with better liaison between units, superior training and more weapons, but it refused to countenance the idea of the organization being turned into a regular army organized along British lines. The result was that in November 1947, at the time of the UN resolution, Haganah was critically unprepared to cope with anything more than a large skirmish. The only Haganah members who could boast any real degree of training and organization were the 3,000 men and girls of Palmach, the defence force's special commando group, and even they had been trained to operate in small inconspicuous units rather than as a unified body.

Nevertheless, in October 1947 Ben-Gurion persuaded the National Council of Palestine to agree to a plan for the mobilization of a Jewish 'citizen army', which would come into effect in a time of crisis. He also received an allocation of three million dollars (he had actually requested five million) from the Jewish Agency Executive for the purchase of more weapons. The following month, a team of four men, Ehud Avriel, Munya Mardor, Yehuda Arazi and Eliahu Sacharoff, left for Europe and the United States on a large-scale arms buying assignment. They were to give special attention to the purchase of artillery and aircraft.

The most important task was assigned to Ehud Avriel, who was sent to Europe with an order to buy 450 machine-guns and 10,000 rifles. After his work with Aliyah Bet, the organization responsible for smuggling illegal Jewish immigrants into Palestine under the noses of the British administration, Avriel had access to a vast network of Haganah agents and other contacts scattered throughout the continent, and through them he now put out a series of exploratory feelers. It was not long before offers of arms began to flow in to his headquarters in

Paris, and an investigation revealed that most of the offers originated in Czechoslovakia.

Armed with forged identity documents that proclaimed him to be an official representative of the Swiss Government, and working under the name of Herr Ueberall (the name means 'everywhere' in German) Avriel arrived in Prague during the second week of December 1947 and contacted two Haganah agents there, Uriel Doron and Pino Ginsburg. After several days of preparation, the trio, still posing as Swiss representatives, made the first direct approach to the Czech Ministries of Defence and Supply. To their surprise and delight, they found the Czechs willing to co-operate. The reason was not hard to find. The Czechs badly needed dollars; the Haganah agents had them to spend. Nevertheless, it was not until the Russians gave their approval that the first arms deal could be concluded, in January 1948.

The problem now was to get this and subsequent consignments of arms from Czechoslovakia to Palestine. The man in charge of the clandestine pipeline was Munya Mardor, who had been given a roving commission in Europe by Ben-Gurion himself. Under his direction, part of the first batch of Czech arms was shipped through the American Zone of Germany to Belgium, and from there transported to Palestine by sea. Like earlier consignments from Italy, the arms were broken down and carefully concealed in crates containing machinery. The route through Belgium, however, was dangerous and unsatisfactory; too much time was being lost in transit while the situation deteriorated steadily in Palestine. But early approaches to Poland, Rumania and Yugoslavia for permission to transport the arms across their territory ended in failure, and for the time being at least the German-Belgian pipeline remained the only possibility.

The initial consignments of Czech arms arrived in Palestine towards the end of March 1948 and were hastily distributed among the Haganah units. It was not a moment too soon; the Jewish population in Jerusalem was virtually surrounded by Arab irregulars, who controlled all but one of the major roads leading into the city, and that, too, fell late in March. The plight of the 100,000 Jews in the ancient city was now desperate; they were isolated and faced with starvation, and the Haganah garrison, although in a position to fight off an Arab attack, had neither the arms nor the personnel to mount a counter-offensive with any hope of success.

There was no longer any time for secrecy. On March 29th, Ben-Gurion cabled Avriel in Prague with an urgent request for rifles, trench mortars and ammunition. The Haganah agent acted quickly; on March 31st, a four-engined Douglas DC-4, chartered from an American company, and with an American crew, took off from an airstrip near Prague, laden with arms and a full load of fuel. Hours later, the aircraft touched down at Beit Daras, in southern Palestine, with only enough fuel left for another five minutes' flying. The weapons were hurriedly unloaded and, still in their crates, rushed to the Palmach troops who were assembling near Tel Aviv in readiness to break the Arab stranglehold on Jerusalem.

The need for an air force had by this time become desperately apparent to Haganah's leaders. First and foremost, Haganah needed transport aircraft, not only to fly in supplies of arms but to ferry combat units rapidly to areas threatened by Arab attack. But transport aircraft would be of little use without a combat element to protect them; the Arab belligerents could muster over 100 aircraft between them, including 40 Spitfires. With no combat aircraft to oppose them, these would be able to rove at will over the whole of Palestine. Fortunately, much of the groundwork for the formation of an air arm had been laid as long ago as 1937, when several gliding clubs had been set up under Haganah direction and a number of small airstrips built throughout the country, many of them on the coast. The result was that, in 1948, Haganah possessed a cadre of personnel who already had some knowledge of flying. More important still, the Jewish community in Palestine included some 2,500 men and women who had served with the Allied air forces during the war, either as aircrew or ground staff.

Plans for an air arm proper had been made in October 1947, when the Haganah leaders had reached a firm decision to expand the defence force. The embryo air force, known as the Sherut Avir, or Military Air Service, still existed only on paper at the beginning of 1948, although its commanders had been designated; the Chief of Air Staff was Aharon Remez, with Heiman Shamir as his deputy.

Then, in January 1948, the Sherut Avir got its first real break when it secretly acquired eight Taylorcraft and Auster J/1 Autocrat light aircraft. These were all 'legitimate' machines, bearing the Palestinian VQ-P registration; one or two were purchased outright by the Sherut Avir, but most were placed at Haganah's disposal by their owners, who volunteered to undertake operational tasks. Then came another windfall; at the end of January, the Royal

Air Force in Palestine offered 25 Auster army co-operation aircraft for sale as scrap, never dreaming that the dealer who bought them was in fact a Haganah agent. The Austers were broken down into their component parts by the Sherut Avir's nucleus of skilled airframe and engine fitters, and 18 complete aircraft were built from them. The 18 machines were given false VQ-P registrations (to avoid suspicion, all of them were identical to those used by other aircraft on the civil register) and were carefully dispersed, either in barns or under camouflage in open country. The main snag with the scheme was that the Austers sometimes had to use airports where the aircraft with identical registrations were based, and the Sherut Avir had to work out a very careful schedule to ward off the possibility of two identically-registered aircraft turning up in the same place at the same time. In spite of these precautions, it did happen on one occasion, when a pilot flying into Lydda for an engine overhaul spotted an Auster bearing the same registration as his own aircraft parked outside a hangar; fortunately, he was able to do a smart about-turn and disappear before anyone noticed the similarity.

These aircraft were pressed into service immediately, carrying food, ammunition and medical supplies to isolated Jewish settlements and patrolling areas where there was a threat of Arab attack. On several occasions, they acted as spotters for convoys carrying food to the Jewish community in besieged Jerusalem before the western road was finally captured by the Arabs in March. The crews of the Austers by no means confined themselves to a passive role; they carried rifles and machine-pistols, and often dived down to take a pot-shot at anything that looked suspicious. A variety of hand-grenades and home-made bombs was also dropped, although some of the latter presented a greater risk to the Austers' crews than they did to the enemy. During the period of tension immediately prior to the end of the British Mandate, the Austers played a significant part that was out of all proportion to their size and scope; apart from their value in giving warning of possible ambushes, their presence alone was a powerful morale-booster to hard-pressed Jews who went in fear of their lives from day to day.

Meanwhile, Haim Slavin's Zionist group in the United States had been far from inactive. In March, a Jewish-American pilot and aero-engineer named Alan W. Schwimmer (who was later to found an aircraft industry in Israel) arranged the purchase of ten Curtiss C-46 Commando and three Lockheed Constellation transport aircraft from USAF surplus stocks, and also recruited crews to fly and maintain them. To turn the operation of these aircraft into a perfectly legitimate venture, Slavin and his colleagues bought up the rights of two defunct air charter firms and then made an agreement with the government of Panama which permitted the aircraft to carry the Panamanian HP- civil registration.

The crews recruited by Al Schwimmer offered their services for a variety of reasons. Most were Jews, who volunteered from idealistic motives; others were men who loved flying, and if the flying went hand-in-hand with some excitement, so much the better. Among them was a man who fell into both categories: Ray Kurtz.

With four other airmen, Kurtz arrived in Panama on May 2nd and immediately set to work to get the transports ready for the long haul to Palestine.

Within a week, five of them were in flying condition
and the necessary paperwork to cover their export
had been completed. On the morning of May 8th
the aircraft—three C-46s and two Constellations—
took off and climbed in formation over the mountains, heading eastwards for Dutch Guiana on the
far side of South America. From there they flew on
to Natal in Brazil, the easternmost airfield in Latin
America and the closest point on the continent to
Africa.

Before taking off on the trans-Atlantic hop that
would take them to Equatorial Africa, Kurtz had
the engines of the five aircraft subjected to thorough
checks. The engines of one of the Constellations
were found to be giving trouble and the aircraft
had to be left behind while repairs were effected;
it would follow on later. The four remaining aircraft staggered into the air, laden with maximum
fuel, and climbed out over the sea; their next land-
fall would be the island of São Tome, nearly 2,800
miles away. Over the Atlantic the aircraft became
split up because of bad weather and engine trouble,
but all of them reached their destination safely. The
wind was in their favour for most of the way, and
they completed this leg of the journey in a little
under 15 hours.

The next leg was a 2,300-mile haul northwards
over the Sahara, ending in Casablanca. By this time,
the number of aircraft had been reduced to three;
yet another had been delayed through technical
trouble, and when the time came to leave Casablanca
it was found that one of the three remaining aircraft
was beset by similar problems. So, with only two
aircraft out of the original five, Ray Kurtz set off on
the last stage but one of the journey, arriving in

Sicily on the night of May 15th. The following day, the two C-46s landed at the former RAF airfield of Ekron, a few miles south of Tel Aviv. Utterly exhausted, their eyes red-rimmed from lack of sleep, Kurtz and his men climbed down and stood in the silence that seemed uncanny after the constant roar of the faithful engines that had carried them over half the world.

It was 20.15 hours on May 16th, 1948. Almost exactly 52 hours earlier, in Tel Aviv, David Ben-Gurion had faced the assembled National Council of Palestinian Jewry and, in tones heavy with the atmosphere of history in the making, proclaimed the establishment of the State of Israel.

CHAPTER TWO

Into Combat

The early morning of Saturday, May 15th, 1948 was bright and clear. The streets of Tel Aviv were deserted; it was as though the city had worn itself out in the celebrations of the previous day, following the proclamation of the State of Israel. The only activity was in the Red House, the seat of government overlooking the Mediterranean, where the Haganah High Command had been discussing the looming threat of war with Prime Minister David Ben-Gurion and members of his cabinet throughout the night. It was strange to see Tel Aviv empty of British soldiers and vehicles; the last British soldiers had already pulled back to Haifa, where they were in the process of embarkation. The Royal Air Force, too, had abandoned all its airfields in the country except one; two squadrons of Spitfires had stayed behind at Ramat David to cover the evacuation of the British forces. The other airfields had already been taken over by Sherut Avir personnel.

At 05.00, the morning silence was split by the distinctive crackle of Rolls-Royce Merlin engines as two Spitfires swept in low towards the capital over the sparkling waters of the Mediterranean. The aircraft were LF. Mk.9s, and they bore the green-and-white roundels of the Royal Egyptian Air Force. At 1,000 feet, the two Spitfires roared over the mouth of the Yarkon River and dived on the Reading

Power Station to the north, sandwiched between the harbour and Tel Aviv's small airstrip. While one of the fighters circled watchfully overhead, the other dropped two 250-pound bombs, causing slight damage to the power station and its surrounds. Both aircraft then made two or three strafing runs over the area.

On the roof of the power station, a lone Haganah machine-gunner let fly at one of the Spitfires as it flashed overhead. Incredibly, his bullets found their mark and the fighter turned away, streaming a white trail of glycol coolant from a punctured tank. A few moments later, it lost height and slid in to make a belly-landing on the nearby beach, slewing to a stop in a cloud of sand. A group of Haganah men ran across the beach towards it; the pilot, unhurt except for a few bruises, climbed from the cockpit and quickly raised his hands.

Less than an hour later, a team of Sherut Avir technicians arrived at the beach and, assisted by an army of willing helpers, salvaged the fighter and took it to the nearby airfield for repairs. The damage was not serious, and within a few days the Spitfire, patched up with spare parts found on an RAF dump, was flying again, its Egyptian roundels painted out and replaced by the six-pointed blue Star of David. It was the Sherut Avir's first combat aircraft.

A matter of minutes after the two Spitfires attacked Tel Aviv, reports came in that several settlements in the Negev had also been strafed, and that other villages along the borders of Syria and Transjordan had been subjected to heavy artillery fire. The Arab offensive against the infant State had begun.

Contrary to claims in subsequent reports on the Arab-Israeli war of 1948, there was no great disparity of numbers between the Arab and Israeli forces engaged at the outbreak of hostilities. On the morning of May 15th, the Arab armies had mustered a total of some 23,000 men on all fronts, against a Haganah total of about 19,000. Another 6,000 Haganah combatants were either held in reserve or in various stages of training. The real disparity came in terms of equipment; the Arabs had an overwhelming artillery superiority, and they possessed something vital to the successful waging of modern warfare—air power. The Royal Egyptian Air Force's operational strength at this time comprised some 40 Spitfires, 25 Dakotas (all of which had been converted to carry small bomb loads), 15 Harvards, 5 Hawker Furies and about 30 miscellaneous trainers, communications and transport machines. A large number were unserviceable at any one time because of the REAF's atrocious standard of servicing and the difficulty in obtaining spare parts; nevertheless, the REAF remained a serious threat as long as it was allowed to operate unopposed in the sky of Israel.

Two months earlier, in March, Ehud Avriel had begun negotiations with the Czechoslovak Defence Ministry for the purchase of a small number of surplus fighter aircraft. The machines involved were Avia C.210s, and they had an interesting history. During the Second World War, several versions of the famous Messerschmitt Bf 109 fighter had been built by the Avia Factory in Czechoslovakia, and when the Germans withdrew from the country one step ahead of the advancing Russians in 1944, the Czechs found themselves with enough component

parts and complete facilities to build 500 of the
fighters. In 1945, production of the Messerschmitt
Bf 109G-14 continued at Avia's Prague-Cakovice
factory, and the aircraft entered service with the
post-war Czech Air Force under the designation of
S-99. Later versions were powered by the Junkers
Jumo 211F engine instead of the usual Daimler-
Benz, supplies of which quickly ran out, and the
Jumo-powered aircraft were designated S-199.

Few aircraft in history can have been as univer-
sally hated by its pilots as the S-199, whose handling
qualities varied from poor to downright vicious. The
accident rate was frightful; the aircraft had a
tendency to swing violently and without warning
on take-off and landing, and many Czech pilots
killed themselves when their S-199s lurched off the
runway and turned over, bursting into flames. The
aircraft was known as 'Mezec' (Mule) by the men
who flew it, and the name accurately summed up
its handling qualities.

In spite of its vices, the S-199 equipped most first-
line Czech fighter squadrons until they began to
receive more modern Russian equipment early in
1948. The surviving S-199s were then allocated to
reserve squadrons or put up for export and it was at
that point that Ehud Avriel arrived on the scene.
The S-199, or C.210, as it was designated for export,
was the answer to the Sherut Avir's prayer. The
fact that the machine had a reputation for being a
killer didn't matter; it was a fighter, and it could
meet the REAF's Spitfires on more or less equal
terms.

Avriel concluded a deal for the supply of an
initial batch of eleven C.210s in April, and these
were transferred to Zatec airfield, Haganah's main

supply base in Czechoslovakia, where they were dismantled in readiness to be airlifted to Israel. By this time the air transport problem was more serious than ever, for diplomatic intervention by the United States Embassy in Prague had effectively put a stop to the use of American charter aircraft and crews. From now on, the Israelis were going to have to rely on their own resources, and on May 17th an Air Transport Command was formed with Munya Mardor as its commander. On paper, the new Command had a total of ten transport aircraft, but only three of these, two C-46s and one four-engined C-54, were operational. Three more aircraft were grounded with technical troubles, one was in Africa and one in Brazil, undergoing repairs, another had been written off in a landing accident en route, and a seventh had been forced down through engine failure on the island of Rhodes, where it had been interned by the Greek Government.

Of the operational aircraft, the two C-46s were fully committed to airlifting supplies to Israeli ground forces and settlements. They had been commandeered for this task a matter of hours after Ray Kurtz arrived with them. Only the C-54 was available to fly arms into the country and on May 20th, this aircraft was used to transport the fuselage of the first C.210 to Israel. The C-54 took off again immediately after it was refuelled, returning to Czechoslovakia to pick up the C.210's wings and engine.

All supplies flown in from Czechoslovakia, including the C.210s, were brought to Ekron, which was Air Transport Command's main base. To safeguard the airfield and the precious cargoes against enemy air attack, every available anti-aircraft gun was brought up and positioned around the perimeter.

As well as the equipment, more volunteers were continuing to arrive from overseas, mainly from North America; among them was Chalmers H. Goodlin, who, a year earlier, as a test-pilot with the Bell Aircraft Company, had made the first unpowered test-flights of the Bell XS-1 research aircraft, the machine that subsequently became the first in the world to exceed the speed of sound in level flight. Other volunteers never reached Israel; some were interned en route by hostile governments, and at least one was killed when the aircraft he was flying to Israel crashed. The latter was a Canadian named George Beurling, who, as an RAF fighter pilot during the Second World War, had destroyed 27 German and Italian aircraft during the defence of Malta. His friends knew him as 'Screwball' Buerling, but there was nothing crazy about him; a superb pilot, he possessed uncanny eyesight and fantastic marksmanship, and had spent most of his spare time during the hectic air battles over Malta working out newer and better air fighting tactics.

Beurling had volunteered to fly one of three Norseman light transport aircraft, acquired by Haim Slavin's organization in North America, on the long journey to Israel. On May 20th, 1948, he supervised the loading of the aircraft with medical supplies at Urbe Airport, Rome, and took off on the last leg of his flight. A few hundred feet off the ground the engine stalled and the overladen aircraft spun to earth. Beurling was killed instantly. He was just 26 years old.

Two days later, on May 22nd, a lone Egyptian Spitfire appeared soon after dawn over Ramat David airfield near Haifa, where the Spitfire FR.18s of the RAF's Nos. 32 and 208 Squadrons were still based. After circling the area several times, the Spitfire swept in and dropped a couple of small bombs, destroying two RAF aircraft, and then making two or three strafing runs over the base. Four of 208 Squadron's Spitfires were hurriedly scrambled, but the intruder had gone. Two hours later, however, Ramat David was again attacked, this time by three Spitfires which bombed a hangar, destroyed a Dakota transport on the ground with cannon and machine-gun fire, and damaged seven more RAF aircraft. This time, the attackers did not escape; they were intercepted by four RAF Spitfires and one of them was shot down within seconds. The second flew into the ground as its pilot tried desperately to get away and exploded, and the third crashed after receiving a direct hit by an anti-aircraft shell.

The following day, the Egyptians apologized to the British authorities for the 'regrettable navigational error' on the part of their pilots. It was clear that the Egyptian pilots had believed Ramat David to be in the hands of the Israelis, and that they had assumed the aircraft on the ground to be Israeli machines. The Israelis learned two things as a result of the attack; first, that the Egyptian intelligence was poor, and second, that it could only be a question of time before the Egyptians made further attacks on airfields in Israel—which underlined the need for an efficient air defence system.

By the end of May, eleven C.210s had reached Israel via the air bridge (code-named 'Operation Balak'), together with spare parts and ammunition, and 15 more aircraft were at Zatec awaiting transportation. The first four aircraft to arrive were hastily assembled and pressed into service in the defence of Tel Aviv, which had already been bombed several times by Egyptian aircraft and which was now being threatened by an Egyptian expeditionary force under Colonel Mohammed Neguib, pushing northwards through the Gaza Strip. The Egyptian force was opposed by an Israeli brigade whose ranks had been seriously depleted as a result of almost continuous action since December 1947, and the situation was desperate.

The first success chalked up by the C.210s fell to a 27-year-old ex-RAF Flight Lieutenant, Moodi Alon. One morning during the second week of June, two Egyptian Dakotas, converted for use as bombers, droned in towards Tel Aviv from the sea, escorted by four Spitfires. Alon, patrolling over the city, was a few thousand feet higher than the enemy and in a good position to attack. The escorting Spitfires scattered in all directions as he dived out of the sun, closing in on one of the Dakotas. A short burst of cannon-fire and the Egyptian aircraft spiralled down to crash in the sea, enveloped in flames. Alon attacked the second Dakota head-on and it went into a spin, its cockpit shattered by cannon-shells. The Spitfires didn't stop to fight; they dived away and headed for home, flying low over the sea.

Three days later Alon once again tangled with the Egyptian Air Force near Tel Aviv. He got on the tail of one Spitfire and shot it down, but cannon-fire from another aircraft riddled his fighter and shell-splinters tore into his body. The wounded pilot managed to land on Tel Aviv airstrip and mechanics lifted him gently from the bloodstained cockpit, but he died later in hospital. He was subsequently fêted as a national hero.

After several days of bitter fighting, the Israeli brigade opposing Neguib's thrust into the heart of

Israel, successfully prevented an Egyptian break-through, but at a cost of over 1,000 casualties, including 200 dead. In the north, the Israelis had also sustained heavy losses in combat against the Syrians, who were pushing on relentlessly towards the Jewish settlements in the Jordan Valley; while in the northwest, two Israeli battalions fought a successful holding action against a Lebanese force that outnumbered them six to one.

When the first United Nations ceasefire came into effect on June 11th, both sides were to all intents and purposes exhausted. The Haganah had no reserves left; every able-bodied man had already been committed to the fighting against the Syrians, Lebanese and Iraqis in the north, the Arab Legion in and around Jerusalem and the Egyptians in the Negev. The ceasefire provided a much-needed respite at a critical moment, enabling both sides to recover and marshal their strength; it had come just in time to ward off a complete Israeli collapse. As it was, the Israelis, although they had held the enemy in check, had lost more than a third of the territory allotted to them by the United Nations.

Apart from satisfying the urgent requirement for equipment, particularly artillery and aircraft, Israel's main hope for survival was to confront the attacking armies with a unified chain of defence. The first step towards this aim had already been taken; on May 31st, eleven days before the ceasefire, the Haganah officially ceased to exist and the Tsvah Haganah le Yisrael, the Israeli Defence Force, came into being. At the same time the Sherut Avir (Military Air Service) became simply Chel Avir, or Air Force.

By the middle of June, the build-up of the Chel

Avir into an operationally effective force was almost complete. Priority had been given to the airlifting of the C.210s from Czechoslovakia, and eleven of the 25 had now reached Israel. By this time, a total of 107 tons of equipment, mainly the C.210s and their spare parts, had been lifted into the country by the hard-worked aircraft of Air Transport Command, which had made some 30 round trips since the operation began. The flights were carefully timed to reach Ekron during the hours of darkness, to reduce the risk of interception by enemy fighters.

In addition to the C.210s, the Israeli Air Force, as a result of the efforts of Ehud Avriel and his team in Prague, could now look forward to the delivery of a number of Spitfires. Early in June, Avriel successfully concluded negotiations for the supply of 50 Spitfire Mk. 9s—all ex-Czech Air Force fighters. The snag was that even with the addition of long-range tanks, it would be almost impossible to make the 1,400-mile flight in one hop; they might just make it, but it would mean stretching them to the limit of their endurance. But there seemed to be no alternative; Air Transport Command was fully committed to airlifting the remaining C.210s and other urgent supplies, and no airfields nearer to Israel were available for staging. In fact, the use by Israeli aircraft of the European airfields that were available was becoming increasingly restricted; in the beginning, for example, the French authorities had freely permitted the Israelis to use Campo Delloro airfield near Ajaccio, in Corsica, for refuelling aircraft en route to Israel, but now this field was being used by growing numbers of American and British aircraft, and the word about the activities of the transports in their Panamanian colours was beginning to spread.

Accordingly, it was decided to attempt the hazardous direct flight to Israel with the first batch of five Spitfires. Fitted with makeshift fuel tanks and stripped of most of their internal equipment, including radios and armament, to reduce their weight, they took off from Zatec early in July and set course south-eastwards. The flight was a disaster. Only three of the fighters reached Israel; the other two were forced to land on Rhodes when the pilots found themselves unable to switch from main to auxiliary fuel tanks because of faulty fuel cocks, and they were interned.

Then, on July 15th, the problem was solved, for the time being at least. Shaul Avigur, who had been one of Haganah's senior planners in Palestine before independence, and who was now responsible for co-ordinating clandestine arms-purchasing activities in Europe, succeeded, after protracted negotiations with the Yugoslav authorities, in obtaining permission to use a deserted airfield on Yugoslavia's Adriatic coast. The field had no hangarage or fuel —in fact, no facilities whatsoever—but it had a 5,000-foot airstrip suitable for the operation of large, heavily-laden transports. Within ten days a team of Air Force technicians had been flown to the spot, and a fuel dump and makeshift servicing facilities were set up with the help of a Yugoslav Army unit. By the end of the month, the Czech Spitfires were staging through the Yugoslav base on their way to Israel, and by the middle of August about 30 had arrived in the country. These formed two squadrons, and were flown by volunteers from half a dozen nations, including Britain, America and South Africa, as well as by Israeli pilots. Among the most notable were Boris Senior, who had served in the South African Air Force during the Second World War; Paul Homeski, a young veteran of the Free French Air Force; and Ezer Weizman, a former RAF pilot who had interrupted his studies at the London School of Economics to offer his services, and who was later to become the Air Force Commander-in-Chief.

The acquisition of the Yugoslav airfield had come only just in time, for a few days later, in response to growing American pressure, the French told the Israelis that they would no longer be allowed to use the airfield on Corsica. Before it was finally closed to them, however, the Israelis used the Corsican staging-point for one last spectacular coup.

For some months, Jewish agents had been operating in Britain, raising funds to obtain arms and recruiting volunteers. In the spring of 1948, one group of agents created a bogus film company under the pretext of making a film about New Zealand's part in the Second World War, and purchased four surplus Bristol Beaufighters with the full approval of the unsuspecting authorities. In any case, the Beaufighters were regarded as being of little more than scrap value, with only a few hours of life left in their engines and airframes. The 'film crews' thought otherwise; some of them were in reality skilled airframe and engine mechanics, and they believed that they could get the Beaufighters into condition to undertake the flight to Israel where they could be thoroughly overhauled and repaired with the abundant spares salvaged from RAF dumps.

One morning in July, the four Beaufighters, with volunteer pilots at the controls, took off from a deserted airfield in southern England which had been taken over for 'filming'. They were next seen in Corsica, where they refuelled before taking off on the last leg of their journey.

With the Yugoslav base at their disposal, the Israelis did not consider the loss of Campo Delloro to be too serious. Late in July, however, came a heavy blow; the Czechs warned the Israelis that (once more because of American diplomatic pressure) they could no longer provide facilities at Zatec. In effect, the Israelis were given 14 days' notice to dismantle their equipment and evacuate all personnel, particularly those who were still American citizens. The Israelis made good use of the period of grace; in one week alone, between August 6th and 12th, Air Transport Command lifted out 40 tons of supplies, including 26 tons of bombs. With this final determined effort, very few supplies had to be left behind. The Israelis saw to it that those which were left did not include spares and ammunition for the C.210s, all 26 of which had by this time reached Israel, or for the 30 Spitfires.

The closure of Zatec, however, although the Czechs insisted that it was only a temporary measure, meant that the Israelis' main source of combat aircraft had now dried up. Avriel thanked providence that Ray Kurtz and his ingenious band of airmen had been able to get the three Flying Fortresses out in time; at least the Chel Avir now had a bombing force of sorts. In spite of the formidable difficulties, however, and thanks to the efforts of Jewish organizations overseas, aircraft of many different types continued to trickle into Israel throughout that summer. There was no attempt at standardization; if an aircraft flew, or had even a remote chance of being restored to flying condition, the Israelis bought it. Bad luck dogged the Israelis on a number of occasions; in addition to the Norseman that crashed at Rome in May, killing 'Screwball' Beurling, two more Norsemen were forced down over Egyptian territory en route and were lost. Three others managed to get through safely, and one of these remained in service until the early 1960s, bearing the registration 4X-ARS. Several aircraft reached Israel via South Africa, where there was a large and active Zionist organization; one of them was a Miles Aerovan light transport (British registration G-AJWI) which was sold to Mayfair Air Services Ltd. by Lees-Hill Aviation (Birmingham) Ltd. in June 1948.

The respite from war lasted only a month. On July 9th, following the rejection by both sides of proposals to prolong the truce and for a lasting settlement of the Arab-Israeli problem made by Count Bernadotte, the United Nations mediator, the Egyptian Army once more went over to the offensive in the Negev. But the situation now was vastly different from that of a month earlier. The Israeli Army was turning into a highly-mobile force, with a nucleus of tanks and artillery, and was well equipped with mortars and small arms; most important of all, it now had air support. Nevertheless, it was the Israelis who had the first taste of the havoc wrought by concentrated air attack, when Israeli and Syrian troops once more joined combat in the north; the Syrians used North American Harvard aircraft fitted with machine-guns and racks for light bombs in the close-support role, and these inflicted heavy casualties on the Israeli infantry. In the renewed fighting against the Transjordanian Arab Legion, however, the Israelis were able to use their own aircraft to good effect; the enemy had none to oppose them. In the Negev, the Israelis recaptured a number of villages from the Egyptians in spite of the fact that the Royal Egyptian Air Force still ruled this part of the sky; the available Israeli fighter aircraft were being held in reserve to defend Tel Aviv against Egyptian bombing attacks. Then, on July 18th, a second truce brought the fighting officially to an end. But not in reality: both sides continued to raid each other's territory.

The truce had left the Egyptians in possession of areas of the Negev where Israeli settlements, although not overrun, were nevertheless completely cut off from outside help except by air. At a meeting of Israeli Chiefs of Staff and senior officers early in August, Prime Minister Ben-Gurion requested the full support of Air Transport Command in lifting supplies to the beleaguered areas. It would be a formidable task. Still desperately short of fuel and aircraft, the Command was required to carry a minimum of 2,000 tons of provisions and equipment to makeshift airstrips in the desert, behind the Egyptian lines, and at night into the bargain.

An initial reconnaissance revealed a reasonably flat stretch of ground near the settlement of Ruhama, and a few days later a small force of Army engineers arrived at the spot with tractors and bulldozers, having made a detour around some Egyptian camps on the way. The engineers immediately set to work levelling out the strip, but they reported that they could not be certain whether it would be firm enough to receive heavily-laden transport aircraft. There was only one way to find out; on August 18th, a C-46, its Panamanian colours now replaced by the six-pointed blue star, landed safely on the strip after flying across the desert in broad daylight, running the risk of interception by Egyptian Spitfires. On board the aircraft was Leo Gardner, Air Transport Command's operations officer, who reported that the airstrip appeared capable of taking aircraft up to a maximum of 30 tons all-up weight, provided the ground was dry. The biggest danger was likely to come from the enormous dust-clouds kicked up by the aircrafts' slipstream; with several aircraft landing one after the other, at night, this would result in bad visibility just before touchdown and might also block the engine filters. However, it was a chance the transport crews would have to take.

After a series of trial flights, which proved entirely successful, the airlift proper, code-named 'Operation Dustbowl', began shortly after dusk on August 23rd. In addition to the C-46s, the Command now had one Lockheed Constellation at its disposal; it was the aircraft that had been grounded in Panama with engine trouble three months earlier, and it provided a valuable boost to the Command's weight-lifting capability.

By August 25th the operation was in full swing, with six aircraft flying supplies into Ruhama almost non-stop. Facilities at the airstrip were primitive, to say the least; runway lighting consisted of sand-filled cans that had been soaked in petrol. Nevertheless, during the first week of the operation the Command flew in 65 tons of supplies nightly, a figure that astonished the planners at GHQ, who had reckoned on a maximum of 15 tons. In addition to cargo, the aircraft also carried infantry reinforcements for the Negev Brigade, and flew out battle-weary troops on the return journey. Women and children from Ruhama settlement were also evacuated.

Surprisingly, it was several days before the Egyptians realized what was happening and even then, the airlift came to their notice purely by chance. It happened when a C-46, whose engines were found to develop insufficient power for take-off after a night flight to Ruhama, had to remain at the airstrip until it could be made serviceable. The following morning, a flight of Egyptian Spitfires appeared over the airfield and circled it a couple of times; they showed no inclination to attack, but the damage had been done. The Israelis took immediate steps to strengthen the airstrip's defences in anticipation of an Egyptian ground attack.

Meanwhile, Air Transport Command made every effort to increase its deliveries. On one night during the first week in September, five aircraft made 13 flights to the airstrip and carried a record 81 tons of freight. For the nine crews engaged in the operation, it was exhausting work; the night-flight across enemy-held territory and the hazardous landings on the makeshift runway, coupled with the fact that no crew enjoyed more than a couple of hours' rest between trips, took its inevitable physical and mental toll. Some crews, almost at breaking-point, approached the Air Transport Command Commander-in-Chief, and requested the temporary suspension of the flights for a day or two; the Commander-in-Chief sympathized, but had no alternative but to turn down the request. By September 9th, the transport aircraft had made 170 flights to the Negev, during which they carried over 1,000 tons of freight and 621 passengers on the way in and 120 tons of cargo and 1,764 passengers on the homeward runs. This marked the end of the first phase of the operation, and for the next two weeks supply flights to the Negev settlements were cut down to two or three a night, allowing Air Transport Command to employ most of its aircraft on arms flights from Europe in readiness for the coming offensive against the Egyptians.

What the Israelis needed now was an excuse to launch their offensive. This came in the afternoon of October 5th, when one of their supply convoys on its way to the Negev settlements with the authorization of the United Nations Observer HQ, was ambushed near Karatiya. Less than an hour later, at about 18.00, six Israeli Air Force Spitfires attacked the Egyptian airfield at El Arish with rockets and cannon-fire, damaging several Egyptian aircraft. Rafah, Gaza, Majdal and Beit Hanun were also attacked just before dark, and a few minutes later the Israeli ground forces went over to the offensive. By daybreak on the 16th, they were within sight of the coastal road at Beit Hanun, the main supply route to the Egyptian camps in the area. A second Israeli thrust, designed to break through to the vital crossroads at Faluja, met with less success; the Egyptians fought back savagely, destroying or damaging all of the few Israeli tanks and cutting one infantry company to pieces. In spite of this set-back, the Israeli offensive was renewed all along the front, and by nightfall on the 17th, Israeli forces were in possession of high ground overlooking the crossroads. During the next 48 hours the Israelis made several attempts to capture the strongly-fortified Egyptian position at Huleiqat, further to the south, as a first step towards encircling the Egyptians at Faluja. The defences were finally overrun on the night of October 20th, at a cost of severe casualties. On the 22nd the Israelis made a determined drive to seal off the coastal road, but the Egyptians anticipated the move and managed to pull out an entire brigade before it was cut off. Nevertheless, the Israelis easily captured the towns of Majdal and Isdud, effectively eliminating the Egyptian threat aimed at the heart of Israel. The breaking of the Egyptian stranglehold on the Negev was completed on October 21st, when the Israeli Negev Brigade took Beersheba.

Meanwhile, a second phase of the offensive had developed with an Israeli thrust towards Jerusalem through Hebron and Bethlehem. There was little resistance; the Egyptian soldiers, demoralized by the defeat in the west, showed scant inclination to fight. However, the Israeli plans to sieze Bethlehem and Hebron were thwarted by a United Nations ceasefire, which brought the fighting almost to a standstill along the whole front.

The part played by the Israeli Air Force throughout this period of fighting had been significant. On September 28th, Air Transport Command had once more stepped up its airlift to the Negev; this time, the cargoes consisted mainly of military equipment and troop reinforcements. The initial flights of the new phase were made to a small airstrip at Urim, on the Gaza Front, which had been levelled out in less than 24 hours by Israeli engineers.

A matter of hours before the Negev offensive was to begin, the Command was ordered to convert some of its aircraft to carry bombs. Half a dozen mechanics performed miracles, fitting four of the C-46s with internal racks along which the crews could manhandle bombs and drop them out of the main hatch. A small number of Dakotas were similarly equipped, but these had been converted as bombers almost from the outset. Together, the C-46s and Dakotas, assisted from time to time by the Air Force's trio of B-17 Flying Fortresses, kept up constant pressure on several major enemy targets during the hours of darkness for four days prior to the offensive. When the ground offensive was well under way, the transport aircraft reverted to their more normal role of airlifting supplies and troops to the forward areas. The operation ended on October 21st; during two months of intensive flying, Air Transport Command had made over 400 flights to and from the Negev, delivering 5,000 tons of equipment and carrying more than 10,000 passengers.

No transport aircraft had been lost through enemy action, even though some flights had from necessity been made in broad daylight. By the middle of September, the 30 ex-Czechoslovak Spitfires introduced into operational service at the beginning of the month had tipped the scales of air superiority in the Israelis' favour; the sporadic air battles that flared up over the Negev in October had been one-sided, and several Egyptian Spitfires had been destroyed. As time went by, the Egyptian aircraft put in fewer appearances, mainly because of the REAF's atrociously low state of serviceability, which resulted in no more than a small proportion of its combat force being in an airworthy condition at any time.

Towards the end of October, fighting broke out again in the north when Lebanese troops attacked and surrounded an Israeli kibbutz. A counter-offensive was immediately launched by three Israeli brigades; in a campaign lasting 60 hours, they killed 400 Arabs and drove the remainder back across the Lebanese border.

On November 11th, the United Nations Security Council passed a resolution calling for the tenuous truce between Israel and Egypt to be replaced by a permanent armistice. The Egyptian Government, however, refused to take part in armistice talks, and Israel's leaders became convinced that they would only be forced into taking such a step by further military defeats. Accordingly, the Israeli GHQ was authorized to go ahead with Operation Horev. This was the final big offensive that was designed not only to drive the last Egyptian forces from the soil of Israel, but also to deal the Egyptian Army such a blow that it would be incapable of undertaking any aggressive action for some time to come.

The opposing forces were more or less evenly matched; five and a half Egyptian brigades against five Israeli. The difference was that the Israeli Air Force now enjoyed complete air superiority; on December 23rd, the date when Operation Horev was launched, there were 113 aircraft on the Chel Avir's inventory, and of these 70 were fighters or bombers. With Egyptian opposition almost completely swept from the sky, the Israeli pilots were now free to fly in support of the ground forces.

The Egyptians were taken completely by surprise. Under cover of a diversionary attack on the Gaza Strip, a strong Israeli force overran the Egyptian defensive positions on the vital Auja el Hafir crossroads on the frontier of Sinai. In the feint attack on the Gaza Strip the Egyptians fought back bitterly and inflicted heavy casualties on the Israelis, but the assault had served its purpose. The road into Sinai was now wide open and the Israelis pushed on, capturing Abu Ageila on December 28th and attacking the desert airfields of Bir Hasana and El Hama, which had already been subjected to heavy strafing attacks by the Israeli Air Force. On the 30th, the main Israeli column captured the airfield of El Arish, and prepared to advance on the town itself, four miles to the north.

The plan was thwarted by a political issue. Almost as soon as the Israeli forces crossed into Sinai, Britain had invoked the twelve-year-old Anglo-Egyptian Treaty of Friendship and had sent an ultimatum to Israel, demanding the withdrawal of all Israeli forces from Egyptian territory. A similar demand was made by the U.S. State Department. Faced with a possible threat of British military intervention, the Israelis called off the attack on El Arish and instead launched an assault on Rafah, beginning on January 3rd, 1949. Three days later, after fierce fighting, the Egyptian forces in the Gaza Strip were completely cut off from outside help; on that same day, the Egyptian Government announced that it was willing to enter into armistice negotiations, and on January 7th the fighting came to an end on the southern front.

It was then that the Israeli Air Force, for some unaccountable reason, precipitated a crisis with Britain that came close to ruining Israel's bargaining position and increased the threat of British intervention.

On the morning of January 7th, the Royal Air Force carried out two photo-reconnaissance missions over Sinai, close to the Israeli border. The first flight, consisting of a Mosquito and four Tempests, carried out its mission without incident and returned safely to base; but with the second flight, made up of four Spitfire FR.18s of No. 208 Squadron, it was a different story. Flying in two pairs, one at 500 feet and the other 1,000 feet above and behind, the Spitfires, following their instructions, sped along the Ismailia-Beersheba road and then turned north to survey the road leading from El Auja to Rafah, which they followed as far as the frontier. At that point, as they turned for home, they ran into heavy anti-aircraft fire; one of them was hit and its occupant, Pilot Officer Close, baled out.

The other three Spitfires circled overhead; they had eyes only for the parachute descending slowly earthwards, and never saw the two Israeli Spitfire Mk.9s arcing towards them out of the sun until it was too late. All three RAF Spitfires were hit almost at once; one went down vertically, its pilot dead at the controls, but the other two pilots, Flying Officers Cooper and McElhew, managed to bale out. McElhew and Close were picked up by Israeli troops and were handed over later, but Cooper, who was injured about the face and head when he hit rocky ground, was the first to get back to base; he landed near a Bedouin encampment, and they brought him in shortly after nightfall.

When the Spitfires became overdue, a flight of four Tempests was sent out to look for them. This time, unlike the Spitfires, the Tempests actually crossed the frontier into Israeli airspace. As they did so, they were attacked by a flight of Israeli Spitfires; one of them was hit and crashed four miles inside Israeli territory. The other three Tempests returned the fire before breaking off the combat and heading for the Canal Zone.

Either by design or mistake, the Israelis had suc-ceeded in destroying five RAF aircraft at no loss to themselves. It was difficult to believe that the Israeli pilots had made a mistake in the first en-counter, and impossible in the second; quite apart from their distinctive RAF roundels and their large 'RG' squadron letters, No. 208 Squadron's Spitfire 18s were distinguishable from the REAF's Spitfire 9s by their bubble cockpit canopies, and in the sec-ond instance the Tempests, with their huge radiator scoops under the nose, bore little or no resemblance to any aircraft serving with the REAF.

The two Spitfires that attacked the 208 Squadron aircraft were camouflaged in dark green and grey, following the RAF pattern, and had red spinners. In addition to their national markings, the six-pointed blue star, they carried diagonal red and white stripes on their rudders. Their serial numbers were 2012 and 2016, the last two digits of which were painted in black on their rear fuselages, aft of the national marking.

The names of their pilots were never revealed.

CHAPTER THREE

The Other Side of the Fence—1

The temporary cessation of hostilities between Egypt and Israel in January 1949, when Israeli troops began their withdrawal from Egyptian soil, found the Royal Egyptian Air Force in a desperate situation. Morale of the Egyptian personnel was at rock-bottom. They had entered the conflict with supreme confidence and a certainty of success; instead, they had received a severe trouncing and a shattering blow to their pride.

Worse still: combat attrition, technical problems and an incredibly low standard of servicing had combined to reduce the operational strength of the REAF to 30 or so first-line aircraft. The majority of these were ex-RAF Spitfire 9s and 22s, but there were also twelve Hawker Furies (at least seven of which were unserviceable at any given time), eight war-surplus Macchi C.205 Veltros and a small number of Fiat G.55 advanced trainers. Because of the embargo on the shipment of war materials to the Middle East, the problem of acquiring spare parts was critical. To overcome it, the Egyptian ground crews had been forced to resort to cannibalization: stripping damaged aircraft that were not immediately repairable in order to maintain at least a token first-line combat force. With the opposing sides in the Middle East held at arm's length by a tenuous armed truce, there was a strong possibility that fighting on a large scale could break out again at any time and in that event, the REAF would have been completely incapable of carrying out even a fraction of its required tasks.

On August 8th, 1948, as part of a plan to expand and modernize the REAF, the Egyptian Government had ordered two Gloster Meteor F.4 jet fighters from Britain at a total cost of £60,000. This order was suspended when the arms embargo was imposed, and authorization for the sale to go ahead was not given until July 1949. Three more Meteor F.4s were ordered in January 1949, while the embargo was still in force; they were to be delivered immediately the arms embargo was lifted or within three months of the signing of the contract, whichever came the sooner. One two-seater Meteor T.7 trainer was also ordered.

The first F.4 was eventually delivered on October 27th, 1949, accompanied by the T.7. The latter aircraft was flown by Gloster test pilot, Squadron Leader Bill Waterton, who remained in Egypt for the best part of a month to assist Egyptian pilots in converting to the new type. The second F.4 was delivered on January 16th, 1950, and the second batch of three F.4s was also delivered between that date and February 17th. A third and final batch of F.4s, seven aircraft in all, arrived in Egypt between March 27th and May 22nd, 1950, as did two more T.7s. A further three T.7s were ordered the following September; this original order was subsequently cancelled, but three refurbished ex-RAF T.7s were eventually delivered in 1955.

While the deliveries of F.4s and T.7s were proceeding, the Egyptian Government ordered 24 of the more modern Meteor F.8s. Hopes of acquiring these, however, were shattered in October 1950,

when the arms embargo was imposed once more. Fourteen of the aircraft went to the RAF, the remaining ten to the Royal Danish Air Force. It was not until December 1952 that the British Government authorized Glosters to proceed with the sale of 12 F.8s to Egypt; four of these reached Egypt in February 1953, the formation staging through Rome and led by Bill Waterton. Then came the Suez Canal crisis, and plans for the delivery of the other eight aircraft were halted. Five of them were sold to Brazil, and the other three—ironically —went to Israel! A further eight Meteor F.8s, all ex-RAF aircraft, were, however, sold to Egypt in 1955, as were six refurbished Meteor NF.13 night fighters.

In December 1949, while Meteor deliveries continued, the Egyptian Air Force received an initial batch of Vampire FB.5 fighter-bombers from Britain. Further deliveries of Vampires continued spasmodically until March 1956. By this time a total of 62, made up of six different Vampire marks, had reached the EAF's first-line squadrons. The Egyptians now had six first-line fighter squadrons, four of them equipped with Vampires, one with Meteor F.4 and F.8 day fighters and the other with Meteor NF.13 night fighters. There was one bomber squadron, equipped with a mixture of elderly Lancaster B.3s and Halifax A.9s, and three transport units equipped with Dakotas, C-46s, de Havilland Devons and Avro Ansons.

By the beginning of 1955 the surviving Spitfires, Fiat G.55s and Hawker Furies had been relegated to the advanced training role, which was now mainly carried out by Meteor T.7s and Vampire T.55s. The basic training squadrons were equipped with de

Havilland Chipmunks and Bucker Bestmanns, 60 of which were built at the Heliopolis Aircraft Factory under German supervision.

The training organization was primitive, to say the least. There was a critical shortage of equipment, including flying clothing; most of it had been turned over by the RAF and was sadly the worse for wear after many years of use. Worse still, the Egyptian instructors themselves had very little flying experience; an instructor with 900 hours to his credit was considered to be an expert. In fact, the whole system was at fault; before they began their air training, all Egyptian Air Force cadets had to spend two years in a military academy, where they were given exactly the same training as army cadets. During this time, they were given two or three air experience flights in a Chipmunk, but they were not allowed to handle the controls, and were given no instruction at all on how an aircraft actually worked. The result was that they were plunged head first into training at the end of their two years in college in a state of ignorance, and yet their instructors classified them below average if they failed to master the intricacies of flying within 20 hours or so.

In an effort to remedy the situation, the Egyptian Government launched a big advertising campaign in Britain to recruit ex-RAF instructors. They wanted twenty; they got six. And when those six arrived in Egypt in October 1955, they were horrified. On reporting for duty at Bilbeis flying college, one of the first people they met was a harassed representative from de Havilland aircraft, who had flown to Cairo at the request of the Egyptians to 'examine and report on the technical defects of the

Chipmunk trainer'. Five Egyptian students had been killed when their Chipmunks had failed to come out of a spin, but the representative had gone over both the wreckage and several intact Chipmunks thoroughly and had found absolutely nothing wrong. There was only one way to find out; he took a Chipmunk up and spun it. The aircraft came out of the spin faultlessly. He spun it several more times, and each time the recovery was perfect. Only then did it dawn on him; the Egyptian students had been taught how to get into a spin all right, but once the spin developed they panicked and completely forgot the correct technique for getting out of it again.

The ex-RAF instructors soon found that although their Egyptian pupils were friendly and anxious to learn, their intelligence, in the majority of cases, was just not sufficient to master flying techniques. It didn't matter what their landings were like, just as long as they walked away from them afterwards. Yet when the British instructors recommended that students be given further preliminary training at the end of their six-month basic flying course, no one took the slightest notice; the authority lay with the Egyptian instructors, who stubbornly refused to listen to sound advice and who passed the students as a matter of routine. After the Chipmunk, the students did a further six months advanced training on the Harvard, then went to Fayid for a jet conversion course on Meteors and Vampires.

In August 1955, two months before the British instructors arrived in Egypt, the Egyptian Government concluded an arms deal with Czechoslovakia for the supply of military equipment, including radar, tanks and aircraft, at a cost of 80 million pounds. The first batch of Czech-built MiG-15 fighters reached Egypt in October aboard the Soviet freighter Stalingrad; they were unloaded at Alami and taken to Almaza for assembly under the direction of a group of Czech Air Force technicians. More MiGs reached Egypt in a steady flow over the weeks that followed, assembly work being taken over gradually by Egyptian technicians who had undergone a crash technical training course in Czechoslovakia. The first two MiG-15 squadrons were formed at Almaza in December 1955, and on January 15th, 1956, these took part in a flypast over Cairo, although at this time they were still flown by Czech pilots. By March 1956, a total of 60 MiGs had been delivered and two more squadrons were forming at Kabrit.

Hard on the heels of the first MiGs, in December 1955, came an initial batch of 12 Ilyushin Il-28 jet bombers: the first of a total order of 60. These were also assembled at Almaza, and by March 1956 29 had been shipped to Egypt. Twenty Ilyushin 14 transports were also acquired to supplement the existing transport force of Dakotas and C-46s. An initial batch of 12 Yak-11 trainers was ordered to replace the ageing Harvards, although the first of these did not arrive until August 1956. Two of the Yaks were destroyed almost immediately, when they were crashed by Egyptian instructors during a flypast for the benefit of President Nasser, and three more were written off accidentally within a month. The Yak was a good deal less responsive than the Harvard, and both instructors and students seemed unable to master it. Conversion to the MiG-15 presented an even bigger problem; the students, used to the lower handling speeds of the Meteor and Vampire, were seldom capable of putting the MiGs down at the right place on the runways, and several were destroyed in landing accidents. The Egyptians' solution to this problem was to make the runways longer.

By the spring of 1956, the Egyptian Air Force possessed some of the finest air bases in the Middle East: airfields which had been evacuated by the Royal Air Force under the terms of the Anglo-Egyptian Treaty of 1954. The largest of these was Abu Sueir, ten miles west of Ismailia on the main Cairo road, with a main runway 9,000 feet in length. The last RAF unit, a Meteor squadron, left on March 10th, 1956, and less than a month later the first MiGs moved in. Although Abu Sueir was used mainly as a storage depot for complete aircraft until sufficient crews had been trained to fly them, an operational MiG-15 squadron, No. 30, with 15 fighters, was formed there in June 1956.

Ranking next to Abu Sueir as a base of first-line importance came Kabrit, at the southernmost point of the Great Bitter Lake. Immediately after the RAF moved out, the Egyptians set to work lengthening the base's 6,000-foot runway by an extra 3,000 feet to make it suitable for the operation of fast jet traffic. When this work was completed, Kabrit was used as a base for the EAF's MiG-15 Operational Conversion Unit, which shared the airfield with two operational MiG squadrons—Nos. 1 and 20. The task of these squadrons, together with No. 30 at Abu Sueir, was to provide fighter cover over the Canal Zone; both bases were defended by a concentration of Czech-built 20 mm anti-air-craft batteries.

Fayid and Kasfareet, further to the north-west, were used by the EAF's Vampire and Meteor squadrons. Fayid was the home of No. 2 Squadron, with 15 Vampires; No. 5, with 12 Meteors (F.4s and F.8s); and No. 40 with 20 Vampires and ten Meteors, six of them tropicalized NF.13s. The EAF's remaining Vampire squadron, No. 31, was based on Kasfareet.

Further inland, Cairo West served as the main EAF bomber base, and by the late summer of 1956 a total of 24 Il-28 jet bombers were deployed there with Nos. 8 and 9 Squadrons. Only 12 of these aircraft were anywhere near an operational standard; the remainder were being used for operational conversion, and other Il-28s which had reached Egypt by this time were held in storage. Almaza, the other military airfield in the vicinity of Cairo, was the EAF's transport base, and the home of No. 3 Squadron with 20 Ilyushin 14s, No. 7 with 20 C-46 Commandos, and No. 11 with 20 Dakotas. The EAF's six recently-acquired Meteor NF.13 night-fighters were also based there, as were eight elderly Hawker Fury piston-engined fighters.

With Czech and Soviet help, the Egyptian Air Force had, on paper at least, been turned into the most formidable striking force in the Middle East almost overnight; the balance of power between Egypt and Israel had been completely reversed, and it was not until the late summer of 1956 that the Israeli Air Force was able to redress the balance to some extent and match the EAF in terms of equipment. In one respect, however, the Israelis enjoyed an overwhelming advantage: the overall ability of its personnel. For in terms of skill and organization, the Israeli Air Force had come a long way from those hectic days six years earlier, when its motley band of volunteers had fought and died in the cockpits of their Avias and Spitfires, setting a tradition as proud as that of any air force in the world.

CHAPTER FOUR

Years of Expansion

During the months which immediately followed the end of the War of Independence, the Israeli Air Force continued to grow in leaps and bounds: there was little attempt at a planned expansion. Early in 1949, the Czech airfield of Zatec was once again made available for a limited period, and this enabled a further 20 Spitfires to be flown out to Israel, together with a substantial quantity of spare parts and ammunition. This brought the total of ex-Czech Air Force Spitfires in Israeli service to 50; all were LF.Mk.9s, although a few were subsequently fitted with high-altitude engines by the Israelis, bringing them up to HF.Mk.9e standard. In 1950-51, 30 more Spitfire 9es were acquired from the Italian Air Force. These served mainly as operational trainers, a role which was also assumed by the ex-Czech aircraft in July 1950.

In 1954, by which time the IAF's first-line air defence units were operating jet aircraft, the Israeli Government authorized the sale of 30 serviceable Spitfire 9s to the Union of Burma, which desperately needed combat aircraft for operations against Communist insurgents. Ferrying the aircraft to Burma, however, presented a problem, as the Arab nations would not allow Israeli aircraft to fly over their territory. To overcome this snag, the Israelis devised an elaborate ruse. The Spitfires, painted in Burmese insignia, were flown out to Catania in Sicily, where they were handed over to the pilots of a British charter company. A false flight plan was then filed to give the impression that the Spitfires had come direct from Britain; the Arabs were completely taken in and gave their permission for the aircraft to stage at various points in the Middle East on their way to Rangoon.

Even so, the flight was a risky venture. Two Spitfires in the first batch came to grief, one when it made a wheels-up landing at Shaibah and another when its undercarriage was badly damaged on a rough airstrip at Calcutta. Then, to complicate matters still further, the newspapers got hold of the story and all hope of maintaining secrecy was shattered. Repercussions followed quickly; the last two Spitfires in the second batch, which had been given permission to stage at Habbaniyah, were pounced on by Iraqi fighters and forced to turn back at gunpoint. They were escorted as far as the Syrian-Lebanese border, when, critically short of fuel they were forced to land at Beirut, where they were impounded and the pilots arrested.

With only seven spitfires successfully delivered so far, the Israelis were now forced to organize an alternative route. The aircraft at Catania were flown back to Israel and fitted with long-range tanks at Lydda; from there, they took off in weekly flights of four and headed out to sea, flying over the Mediterranean for ten minutes before turning northwards and heading for Turkey. After landing and refuelling at the Turkish airfield of Digarbekir, not far from the Soviet border, they flew on to Burma via Iran, India and East Pakistan. Several Spitfires were forced down on old wartime airstrips in Burma when they ran into tropical storms, but all took off again safely and reached their destination.

In 1950, when the Spitfires and Avia C.210s still formed the Chel Avir's first line of defence, the Israelis had some 30 different types of aircraft on their inventory. These included 50 de Havilland Mosquitoes of various marks, which were either purchased direct from Britain or rebuilt from scrapped machines found on RAF dumps. They formed the backbone of the IAF's bombing force, which had hitherto consisted solely of the three elderly B-17 Fortresses. Other aircraft in service were three de Havilland DH.89A Rapides (one of which had been captured from the Egyptians during the Israeli thrust into Sinai); Auster AOP.3s, rebuilt from scrapped RAF machines; five Taylorcraft Plus Cs; two Auster J/1s, formerly used by the Palestine Aero Club; one Aeronca L-18; Canadian-built DH.82C Tiger Moths; two pre-war Polish RWD-13s, also from the Palestine Aero Club; one Republic Seabee; two Grumman Widgeons; three Beechcraft Bonanzas; three Noorduyn Norsemen; one Miles Aerovan; one Fairchild F-24 Argus; two Nord Norecrins; one Douglas DC-5; Boeing PT-17 Kaydets and North American Harvards, used as primary and advanced trainers; Avro Anson Mk.1; Airspeed Consul; one Lockheed Hudson; Douglas C-54; Lockheed Constellation; Douglas C-47 Dakota; Curtiss C-46 Commando; Miles Gemini; and a few new de Havilland Chipmunks, purchased from Canada.

This motley collection of aircraft created an enormous spares problem, and in 1951, when the IAF became fully integrated with the Israeli Army and Navy under the title Israel Defence Forces Air

Force, it was decided to standardize on a limited number of types. The Israelis were anxious to obtain jet equipment for their first-line squadrons, but in the meantime a contract was signed with the Swedish Government for the purchase of 25 ex-Swedish Air Force P-51D Mustangs, for use as an interim replacement for some of the Spitfires and Avias. These aircraft were delivered over a period of eight months, between November 1952 and June 1953.

Meanwhile, on February 10th, 1953, the Israeli Government placed an order with Gloster Aircraft Ltd. for the supply of eleven Meteor F.Mk.8 jet interceptors and four T.7 trainers, at a total cost of £420,000. Before leaving Britain, the aircraft were fitted with Martin-Baker M2E ejection seats and equipped with winches, enabling them to act as fast target-tugs as well as interceptors. The aircraft, three of which had originally been destined for sale to the Egyptian Air Force, were delivered to Israel between August 21st, 1953 and January 17th, 1954. On their arrival, the Israelis fitted them with cannon armament and underwing points for the installation of American HVAR air-to-ground rockets. The four Meteor T.7s were delivered in June 1953; they were followed shortly afterwards by two more, both ex-Belgian Air Force machines, which were delivered by Avions Fairey. The final Meteor variants supplied to the Israeli Air Force were the NF.13, six of which were delivered between September 1956 and March 1958, and the FR.9, seven of which were supplied by Flight Refuelling Ltd. in 1954/55.

Throughout the early 1950s, David Ben-Gurion had campaigned vigorously for the creation of

Israel's own aircraft industry. Initially, his plans met with a great deal of political opposition; many of the Prime Minister's colleagues believed that an indigenous aircraft industry was a luxury that Israel could ill afford. By 1953, however, with the Israeli Air Force squadrons on the point of receiving modern jet equipment, the need for an advanced overhaul and maintenance organization had become apparent and Ben-Gurion received enough support to overrule those who opposed him on the grounds of the country's shaky economic position. That was not the only objection; in 1953, after five years of independence, Israel was still emerging from the embryo stage and skilled technical personnel were fully committed to projects that would directly aid the country's economy. Even if an aircraft industry was founded, it might well be crippled from the outset by a shortage of engineers and technicians.

Nevertheless, Ben-Gurion had his way and Bedek Aircraft, Ltd came into being, directed by the redoubtable Al Schwimmer, who had played such a leading part in acquiring the first transport aircraft for Israel from the United States in 1948, and backed financially by the Ministry of Defence. Based on Lydda Airport, Bedek's main task was to maintain the Israeli Air Force's equipment, but as time went by it gradually assumed more responsibilities such as the overhaul, maintenance and inspection of aircraft operated by foreign charter firms. The 70 technicians originally employed by the firm grew rapidly into 4,000, an increase made necessary as the firm branched out into new fields of aviation technology. A research centre was set up and a number of advanced projects were launched; these included electronic research and experiments with the use of new materials, including plastics, in airframe construction.

By the middle of 1954 the aircraft industry, although small as yet, was on a sound footing, and as Bedek's personnel grew more experienced in maintenance techniques the technical efficiency of the Israeli Air Force increased in direct proportion. The IAF was now commanded by Brigadier Dan Tolkowsky, an extremely able officer who had flown with the RAF's 94 and 238 Squadrons during the Second World War, and under his leadership it lost the haphazard, makeshift atmosphere that had characterized it during previous years. Tolkowsky recognized the need for discipline, which was something the Israeli Defence Forces had lacked until now: because of their 'citizen soldier' status, it had been common for Israeli enlisted men to address their officers by their first names and even to argue about the orders they received. This had been true in particular of the Air Force, many of whose pilots in the beginning had not been Israeli citizens and who had seen no reason why they should be ordered about by men often younger than themselves and lacking experience. They had fought well to a man, but had often preferred to fight in their own individual style rather than as part of a team.

Those days were gone, but in the early 1950s the Israeli Defence Forces still suffered from a hangover of nonchalance. This disappeared gradually as the organization was tightened up. As far as the Air Force was concerned, the first step was to institute a rigorous selection procedure followed by a training programme based on Royal Air Force Lines. From 1951-56, many Israeli students won their pilot's 'wings' after successfully completing a course at a British flying training school; others were attached to RAF advanced flying training schools after carrying out their basic flying in Israel, and when the first IAF Meteors arrived still more Israeli pilots arrived in Britain, either to go through a course at an Operational Conversion Unit or to obtain their Instructor's Rating on jet aircraft at the Central Flying School.

Often, during these courses, the Israelis rubbed shoulders with students from the Arab air forces—and got on well with them. Antagonism between their respective nations was something manufactured by the politicians; these young men, whether Israelis, Egyptians or Iraqis, found no room in the life of an RAF Mess for politics. They had come to Britain to learn how to do the thing they wanted most in the world—to fly. The difference was that the Israelis generally turned out to be the better pilots; in terms of intelligence, initiative and general ability they came much closer to their RAF counterparts than did the Arabs, and they consistently achieved 'above average' ratings.

By 1953, basic flying training in Israel had become standardized on two types of aircraft; the PT-17 Kaydet and the Fokker S.11 Instructor. Forty-one of the latter aircraft were obtained from Holland in 1951-52, but the type proved generally unpopular with instructors and students alike and some of those originally destined for the Air Force were sold to various flying clubs at cut prices. The average IAF student completed between 40 and 60 hours on either or both of these aircraft before being awarded his 'wings' at the end of an initial course lasting from 15 to 17 months. Student pilots then went on to complete 160 hours on Harvards before reaching an operational training unit, where, in the case of future single-seat fighter pilots, they converted on to Spitfires. Pilots destined for Air Transport Command or the IAF's Maritime Flight (which was equipped with two PBY-5A Catalinas for patrol and air-sea rescue duties) underwent twin-engined conversion on Ansons and Consuls, which were also used to train other aircrew branches such as navigators. Future bomber pilots completed their OCU training on the Mosquito T.Mk.3.

Although the majority of IAF ground personnel were conscripts called up at the age of 18 to serve for two and a half years (two years in the case of women), all aircrew were volunteers, signing on for a minimum of five years. Many received preliminary training in the Gadna, an organization similar to the British Air Training Corps or Combined Cadet Force and the American Civil Air Patrol.

Israeli Air Force personnel wore blue uniforms similar in style to those of the RAF, but with American-style badges of rank which were common to all three branches of the Israeli Defence Forces. In spirit and temperament, however, the IAF was noticeably closer to the RAF than to the United States Air Force; it compared particularly with the wartime RAF, when young officers had risen to senior positions of command by virtue of drive and enthusiasm backed up by combat experience.

In the space of five years, the Israeli Air Force's young commanders succeeded in forging a high-class weapon out of what had been little more than a nondescript collection of flying museum pieces, and had instilled the IAF's personnel with a deep sense of discipline, teamwork and tradition. All three attributes would be sorely needed in the months to come, as the Arab nations sharpened their weapons and showed every sign of getting ready for a new trial of strength.

CHAPTER FIVE

The Approach to War

On the morning of August 29th 1955, the pilots of two Israeli Air Force Meteor F.8 fighters patrolling at 20,000 feet over the Negev, 30 miles inside Israel, spotted four suspicious aircraft heading northwards in tight formation some 5,000 feet below. Going down to investigate, the Israeli pilots identified the aircraft as Vampire FB.52s which bore the green and white roundels of the Egyptian Air Force.

The Israelis at once flicked off the safety-catches of their cannon and went in to the attack. Seconds later one of the Vampires was spinning down in flames, shattered by 20 mm shells. The other three aircraft turned westwards and headed out over the Mediterranean in a long dive, shaking off the Meteors.

Less than 48 hours later, two more Meteors surprised a second formation of four Egyptian Vampires near Ashkelon, north of the Gaza Strip. After a twisting dog-fight that lasted five minutes, two of the Vampires crashed in flames; the other two broke off the combat and escaped.

The two incidents brought to a new peak the tension that had been steadily mounting between Egypt and Israel since 1952, when, in addition to their refusal to allow Israeli shipping or vessels destined for Israel to use the Suez Canal, the Egyptians had occupied the island of Tiran at the entrance to the Gulf of Akaba and set up a battery of naval

guns across the straits at Ras Nasrini on the mainland, effectively denying access to the new Israeli port
of Eilat. Then, in September 1954, the Egyptians had
seized the Israeli vessel *Bat Galim* as she attempted
to pass through the Suez Canal, and soon afterwards
a number of Israelis had been executed following
an espionage trial in Cairo. The Israelis had retaliated almost immediately by attacking the Egyptian
town of Gaza in February 1955, killing or wounding
70 Egyptians before withdrawing.

Nasser's reaction to this Israeli counter-stroke
was to reinforce the garrison in the Gaza Strip and
to set up a secret commando school at Khan Yunis,
where Arab volunteers received intensive training
in the art of guerrilla warfare and terrorist operations. During the months that followed, these
terrorists, known as Fedayeen, or 'those who sacrifice
themselves', were a constant thorn in the Israelis'
flesh, slipping over the border to carry out their
work of murder and destruction with the open
acclaim and approval of Nasser's government. In
August 1955, the Israelis launched a punitive strike
against Khan Yunis, killing 40 Egyptians and wounding as many more; this was followed by a period of
fighting along the frontier of the Gaza Strip, which
lasted until the United Nations Truce Commission
was able to impose an uneasy ceasefire.

The Egyptians, however, showed every sign of
becoming more aggressive. In defiance of the armistice agreement of 1949, they now moved a strong
concentration of troops into the demilitarized zone
around Nizana, where a strategic road junction
branched off towards Gaza, Rafah and Abu Ageila,
the main Egyptian base in Sinai. This led to a clash
between Egyptian and Israeli forces in November,

and although the result was indecisive as both sides broke off the engagement after sustaining only light casualties, it strengthened Nasser's resolve to build up his armed forces to the point where they would be capable of overcoming any further Israeli opposition. The result, as we have seen in the preceding chapters, was an approach to the countries of the communist bloc with a request for the supply of substantial quantities of arms.

In August 1955, when Nasser concluded his arms deal with Czechoslovakia, the Israeli Air Force possessed only one combat aircraft that could be classed as anywhere near modern: the Gloster Meteor F.Mk.8. It was apparent that this, the IAF's only jet type, would be no real match for the much faster MiG-15. There was only one aircraft in service anywhere in the world at that time which, in the skies over Korea three years earlier, had proved consistently capable of establishing air superiority over the MiG time after time: the North American F-86 Sabre. Realizing that a direct approach to the United States for help would be out of the question because of that country's embargo on the supply of arms to the Middle East, the Israeli Government turned to Canada, where the American fighter was being licence-manufactured by Canadair under the designation of Sabre Mk.6. Moreover, the Sabre 6 was a superlative aircraft; its Canadair-designed Orenda turbojet gave it a much improved performance over the basic F-86F from which it was developed, enabling it to climb to 40,000 feet in nine minutes flat. With the potential enemy's air bases only a few minutes' flying time away, the Israeli defences would have to be in a position to react quickly if attacking aircraft were to be intercepted, and there was a pressing requirement for a fighter with a fast rate of climb.

The Canadian Government showed every sign of being co-operative and the Israelis immediately placed an order for 24 aircraft. Meanwhile, negotiations had also begun with the French Government and Avions Marcel Dassault for the supply of two principal types of combat aircraft: the Ouragan and the Mystère IIC. The negotiations dragged on for weeks, but in the end the French agreed to supply 30 Ouragans and 24 Mystères. The first 15 Ouragans, part of a batch originally destined for the French Air Force, arrived in Israel in November 1955. They were ferried out by French pilots, some of whom stayed behind to assist the Israelis in the task of converting to the new type.

Soon after this, the Sabre order fell through when the Canadian Government also placed an embargo on the supply of arms to the Middle East. The Israelis were now totally dependent on France, but there was every hope that the French would continue to co-operate. After all, the French Government had an interest in helping Israel to build up a strong defence force; the more Nasser became preoccupied with the Israeli problem, the less likely he would be to step up his aid to the terrorists in Agleria, where French forces were at that time heavily committed.

The Dassault Ouragan, although a distinct improvement on the Meteor, would nevertheless be no answer to the threat posed by Nasser's MiG-15s. Designed as an interceptor, the French fighter was rapidly approaching obsolescence in that role; its maximum speed at sea level was a hard-earned 530 knots and its initial rate of climb 7,800 feet per minute, compared with the MiG-15's 570 knots and 10,000 feet per minute. In view of the MiG's definite overall superiority, the Israeli Air Force decided to use the Ouragan in the fast ground-attack role; its endurance of 70 minutes or so, which severely limited its combat radius, could be greatly extended by the use of wingtip fuel tanks; it was an extremely stable gun platform and well-armed with four 20 mm cannon, which could be quite devastating against most types of ground target; and there was provision for a variety of underwing stores, including 16 rockets.

Meanwhile, the Israelis were beginning to have serious doubts about the potential of the other French aircraft they had ordered, the Mystère IIC. This type was a straightforward swept-wing development of the Ouragan, powered by an Atar 101 turbojet that gave it a maximum speed of 560 knots at sea level and an initial rate of climb of 8,000 feet per minute. Like the Ouragan, the Mystère IIC had four 20 mm cannon in the nose. There was no doubt that the Mystère came much closer to matching the MiG-15 in terms of performance than did the Ouragan, but even so, the Russian fighter was still superior on most counts. What was more, the French Air Force, which had ordered 150 Mystère IICs in 1953, was far from satisfied with the aircraft. Several had been lost through structural failure, and further orders were held up while Dassault's designers tried to sort out the trouble.

In fact, the delay proved to be so serious that the Mystère IIC was overtaken on the production line by another Dassault fighter: the Mystère IVA. Although it bore the same name as the earlier aircraft, the Mystère IVA was in fact a completely new design and was so promising that the French Air Force had placed an order for 325 in April 1953, six months after the prototype had flown. Powered by a Hispano Suiza Verdon engine, the Mystère IVA had a maximum sea level speed of 600 knots and was armed with two 30 mm DEFA cannon. In addition, the aircraft could carry two underwing Matra rocket pods, each containing 19 air-to-ground projectiles, with two 1,000-pound bombs or twelve T-10 missiles as alternative loads. Production aircraft were fitted with a new type of gun-ranging radar, which ensured a high degree of accuracy.

In October 1955, a group of senior Israeli Air Force officers visited the French air base at Cambrai, where the first FAF Mystère IVA unit had formed the previous August. They saw the aircraft put through its paces, and learned that the French pilots were delighted with the way it handled. A fortnight later, on the recommendation of the IAF delegation, the Israeli Government cancelled the Mystère IIC order and changed it to 24 Mystère IVAs. This was quickly followed by a second order for an additional 36 machines, which the French Government agreed to supply in spite of the arms embargo.

The nucleus of the first Israeli Air Force Mystère squadron was formed in April 1956, with the arrival of an initial batch of eight aircraft. The pilots were all experienced on the Meteor and found the French machine a much simpler and more pleasant aircraft to fly. In an effort to expand its limited number of qualified jet pilots, the IAF completely revised its flying training programme; previously, pilots destined for the Meteor had been required to put in a considerable amount of flying on the Mustang before converting to the jet type, but now this requirement was scrapped and pilots went straight on to the Meteor after completing their advanced flying training on Harvards. Jet conversion was carried out on the IAF's four Meteor T.7 two-seat trainers, and this was followed by some 30 hours' familiarization on the single-seat Meteor F.8 before converting to the Mystère or Ouragan.

There was little need to underline the urgency behind this crash training programme. During the first months of 1956 Soviet war material continued to flow into Egypt at an increasing rate and the Fedayeen raids into Israeli territory increased. In April, Egyptian artillery shelled Israeli settlements across the border of the Gaza Strip; the Israelis retaliated by shelling Gaza itself, and an artillery duel developed that lasted almost half a day. A ceasefire was arranged by the United Nations observers, but the echoes of the gunfire had hardly died away when fighting broke out once again. During the second week of April there was a spate of Fedayeen raids, which resulted in many Israeli casualties—mostly civilians. On April 12th, an Egyptian Vampire FB.52 was shot down by a Meteor inside Israeli territory in yet another clash.

It was then that international politics caused Egypt to shift her attention from Israel—for the time being at least. For various reasons, but mainly because of the acquisition of arms from the communist countries by Nasser and his recognition of Communist China, the United States refused to finance the building of the Aswan High Dam, a project on which much of Egypt's future economic development depended. Infuriated by this rebuttal, Nasser retaliated by announcing on July 26th, 1956, that he intended to nationalize the Suez Canal Company. During the first week of August, several reserve formations of the Egyptian armed forces were mobilized.

On September 1st, a full meeting of the Israeli General Staff was convened to discuss the implications of a top secret intelligence report that had just arrived from the Military Attaché in Paris, who had been in consultation with French Defence Ministry officials. The report contained full details of 'Operation Musketeer', the Anglo-French plan to seize key points in the Canal Zone, and even included the names of the British and French officers who had been chosen to command it. The report also indicated that the invasion would probably take place during the last week of October at the very latest. For Israel, the implications of the report were enormous. Not only would an Anglo-French invasion mean the possible opening of the Suez Canal to Israeli shipping, but it would present the Israelis with an unparalleled opportunity to remove the threat of an Egyptian attack from their doorstep. They could mount a lightning thrust into Sinai while Nasser's attention was occupied elsewhere.

During the week that followed, the Israeli Chief of Staff, Moshe Dayan, made a personal tour of Army and Air Force establishments to assess their combat potential. During his visit to the IAF Headquarters and first-line air bases, he was impressed by the tremendous enthusiasm of both air and ground crews. Without exception, they were keen to 'have a go' at the enemy. Nevertheless, the process of conversion to jets, although going ahead smoothly and as quickly as possible without cutting too many corners, was not producing fully qualified jet pilots

at as fast a rate as had been anticipated, and there
was no escaping the fact that if Israel was going to
find herself involved in a war within the next two
months, the IAF would be caught in mid-stride.

Apart from the pilot problem, there was also a
shortage of skilled technical personnel. Ground
crews were still in the process of mastering the tech-
nicalities of the new jet aircraft, and the unservice-
ability rate was high. By the middle of September
the IAF possessed 37 Mystère IVAs, but not more
than 16 of them were serviceable at any one time.
As far as the Ouragans were concerned, the picture
was more optimistic: of the 30 then in service,
between 22 and 25 were fully serviceable.

To add to the difficulties, there was also a shortage
of bombs and rockets. Supplies which had trickled
through from France had immediately been allocated to the Mustang and Ouragan units, but the
2 cm rockets for the Mystères had not yet materialized. There was even a shortage of 30 mm shells for
the Mystères' cannon. In an effort to remedy the
situation, representatives of the Israeli General Staff,
including Moshe Dayan and Shimon Peres, the
Director General of the Ministry of Defence, flew to
Paris on September 28th for urgent talks with
officials of the French General Staff and Defence
Ministry. As well as the bombs, rockets and ammunition, the Israelis' last-minute 'shopping list' included
100 tanks, 300 half-tracks, 300 trucks, 50 tank-
transporters, 1,000 bazookas and 12 transport aircraft. The Israeli Air Force badly needed the latter
to supplement its one squadron of ageing Dakotas.
The aircraft it wanted from the French was the
Nord 2501 Noratlas, a medium transport capable of
carrying 45 fully equipped paratroops.

The French Chief of Staff, General Ely, promised to do everything in his power to help, although there were many difficulties to be overcome, notably the growing commitment of the French armed forces in Algeria and a large amount of material destined for the projected invasion of Suez. It was General Maurice Challe, the French Air Force Commander-in-Chief, who came up with an answer of sorts to the aircraft problem; he indicated that a number of French fighter aircraft and one squadron of transports might be temporarily stationed on Israeli bases. The fighters would have the task of protecting Israel's cities against a possible attack by Nasser's jet bombers, thus releasing the whole of the Israeli Air Force for offensive operations. The transports would be placed at the disposal of the Israelis on the understanding that they could be recalled at short notice when the Suez invasion developed. As Challe pointed out, the French aircraft could operate against the Egyptians just as easily from their Israeli bases as they could from Cyprus, where the French squadrons taking part in 'Operation Musketeer' would be based.

The Israelis flew home on October 2nd, and Dayan immediately called a meeting of the General Staff and told them to prepare for a decisive campaign against Egypt aimed at smashing the Egyptian forces in the Sinai Peninsula and breaking Nasser's stranglehold on the Gulf of Akaba. The Chief of Staff told his commanders to have all their preparations ready by October 20th, and to reckon on the campaign lasting about three weeks. For the sake of secrecy, the Israeli reserves were not to mobilized until four or five days before D-Day.

While preparations for a blitzkrieg against the

Egyptian forces in Sinai were under way, Israeli
forces found themselves engaged in yet another
reprisal action. This time, the target was a police
fort situated just north of the small town of Kalki-
liah, in Jordan. The attack, which took place on the
night of October 10th/11th, was retaliation for a
recent spate of terrorists raids across the Jordanian
frontier with Israel. In spite of the urgent need to
keep all Israeli units rested and at full strength to
meet the demands of the impending Sinai campaign, the threat presented by these terrorist raids
into the very heart of Israel, a mere 12 miles or so
from Tel Aviv, could no longer be ignored.

The assault was launched at 20.00 hours on
October 10th, and the fort fell to the Israelis shortly
before midnight, but the Jordanians put up a much
stiffer fight than had been anticipated. One unit
of 50 men was completely cut off inside enemy
territory and subjected to heavy artillery fire and
determined attacks by troops of the Arab Legion.
Three Israeli companies, one of paratroops and two
of infantry, mounted on half-tracks, raced to the
besieged unit's assistance. The first company broke
through at 02.30, in time to repel another Jordanian
attack. Overhead, two IAF Mosquitoes and two
Harvards circled the area and dropped flares; they
could not be of much assistance while it was still
dark, but their presence might lead the enemy to
believe that much stronger reinforcements were on
their way with air support. Two more flights of
Mosquitoes were, in fact, on the alert in case a
second Israeli force had to be sent in at daybreak,
but they were not needed. The hard-pressed Israelis
succeeded in fighting their way out at 03.30; they
came under heavy fire all the way back to Israeli

territory and sustained many casualties.

On October 22nd, the air support promised by General Challe arrived in the shape of 36 Mystère IVAs of the French Air Force's 2e Escadre, which had flown out to Israel from their base at St Dizier by way of Algeria. They were followed, two days later, by 36 F-84F Thunderstreak fighter-bombers of the 1e Escadre, which was normally based on Dijon, and by eight Noratlas transports of the 64e Escadre. The transports, in addition to ground crews and supporting equipment for the fighter squadrons, also brought some of the desperately needed supplies of ammunition for the Israeli Air Force. The two French fighter wings. were in fact part of the 1st Tactical Air Force (1erCATAC), which was the part of the French Air Force assigned to NATO. During their stay in Israel the Thunderstreaks were to be based on Lydda, while the Mystères and Noratlases shared Haifa between them. The original plan was for the Mystères eventually to be handed over to the Israeli Air Force, but in the event both the 1e and 2e Escadres flew home to Dijon and St Dizier in December 1956. The Noratlases returned to France in November.

On October 25th and 26th, the final operational plans for 'Operation Kadesh'—the invasion of Sinai, now scheduled to begin at dusk on October 29th —were drawn up. As far as the Air Force was concerned, there was a last-minute change of task: the original plan had called for an intensive air strike against the main Egyptian airfields, particularly Cairo West where the Ilyushin 28 jet bombers were based, in the two hours before the scheduled time of the invasion. It was decided, however, to abandon this scheme and to hold the IAF in readiness to support the Israeli ground forces after the attack was launched. There were two main reasons for cancelling the plan to knock out the Egyptian Air Force on the ground; first, it would have meant that the IAF would be stretched to the absolute limit, with no aircraft held in reserve for ground-support duties; and, second, it would have alerted the Egyptians to the fact that Israel was mounting a large-scale offensive in Sinai and not simply carrying out another series of reprisal attacks.

The Israeli decision involved a calculated risk: a gamble based on the assumption that Nasser's two squadrons of Il-28s would not be in a sufficient state of operational readiness to carry out serious attacks on Israel's cities. The latest intelligence indicated that only a handful of Egyptian crews had so far converted to the new aircraft, and the Ilyushins were not likely to be flown in combat by Russian or Czech instructors.

During the first phase of the invasion, therefore, from D-Day up to the end of D-Day plus Two, the part played by the Israeli Air Force would be flexible and would depend to a great extent on how far the enemy committed his air force. The orders were to avoid air combat as much as possible, but at the same time all necessary steps were to be taken to prevent Egyptian aircraft from interfering with the Israeli operations in the Peninsula. To this end, a standing fighter patrol was to be maintained over the east bank of the Suez Canal, but under no circumstances were Israeli aircraft to cross over into the airspace of Egypt proper. Patrolling the Canal was to be the task of the Mystères. In the event of the Egyptian Air Force attacking targets in Israel, the IAF was to mount an all-out offensive against the Egyptian airfields across the Canal, using all available jet aircraft.

From D-Day plus Two onwards, by which time the intentions of the Egyptian Air Force would have been made clear, the IAF's main task would be to provide effective support for the ground forces. The Ouragans, Mustangs and Harvards were earmarked for this purpose, with Mystères and Meteors providing top cover. The IAF was also to be prepared to switch part of its effort rapidly in case other Arab countries entered the war on the side of Egypt. By far the greatest burden was to fall on the shoulders of Air Transport Command: as well as dropping para-troops and landing infantry reinforcements in the forward area, the Dakotas and Noratlases would share much of the responsibility for sustaining the offensive by dropping supplies and evacuating the wounded.

Meanwhile, mobilization had been going ahead smoothly and in strict secrecy. This was achieved in the simplest possible way: after a secret briefing, the commanders of the Israeli units visited certain 'key men', who went from door to door passing on the mobilization order to the reservists by word of mouth. This procedure was so effective that the mobilization of all Air Force reservists was completed in only 43 hours. Under the supervision of their regular officers and NCOs, these men worked like slaves so that as many aircraft as possible were serviceable within the two days or so before the start of the invasion.

Nevertheless, in spite of all their efforts the operational strength of the IAF could not be stretched beyond 136 aircraft by noon on October 29th. The total included 16 Mystères, 22 Ouragans, 15 Meteors,

29 Mustangs, 17 Harvards, 16 Mosquitoes, 16 Dakotas, three Noratlases and two ageing B-17 Flying Fortresses. Compared with the Egyptian Air Force's first-line strength of 245 aircraft, the Israeli line-up hardly seemed impressive, particularly since most of the EAF's aircraft were jet types. However, Israeli intelligence sources on the eve of the invasion indicated that only about half the Egyptian squadrons were operational, and that half the operational total was made up of transport aircraft. These were Dakotas, C-46 Commandos and Russian-built Ilyushin Il-14s—about 20 of each. The remainder of the operational strength consisted of 12 Ilyushin 28 jet bombers, 12 Meteor F.8s, 15 Vampire FB.52s and 30 MiG-15s. The Israeli Air Force therefore possessed 53 operational jet aircraft against Egypt's 69: a ratio that was far from disadvantageous, taking into account factors such as morale and general flying ability. The Israeli pilots, however, were under no illusion that their task would be easy if the Egyptians chose to come up in force to oppose them. Firstly, there was the problem of range as the areas over which most of the air fighting was likely to take place lay at the limit of the Israeli fighters' combat radius, which would mean that they could spend only a very limited amount of time actually engaged in combat before shortage of fuel compelled them to break off and return home. The Egyptian aircraft, on the other hand, would be fighting only a few minutes' flying time away from their bases, and if necessary they could make a short dash to safety across the Suez Canal, where the Israeli pilots, because of their orders, would be unable to follow them.

Secondly, because the IAF would be called upon

to provide air cover for the paratroop units which were to be dropped within striking distance of the Canal, most of the air combats, at least during the first forty-eight hours, would almost certainly take place over enemy territory, with the risk of shot-down pilots being taken prisoner or summarily executed. Egyptian pilots forced to take to their parachutes could be in the air again within a matter of hours which was a vital consideration where pilots were in short supply. It was the kind of situation which had contributed to the RAF's victory during the Battle of Britain, and which had enabled the Luftwaffe to keep going so long in the face of formidable odds during the Battle of Germany four years later.

Finally, there was still a serious shortage of fuel and ammunition. This was rectified to a certain extent when a number of French freighters arrived at Haifa in the afternoon of October 27th, only two days before the invasion, carrying supplies of muni-tions as well as trucks and other equipment. Priority was given to unloading the equipment destined for the army units that would carry out the initial thrust into Sinai, but it was not until the early afternoon of October 29th that supplies of bombs, rockets and cannon-shells reached the IAF's front-line squadrons.

On the eve of the campaign, the Israeli pilots were far from enthusiastic over the directive that their primary task was to be to provide close support for the ground forces. Until then, all their training and planning had been aimed at one goal: the rapid destruction of the Egyptian Air Force, should it be made necessary by the course of events. The IAF's strategists had worked out that, with surprise on Israel's side, it would take three days to shatter the EAF and now the initiative was being handed to the Egyptians on a plate. Most of the pilots would also have preferred if the campaign could have been delayed for another two or three months, by when more of them would have qualified on jets; as it was, there was such a shortage of jet pilots that the flying schools had to be closed so that the instructors could be released for operational flying. This meant that if the Egyptian Air Force did choose to fight and inflicted losses on the IAF's jet squadrons, there would be no replacement pilots.

Nevertheless, the pilots were supremely confident that they could do the job required of them with the equipment available. It was a case of necessity; the war would have to be fought and won with the aircraft that went into action at the start of hostilities. Losses could not be made good with factory-fresh machines, and there were enough spare parts to cope with only a limited number of battle casualties. The IAF's commanders knew that there were maintenance facilities and sufficient stores of fuel and ammunition to keep up a sustained air commitment for about a week; what might happen after that, if the Israeli ground forces had not secured their objectives and were still involved in heavy fighting, was anybody's guess.

CHAPTER SIX

The Sinai Campaign

October 29, 1956. 16.30 hours.

The 16 aircraft hung in the sky like a swarm of locusts, silhouetted against the red glow of the dying sun. They were C-47 Dakota transports, their twin Wright Cyclones throbbing out a steady beat over the desert.

The Dakotas flew on at 500 feet above the sand and rock of Sinai, keeping low to avoid detection by the enemy radar. It was now 70 minutes since they had formed up over their base at Ekron and set course south-westwards, heading out over the Peninsula. In each vibrating metal fuselage sat 25 men of the Israeli 202nd Airborne Brigade. For the most part, they sat in silence on their uncomfortable benches; the dropping-zone was only 29 minutes away, and each man was busy with his own thoughts.

The objective for which the Dakotas were now heading with the 400 men of the Airborne Brigade's 1st Battalion was Mitla, some 40 miles east of Port Suez. The paratroops were to be dropped at the Parker Monument (a memorial to Colonel Parker, the British officer who was twice Governor of Sinai) at the eastern end of the Mitla Pass, the vital defile that controlled the Ismailia-Tor and Nakhel-Suez crossroads. The 20-mile-long pass was murderous territory; the road that wound through it was flanked by steep, scorching rock walls, pitted with fissures and caves that formed natural defensive positions.

The task of the 1st Battalion was to secure the pass before the Egyptians had time to rush up reinforcements, and to hold it until the remainder of the Airborne Brigade arrived. The rest of the Brigade, some 2,000 men, spearheaded by two Companies mounted in half-tracks, had already crossed the start-line on the first stage of the 130-mile dash across enemy territory to Mitla. On the way, the mobile force would have to smash three fortified Egyptian positions at Kuntilla, El Thamad and Nakhel. As an insurance against the mobile force being held up by exceptionally strong enemy resistance, a central task force had also crossed into enemy territory at Sabha, 60 miles further north, and was now racing on towards the strategic crossroads at Kuseima. With this objective captured, the task force would be in a position to go to the aid of the paratroops at Mitla either through Nakhel or Bir Hasana. The whole operation depended on three things: complete surprise, the inability of the Egyptian Military Staff to react quickly to the situation and the ability of the paratroop battalion to deny the Egyptians access to Central Sinai through the Mitla Pass.

Now, the dropping-zone only minutes away, the Dakota pilots took their aircraft up to 1,500 feet. From now on, they kept a watchful eye on the sky over the western horizon, screwing up their eyes against the dazzling glare of the setting sun. The enemy radar was bound to pick them up at any moment and the Egyptian air base at Kabrit, with its two squadrons of fast MiG-15 interceptors, was only 45 miles away on the other side of the Great Bitter Lake.

A thousand feet above the Dakotas, ten more pairs

of eyes also scanned the western sky. They belonged to the pilots of ten twin-jet Gloster Meteor F.Mk.8s, which had come whistling up from the north ten minutes earlier to provide close escort for the Dakotas on the approach to the dropping-zone. The fighter pilots had only located the transports after some difficulty as the Dakotas' light-and-dark brown camouflage mingled well with the hues of the desert, and it was hard to see anything in the glare from the west. In two flights of three and one of four, the Meteors now weaved from side to side above their lumbering charges in a series of gentle turns.

Behind the Meteors, and still 15 minutes' flying time away, came eight more fighters: Dassault Ouragans. They were faster than the Meteors, but possessed a more limited combat radius. Their task would be to cover the Dakotas on the way out while the Meteors headed for home, short of fuel. If it came to a scrap, the Ouragans would stand a better chance against the MiGs than would the British fighters; all the pilots had read the reports on the air fighting over Korea four years earlier, when Meteors, operated by a RAAF squadron, had had to be withdrawn from the interceptor role and turned over to ground-attack duties because of the mauling they received at the hands of communist MiG-15s.

Other Israeli fighters were in the air; as the Dakotas approached their objective, a pair of Dassault Mystères cruised at 10,000 feet over the east bank of the Great Bitter Lake, within sight of Kabrit, ready to engage any Egyptian aircraft that attempted to interfere with the paradrop. Operating at the extreme limit of their range, the Mystères could spend no longer than ten minutes in the patrol area; while two patrolled, two more were on their way out to relieve them and the previous two were on their way home. Twelve Mystères were being used to maintain this standing patrol along the length of the Canal; six were airborne at any given time.

The Mystère pilots experienced a good deal of frustration as they saw the Egyptian MiGs parked in full view on Kabrit airfield, and prayed that the enemy would come up to engage in combat. But the Egyptians' only reaction was to disperse their fighters hurriedly, and the Israeli pilots were powerless to intervene. They were under the strictest orders to remain over the east bank. Sabre pilots over Korea had been faced with a similar problem, being forced to watch the enemy MiGs taking off unmolested from their bases across the Yalu and climbing until height and sun were in their favour.

16.57 hours. Air roared into the bellies of the Dakotas as the despatchers opened the hatches. The paratroops made last-minute adjustments to their equipment, rose from their places and clipped their static lines to the rail running the length of the fuselage above their heads.

Fifteen seconds to go. The first man tensed himself in the doorway of the leading aircraft, watching the scrub and sand crawl past far below. In the cockpit, the pilot sweated as he held the Dakota rock-steady, blinded by the sun's rays. Then the green light flashed on and the leading paratrooper launched himself into space.

As the last stick of parachutes drifted down towards the desert, the Dakotas and their escorts turned for home. Two hours later, all the transports landed safely at their base. But for the air and

ground crews on the Dakota squadron, there was to
be no respite; all that night, and for the three nights
that followed, the transports were to fly sortie after
sortie, dropping supplies to the Israeli forces advancing through Sinai.

The paradrop had gone off with only one hitch:
the Battalion had gone down three miles east of the
actual dropping-zone. The Dakota crews could
hardly be blamed for the error; the glare of the
sun had made visual navigation extremely difficult.
There were 13 casualties, mostly with sprained
ankles.

After a two-hour march, the paratroops reached
the Memorial and dug themselves in close to the
mouth of the Pass. At 21.00, a flight of Dakotas
dropped eight jeeps, four 106 mm recoilless guns,
two 120 mm mortars, ammunition and food. There
was one brief skirmish in the darkness, when an
advance party of paratroops ran into two Egyptian
light reconnaissance vehicles and opened fire on
them. The crew of one was killed immediately and
the vehicle slewed off the road, but the other
escaped. During the hours that followed, the paratroops set up road blocks and cleared a dropping-
zone to the east of their position.

Meanwhile, the main body of the Airborne
Brigade had successfully overrun the frontier post
at Kuntilla, attacking from the west. The same sun
that had blinded the Air Force pilots during the
Mitla drop now blinded the Kuntilla defenders;
although they fought back valiantly, they were
unable to prevent the Israelis breaking through the
perimeter in a crazy charge with their half-tracks,
and the Egyptians retreated into the desert. Two
Israeli half-tracks were blown up by mines and one

1an was wounded; he was later evacuated aboard a
'iper Cub liaison aircraft. Many of the Brigade's
ehicles were already suffering badly from a shortage
f fuel and lack of spares, even spare tyres were in
hort supply. It was not until 22.30, five hours after
he attack on Kuntilla, that the paratroops were able
o push on towards their next objective, El Thamad,
0 miles further on.

The attack on Thamad was launched at 06.00
1ours, about an hour after sunrise, which the
sraelis used to their advantage by launching their
ssault from the east. El Thamad was defended by
wo companies of the Egyptian Frontier Force and
. National Guard detachment; they put up a short
nd spirited fight, but pulled out as soon as the
sraelis broke through into their positions, leaving
0 dead behind them. The Israelis lost three killed
nd ten wounded.

While the remainder of the Brigade regrouped
it an oasis a few miles from El Thamad and strove
o get as many vehicles serviceable as possible, the
2nd Battalion moved into the positions they had
ust captured from the Egyptians. At 07.00, the
oldiers were having their breakfast when the alarm
vent up and two Egyptian MiG-15s swept in low
'rom the west, cannon blazing. The MiGs made
hree or four passes before flying away, leaving the
2nd Battalion with 40 casualties and six vehicles
n flames. Fortunately, the Egyptian pilots had not
:potted the remainder of the Brigade.

During the course of the morning, the Brigade
·eceived an airdrop of supplies from two Noratlas
ransports. The next objective was Nakhel, another
10 miles to the west, but the Brigade Commander
1ad no intention of pressing on without his support-

ing artillery, which had got bogged down in so
sand on the drive from Kuntilla. The artiller
turned up at 11.00, but the Commander still had
difficult decision to make. He could push straigh
on in broad daylight, trusting that the patrollin
Mystères and Ouragans would be able to disrup
any concentrated attacks by Egyptian fighter
bombers; or he could disperse the Brigade and wa
until nightfall, pushing on to the next objectiv
under cover of darkness.

The decision rested on how the battalion at Mit
was faring. More supplies had been dropped to th
battalion at sunrise by Noratlases of the French Ai
Force's 64e Escadre, operating out of Haifa, and nc
long afterwards a Piper Cub, one of five attached t
the Airborne Brigade, had been sent up to mak
contact with the forward element. It had still nc
returned at 10.00, and the Brigade Commander
seriously worried by this time, decided to send ou
a second liaison aircraft.

The fact was that the forward battalion ha
been under attack since dawn. The battalion com
mander had made a serious mistake; in the darknes
of the previous night, he had not realized that th
positions occupied by his men were overlooked b
high ground on the southern side of the Pass. Thi
high ground was occupied by a company of Egyptia
troops who had reacted with amazing speed and
hurried to the spot from the western end of the Pass
and now, even before the sun had begun to dispers
the thick dawn mist that hung over the area, the
began to lob mortar bombs on to the Israeli posi
tions around the Monument. A probing attack b
the enemy was successfully beaten off by the para
troops, but it could only be a matter of time befor

heavier assault developed. In fact, units of the
gyptian 2nd Brigade were at that moment crossing
1e Canal and preparing to race to the scene of the
ction.

The first Egyptian air attack on the battalion at
1itla came at 07.30 on that morning of Tuesday,
)ctober 30th, about half an hour after the Israelis
ccupying El Thamad were strafed. A pair of MiG-
5s raced up and hammered the Israeli positions
vith cannon-fire for five minutes, causing some
asualties and knocking out the Piper Cub that was
tanding, fortunately unoccupied, on the improvised
anding-ground to the east of the Monument. A sec-
nd attack was made an hour later by four Vampire
'B.52s of No. 2 Squadron, Egyptian Air Force,
perating out of Fayid. They were escorted by
another pair of MiGs, but no Israeli aircraft
appeared to challenge them.

At 12.00, the second Piper Cub, which luckily
1ad missed the latest strafing attack, returned to El
Thamad with news of what was happening, and also
vith an urgent request for air support. Until this
noment, the Israeli Air Force had not been author-
zed to undertake close-support operations; but
1ow, as the Egyptian Air Force had struck the first
)low, the Airborne Brigade Commander was in-
ormed by GHQ that air support would be available from 13.00 onwards. The Commander requested
immediate air cover for the battalion at Mitla, and
air support for his own column on the drive to
Nakhel from 16.00 hours. The main body of the
Brigade moved off at 13.30, with Mystères and
)uragans maintaining standing patrols overhead.
No further attacks were made on it by the EAF
during this phase. The Brigade reached Nakhel at

16.00 and the assault began almost immediately
spearheaded by AMX tanks and half-tracks unde
cover of an artillery barrage. After 30 minutes c
stiff fighting the main defensive position had bee
overrun, and by nightfall the whole area was i
Israeli hands. The Israelis found a number of Sovie
half-tracks intact, and at once pressed them int
service alongside their own vehicles. Leaving on
battalion at Nakhel, the Brigade pressed on to Mitl
under cover of darkness. The drive was not oppose
and the link-up with the men of the 1st Battalio
was made at 22.30 hours. Supply-drops were mad
by Dakotas and Noratlases, and the Airborne Brigad
dug itself in and prepared to withstand the Egyptia
counter-attack that was certain to come either late
that night or at dawn.

In fact, Egyptian plans to rush up reinforcement
from the Canal Zone had been seriously hampere
during the afternoon of the 30th by the Israeli Ai
Force, which had really begun its attacks after abou
14.00. Twenty of the IAF's F-51D Mustangs ha
been in action almost continuously, attacking th
Egyptian columns on the east bank of the Sue
Canal; some Mustangs had also provided close sur
port for the Central Task Force's assault on Kuseima
where some of the heaviest fighting of the initia
phase had taken place. The piston-engined Mustang
took severe punishment from ground fire, and two o
them were shot down.

It was the Mustangs which had flown the IAF'
first mission of the campaign. At 15.00 hours or
D-Day, a couple of hours before the paratroop bat
talion went down over Mitla, two Mustang pilot
had taken off on an important detail: to cut th
Egyptian telephone wires running from Kuseima t

Nakhel and from El Thamad to Mitla. The aircraft were fitted with special hooks, trailing at the end of a cable which ran from a small winch under their wings. It was designed to catch the wires and tear them apart: unfortunately, it didn't work. The Mustang pilots were undeterred; swooping down a few feet over the desert, they sliced through the wires with the leading-edges of their wings and flew back to base with no worse damage than a few dents, their job well done.

Throughout the first day, the Egyptian Air Force made only 40 sorties. The last attack of the day was carried out on the Mitla positions by two EAF Meteors, escorted by six MiGs; the latter tangled with the standing patrol of IAF Mystères, and while the skirmish was going on above the Meteors slipped in unnoticed and made their attack. Another dog-fight developed at about the same time over the east bank of the Canal, not far from Kabrit, between eight Mystères and twelve MiG-15s. The Israeli pilots claimed the destruction of two MiGs with two more 'probables'. One Mystère was badly damaged, but the pilot flew back to base and made a safe landing.

By midnight on October 30th, the Egyptian defenders of the Mitla area had received reinforcements in the shape of two more battalions, which took up position at either end of the Pass. Worse still, the bulk of the Egyptian 2nd Armoured Brigade had now been ferried over to the east bank of the Canal and was on its way to the scene. There seemed to be little doubt that the Egyptians were recovering well from the initial surprise of the Israeli blitz-krieg.

At 05.45 the following morning, D-Day plus Two, the

first daylight patrol of IAF Mystères had just arrived over Mitla when four Egyptian Vampire FB.52s whistled up from the west. It was No. 2 Squadron from Fayid again, and this time the ground-attack pilots were operating without the benefit of their usual MiG escort. The two Israeli Mystère pilots curved down to the attack; either the Vampire pilots hadn't spotted them or else they were particularly determined. They carried on with their strafing run, and what followed was a massacre. In less than 60 seconds, the burning wreckage of three of the Vampires, torn apart by a hail of 30 mm cannon shells, was scattered over the rocky defiles. The fourth aircraft made one firing pass and escaped in the direction of the Canal.

The Mystères climbed up again and resumed their patrol; they had not even bothered to jettison their long-range fuel tanks before diving to the attack.

Meanwhile, the Central Task Force, which had captured Kuseima the day before, was finding itself involved in still more heavy fighting. The 7th Armoured Brigade, which had been fighting continuously for two days without rest, encircled Abu Ageila in north-eastern Sinai soon after dawn on the 31st, but came under heavy fire almost immediately from Egyptian artillery positions at Um Shihan. An Egyptian armoured unit made two attempts to dislodge the Israelis from their new positions, but each time it was beaten off with heavy losses. It was here, for the first time in history, that guided weapons were used against tanks. The Israelis had acquired a small stock of Nord SS.10 wire-guided anti-tank missiles shortly before the campaign started, and now they used them to good effect. The SS.10 proved to be a devastating weapon, even against the heavily-armoured Soviet T-34 tanks.

The third Egyptian attempt to overrun the Israeli positions was smashed by air attacks. It was a different story from the previous day, when there had been an almost complete lack of contact between the units on the ground and the Air Force. Vital signalling equipment had been put out of action during the Kuseima battle, with the result that the Central Task Force had been unable to request air support when it needed it most. Worse still, Israeli units had been attacked three times by their own aircraft, which had knocked out one half-track and damaged several other vehicles. Fortunately, there had been no serious casualties. On this morning of D-Day plus Two, however, there was complete liaison between air and ground units and the Egyptians were subjected to a well-directed onslaught by Ouragans, Mustangs and Harvards. Two Mustangs and one Harvard were damaged, but they all returned safely to base. The fact that the Harvards came through unscathed for the most part was little short of miraculous; they were slow training aircraft which had been specially adapted for the ground-attack role, and they were extremely vulnerable to ground fire.

During the morning, the pilot of a reconnaissance Harvard returned with some disturbing news: he reported that he had located large Egyptian reinforcements, including armour, driving eastwards through El Hama, only 25 miles to the west of Abu Ageila. The Brigade Commander immediately ordered two companies of infantry in half-tracks and a squadron of Sherman tanks to set up a road-block 15 miles to the west, and to pin down the advancing

Egyptians for as long as possible. He also asked the Air Force to attack the enemy column.

At 11.50 hours, two Ouragans which were on their way to support the paratroops at Mitla were suddenly diverted to search for the new target. The aircraft arrived over El Hama at 8,000 feet, but the pilots could see no sign of the enemy. They flew southwards for a couple of minutes, then turned east, searching for something—anything—that would present a suitable target for their load of rockets and napalm bombs. The leader, Captain Ladia, had just begun a shallow descent to take a closer look at the terrain when a warning shout from his number two sounded in his earphones.

At once, Ladia shoved open the throttle and broke hard to port, peering over his shoulder as he did so. The sky seemed to be full of MiGs; his No. 2, turning desperately, was being harried by three of them and there were two on his own tail. Shouting to his wingman to keep on turning, Ladia jabbed a button in the cockpit and his auxiliary fuel tanks and underwing stores dropped away to smash uselessly into the desert below. Tracers flickered over his wing and gravity pressed him into his seat as he flung the Ouragan into a tight turn. His No. 2 shot past, blazing away at a MiG and being pursued in turn by two more MiGs. Ladia reversed his turn and broke to starboard, missing another MiG by 30 feet. Turning again, he found himself on the enemy aircraft's tail; the MiG twisted and turned desperately but Ladia clung to him grimly. The Egyptian fighter danced and jinked between the luminous diamonds of the reflector sight. Any moment now. Ladia's thumb caressed the firing-button.

At that moment a stream of glowing shells

twinkled over his cockpit, unpleasantly close. Glancing in his mirror, Ladia nearly had a heart attack: it was filled by the gaping air intake of a MiG-15, its nose lit up with wicked flashes from its twin 23 mm Nudelmann-Suranov cannon.

Almost without thinking, Ladia half-rolled the Ouragan on to its back and hauled on the stick, sending the fighter-bomber plummeting towards the desert a mere 3,000 feet below. He kept the stick in the pit of his stomach and the Ouragan shuddered alarmingly as it came out of its dive. Streaking low over the desert he glanced back, hoping that his manoeuvre had shaken off the MiG; but it was still with him and closing fast.

Suddenly, the MiG broke away sharply and disappeared. Thankfully, Ladia looked around and saw the reason why: two Mystères had come whistling up in the nick of time and the MiGs were heading for home, probably short of fuel. Ladia's own fuel state was perilous; a quick glance at the gauge showed that he had just enough left to get him over Israeli-held territory. Five minutes later, having radioed for assistance, he pulled off a perfect forced landing in the desert, the Ouragan coming to a stop undamaged in a cloud of dust. The pilot sat in the shade of the wing and waited for help; it arrived within half an hour in the shape of a little Piper Cub, which landed and took him on board. When he arrived back at Lydda, his base, he found that his number two had returned safely, although his aircraft was riddled with shell-fragments. Ladia's own aircraft was salvaged quickly by an Air Force maintenance team. It was flying again within 24 hours.

Later, it was established that there had been a

total of eight MiGs. They had been flying top cover for four Meteors which had been attacking Israeli units at Bir Hasana, southeast of El Hama. It was sheer bad luck that the two Ouragans had run into them. The reason why the Ouragan pilots had failed to locate the Egyptian column was simply that the Israeli reconnaissance pilot had made a mistake. The enemy reinforcements were not at El Hama but at Bir Gifgafa, a good 50 miles from Abu Ageila.

The mistake was quickly discovered and a second pair of Ouragans was detailed to attack the Egyptian concentration at its new-found location. Once again, the attack was frustrated by the appearance of enemy aircraft, but this time the odds were even.

The two Ouragans, led by Captain Yankel, were heading towards their objective at 13,000 feet over Jebel Libni when they spotted two aircraft a few thousand feet higher up, flying on a reciprocal heading. The aircraft, swept-wing jets which could have been either Mystères or MiGs, disappeared in the sun a moment later and the Ouragan pilots continued on their mission. Then, suddenly, an aircraft came arcing down behind the Israeli fighter-bombers. There was no doubt about its identity this time: it was a MiG-15. Yankel quickly ordered his wingman to drop all external stores and break hard to port. The MiG started to turn as well, trying to get on the No. 2's tail and firing as he came. Yankel turned even tighter and found himself in a position to get off a deflection shot; he fired a short burst, which missed. Then the No. 2 reversed his turn, which was immediately followed by the MiG, presenting Yankel with the chance he had been waiting for. There were more MiGs above, but there was no time to worry about that now The MiG hung poised in his sights and he loosed off a two-second burst with his four 22 mm cannon. There was a puff of white smoke at the enemy fighter's wing root and pieces flew off him, whirling past Yankel's Ouragan. The MiG came out of its turn and wavered, as though the pilot didn't know where the attack was coming from. The Israeli pilot fired again and saw his shell punch holes in the MiG's port wing. A dense cloud of white smoke, possibly vapour from a ruptured fuel tank, streamed back. Another burst, more hits, and the MiG began to trail black smoke from its jet pipe. A second later it rolled over on its back and went down vertically, hitting the desert in a mushroom of burning fuel and debris. The two Ouragans formed up and headed for home, their original mission unfulfilled.

At 14.00, the Egyptian column at Bir Gifgafa was attacked by a flight of three Mustangs, led by Captain Paz. They dropped their napalm and returned to make several strafing runs; their .5 machine guns were ineffective against the armour of the enemy tanks, but they caused considerable damage to a group of soft-skinned vehicles. The attacks were made through heavy fire and after the third pass Captain Paz felt his aircraft shudder as bullets tore into it. Immediately, black smoke enveloped the cockpit and the oil pressure dropped rapidly. Paz climbed to 1,000 feet and headed away from the enemy concentration. He had to get down quickly; his aircraft might burst into flames at any moment. He decided to make a belly-landing; if he baled out, any Egyptian troops in the vicinity might open fire on him as he hung under his parachute.

Side-slipping to clear away some of the smoke, he

picked out a patch of ground that was relatively free from large boulders and slid down towards it. Seconds later, his aircraft hit the ground with a bone-jarring crash, slewing along in a screech of tearing metal. It came to a stop the right way up and Paz climbed thankfully from the cockpit, shaken but unhurt. Belly-landing a Mustang, with its big radiator under the fuselage, was a hazardous business: there was always a danger that the aircraft might flip over on to its back and catch fire, roasting the pilot alive in the cockpit.

He crawled along a nearby Wadi and hid among some bushes, where he lay low until nightfall. There was no sign of any Egyptians, but he decided not to take any risks. As soon as darkness fell, he set off to walk in the direction of Bir Hasana, where he hoped to join up with Israeli forces. He reached it just before first light, only to find the place still occupied by the Egyptians. There was no alternative but to continue walking, north-eastwards this time, towards Jebel Libni. He eventually reached his goal after marching for 30 hours through the desert, collapsing exhausted into the arms of an Israeli patrol. During his marathon trek, his rations had consisted of less than a pint of water and a few sweets, together with a few desert leaves which he had chewed for their juices.

The pilots of the Israeli Mystère squadron were also engaged in several skirmishes on October 31st, and one pilot achieved the distinction of shooting down two MiGs in the course of a single patrol. The pilot, Captain Yac, was leading a pair of Mystères on a routine patrol at 20,000 feet over El Arish when his wingman spotted a trio of bandits at ten o'clock, 2,000 feet below. With the second

Mystère covering his tail, Yac dived down to the attack. As the distance narrowed, he noticed that there was something unfamiliar about the MiGs. Their wing-shape was more graceful than that of the MiG-15, with rounded wingtips, and their fuselages seemed to be longer. A moment later he identified them as MiG-17s; they were newer and more potent than the MiG-15, and it was known that the Egyptian Air Force had recently received a small number.

The MiG pilots saw the Mystères curving down on them and black trails streamed from the jet-pipes of their aircraft as they pushed the throttles wide open, heading for safety on the other side of the Canal. Two of them made it, but the third was hopelessly cornered by the two Mystères. Whichever way the unfortunate pilot turned, he ran into a stream of glowing 30 mm cannon shells. Finally, Yac put a solid burst into the Egyptian's fuselage, just in front of the green-and-white roundel. There was no fire or debris, but the MiG immediately flicked into a fast spin. Yac followed him down; the MiG came out of the spin and levelled at 10,000 feet. Yac closed in fast to give him the coup de grace, but it was not necessary. The MiG's glittering cockpit flew off and a black bundle shot from the cockpit as the pilot used his ejection seat. Yac watched for the parachute opening, but nothing happened. Pilot and seat went on tumbling towards the desert two miles below, dwindling to a black speck that finally vanished against the tawny background. The abandoned MiG-17 went into a shallow dive and exploded on the ground a few seconds later.

The whole action had lasted less than a minute and a half. In spite of the fact that one of his twin 30 mm cannon had jammed, Yac decided to continue

with the patrol. Climbing up to 20,000 feet once
more, the two Mystères described a lazy circle over
the coast to the east of El Arish. A couple of minutes later, the Israeli pilots noticed a tight formation
of seven jets heading northwards in the direction of
Gaza. Leaving his wingman to provide top cover at
20,000 feet, Yac dived down to investigate. This
time, they were MiG-15s; as soon as they saw him
coming, four of the Egyptians disappeared into a
thin layer of cloud. Two of the remaining three
broke away and curved round to get on Yac's tail
as he closed in on their leader, and for the next
half-minute the four aircraft followed each other in
a crazy tail-chase across the sky. Yac got off a quick
deflection shot at the MiG ahead of him, but his
shells missed by several feet. With the other two
MiGs boring in rapidly for the kill and only one
cannon working, the Israeli pilot decided to break
off the combat and headed eastwards, shaking off the
MiGs with a long, fast dive over the desert.

Spotting two more Mystères a few thousand feet
above, Yac climbed up to join them. Suddenly, a
warning shout echoed over the R/T; it took him a
couple of seconds before he realized that it came
from one of the Mystères up above and that it was
directed at him. At that same moment, he saw the
aggressive shapes of two MiGs hurtling towards
him, straight out of the glare of the sun. He broke
into them and they shot past, one on either side,
taken unawares by his manoeuvre. Yac turned
steeply and found himself on the tail of one of the
MiGs. The Egyptian pilot was no novice; he handled
his aircraft expertly and the Israeli needed all his
skill to keep on the enemy's tail. Yac fired twice
with his one remaining cannon, first at 600 yards

and then at 400; he missed both times. The MiG turned in the direction of El Arish, losing height gradually. It was a fatal mistake. Yac fired again from 200 yards and a large chunk of metal whipped back from the MiG's green-and-white striped wing-tip. The Israeli pilot closed in for the kill and jabbed the firing-button as the MiG filled his sights less than 100 yards away. Nothing happened: he was out of ammunition.

The MiG's speed was down to about 250 knots now and he was obviously in trouble. Yac made two more passes, pulling up steeply in front of the MiG and causing the enemy aircraft to take evasive action; he hoped that the MiG might stall and spin into the ground, but it didn't happen. A few moments later, his fuel tanks almost empty, Yac was forced to abandon the chase. Swearing, he headed back to base, leaving the MiG staggering on its way.

Two days later, an Israeli Army pilot flying a Piper Cub in the area spotted something glittering in the shallow waters of Lake Sirdon, to the west of El Arish. Going down to investigate, he saw that it was an aircraft, apparently intact, lying in the shallows. Later, the pilot returned with a Navy salvage expert on board and landed near the village of Bardarwil to make a closer reconnaissance. They returned with the news that the aircraft was a brand-new, almost undamaged MiG-15bis, lying under half a fathom of water. Captain Yac's victim had not got away, after all; the Egyptian pilot had apparently made a good belly-landing in the lake and then got out safely.

A combined salvage operation involving the Israeli Air Force and Navy was immediately

launched. It was to last four days. An IAF Dakota
dropped inflatable life rafts, and the salvage crew
commandeered some flat-bottomed fishing boats.
Working from a motorized landing craft supplied
by the Navy, the salvage crew succeeded in raising the MiG and getting it in position on top of the
fishing boats. A Nile barge then towed the whole
assembly 14 laborious miles along the channel that
led to the open sea to the west of El Arish, where
the Israeli motor ship M. S. *Rimon* was waiting. The
vessel was finally reached after a twelve-hour battle
against strong headwinds, and the *Rimon* towed the
MiG to the port of Haifa.

The Egyptian fighter was at once taken to the
nearby IAF base, where it was subjected to a thorough investigation by Air Force Intelligence officers.
The MiG was then restored to flying condition by
IAF maintenance crews and, with its Egyptian
roundels painted out and replaced by the blue star
of Israel, was later tested to exhaustion by IAF
pilots, who matched it in mock combat against their
own Mystères and Ouragans. By the time they had
finished, they knew as much about the MiG's technical aspects and probably a good deal more about
its combat potential, than did its rightful owners.

The salvage of the MiG had been made possible
by close co-operation between the Israeli Navy and
Air Force: it was the same kind of co-operation that
had resulted in an important Israeli naval victory on
October 31st. At 03.30 that morning, the Egyptian
warship *Ibrahim el Awal*, an ex-British 'Hunt' class
destroyer, HMS *Mendip*, appeared off Haifa and
lobbed 220 four-inch shells into the port from a
range of six miles. Eight minutes later she was fired
on by the French destroyer *Crescent*, cruising in the

vicinity, but contact was lost in the darkness.

At 03.56 the two Israeli destroyers *Jaffa* and *Eilat*, then at sea some 30 miles west of Haifa, set off to intercept the enemy vessel, which was withdrawing towards Port Said. They sighted her at 05.27 and opened fire at a range of 9,000 yards. The Egyptian captain, seeing that his escape route to Port Said was effectively blocked, at once turned northwards and headed at full speed for the Lebanese port of Beirut. The *Ibrahim el Awal* was pursued by the two Israeli warships, which continued to fire on her and caused some light damage.

At dawn, Naval HQ requested air support. At 05.46 a Dakota arrived on the scene and circled the enemy ship, which was now 37 miles from the coast of Israel. The crew of the Dakota made a positive identification and vectored two Ouragans on to the Egyptian destroyer. The Ouragans dived through a heavy flak barrage and launched 32 armour-piercing rockets at the warship, badly damaging her forward part, knocking out her steering gear and electrics and jamming her ammunition lifts. They then returned to make several firing passes, sweeping the destroyer's decks with a hail of 20 mm cannon shells.

At 07.10, after radio consultation with his base at Alexandria, the Egyptian captain ran up the white flag and ordered his men to abandon ship. The crew was taken off by the *Jaffa* and *Eilat*, which had come up alongside. An attempt by the Egyptians to scuttle the vessel was frustrated by the seacocks, which were so badly rusted that they could not be turned, and the destroyer was taken in tow by the *Eilat*. She was later repaired and pressed into service in the Israeli Navy under the new name of *Haifa*.

The brunt of the day's air operations, however,

was undoubtedly borne by the Israeli ground-attack pilots. They made sortie after sortie against the Egyptian column at Bir Gifgafa on the central road into Sinai, and inflicted considerable damage. The Egyptian 1st Armoured Brigade in particular came in for a good deal of their attention; it was known as the 'Soviet Brigade' because it was equipped exclusively with Russian tanks and self-propelled guns. It was also well defended by mobile 40 mm anti-aircraft cannon, and every attack had to be pressed home through a barrage of shellfire. The Mustangs, because of their lower speed and the vital radiator positioned under the fuselage, were particularly vulnerable; most of them returned to base with varying degrees of battle damage. The Israeli aircraft, however, made their runs at very low altitude and the Egyptian anti-aircraft gunners appeared to have difficulty in depressing their weapons sufficiently. On at least two occasions, while pumping shells horizontally across the desert after an attacking aircraft, they pulverized some of their own vehicles that got in the line of fire. During strikes flown later in the day, the Mustang pilots went after the guns and put several out of action, tearing the crews apart with well-aimed bursts from their six .5 Brownings. The Israeli pilots took incredible risks to make sure of destroying their objectives, and a strafing run nearly ended in disaster for one Mustang pilot when he dived on an ammunition truck, blazing away and breaking off the attack only at point-blank range. The ammunition truck blew up with a roar and he found himself flying into a cauldron of smoke and debris. The blast picked up his aircraft and tossed it upwards like a leaf; it missed colliding with a second Mustang by a matter of inches.

Of the six Mustangs that took part in that particular attack, five were damaged by ground fire or
flying debris, but all of them were patched up within
an hour of their return to base. By the end of the
day, the forward elements of the 1st Armoured
Brigade had reached Bir Rod Salim, 20 miles west
of Jebel Libni, but the main body was still in the
Gifgafa area. The road between was littered with
the burnt-out wrecks of 90 tanks, trucks and personnel carriers: grim testimony to the efficiency of
the Israeli pilots.

A second Egyptian column, consisting of two small
mixed brigades of armour and infantry, was also
attacked from the air as it advanced along the coastal
road that led eastwards. Preliminary attacks on this
column were made by F-84F Thunderstreaks of the
French Air Force's 1e Escadre, operating out of
Lydda. Together with the Mystères of the 2e
Escadre, based on Haifa, these aircraft had been
held in reserve as an insurance against possible
attacks on Israeli cities by Egyptian Ilyushin 28 jet
bombers. However, the Anglo-French ultimatum to
Egypt and Israel of October 30th (stating that British
and French forces would occupy key positions on
the Suez Canal unless both sides ceased hostilities
and withdrew ten miles either side of the Canal,
and to which the Israelis had readily agreed since
they were still a long way east of their main
objective) had been rejected outright by the Egyptians in the early hours of the 31st. The two French
fighter squadrons based in Israel were now released
for offensive operations. The other French Air
Force unit in Israel, the 64e Escadre, had been
operating its Noratlas transports in support of the
Israeli advance since the first night of the campaign.

All the French aircraft in Israel had had their
blue, white and red roundels obliterated; some carried hastily-stencilled Israeli stars, but many bore
no insignia at all. The majority of the F-84Fs and a
few of the Mystères also carried black and yellow
'invasion stripes' similar to those painted on Allied
aircraft during the subsequent attacks on the Canal
Zone; this led to some confusion and angry Egyptian
claims that there had been collusion between the
British, French and Israelis over plans for the invasion. The Egyptians had no idea at the time, however, that French combat aircraft were operating
from bases in Israel.

In the Mitla Pass, meanwhile, the men of the
Airborne Brigade had been involved in a day of
heavy fighting as they tried to dislodge the Egyptians
from their positions. Throughout the day, in spite
of the air patrols, the Brigade had been subjected
to sporadic air attacks by Vampires and Meteors.
One of the most intense came at 16.00, when part
of the attacking Israeli force was cut off in one of
the defiles and strafed by four Meteors, which caused
heavy casualties. One Meteor was destroyed by the
Mystères, the second to fall victim to their guns that
day. The air attacks ceased at dusk, by which time
the paratroops had seized the high ground on both
sides of the Pass. Three hours of desperate hand-to-
hand fighting followed as the Israelis winkled out
the Egyptians from their positions in the caves below.
Once they realized there was no escape, the enemy
fought bravely to the end. By 19.00, the exhausted
paratroops had almost completed their mopping-up
operation, at a cost of 50 killed and three times that
many wounded.

An hour later, three IAF Dakotas appeared overhead and dropped supplies. On learning that there were many casualties below, the transport pilots ignoring strict orders not to land under any circumstances, dropped flares and came sliding down towards the primitive airstrip. All the Dakotas made a safe landing, and about a hundred wounded were hurriedly embarked. It was as many as the aircraft could hold; they were already dangerously overladen, and take-off was hazardous in the extreme. But the three aircraft lurched into the night sky without mishap, their wheels bumping over the rough ground at the far end of the airstrip. As they droned away, the paratroops consolidated their new positions and prepared to withstand a possible counter-attack by Egyptian armour approaching from Bir Gifgafa.

So D-Day plus Two came to an end. In the northeast, the key Abu Ageila position had been captured by the Israelis, who were now getting ready to launch an assault on Um Gatef; the 7th Armoured Brigade was at Bir Hasana and ready, if necessary, to go to the help of the paratroops at Mitla; and Rafah, the key to the Gaza Strip, was also under heavy Israeli attack. Air power had been used extensively, but had not played a decisive part in the day's operations; the IAF had destroyed three Vampires, three MiGs and two Meteors, but they had not prevented the EAF from pressing home its attacks. On the other hand, some Israeli ground support missions had been frustrated by the appearance of Egyptian fighters overhead. The ground-attack missions which had hit their objectives had undoubtedly hampered the Egyptian columns, but had not succeeded in distrupting them. Nevertheless, when darkness fell on October 31st, there was no disputing that the

Israeli Air Force enjoyed a considerable measure of air superiority.

It was then that events took a completely new turn. Shortly before midnight, Canberras of Nos. 10 and 12 Squadrons and Valiants of No. 148 Squadron, Royal Air Force, operating out of Malta, bombed the Egyptian airfields of Almaza, Bilbeis, Cairo West and Inchass. The bombers, which had not been sent on their mission until after dark because of the risk of daylight interception by Egyptian MiGs, bombed their targets from over 40,000 feet, and the damage they inflicted was negligible. However, the raids had one immediate result: later that night, 30 MiG-15s and MiG-17s (some Egyptian, some on their way to join the Syrian Air Force, and all flown by Soviet and Czech instructors) flew out of the danger area to bases in Syria and Saudi Arabia. Twenty Ilyushin 28s were also flown from Cairo West to Luxor in the south, where it was believed they would be safe from air attack. These aircraft all belonged to Nos. 8 and 9 Squadrons, Egyptian Air Force, and were not yet fully operational: the Egyptian crews were converting to the new type.

On the morning of November 1st, President Nasser, now faced with the threat of invasion by British and French forces, ordered all Egyptian units not actually engaging the Israelis in the Sinai Peninsula and the Gaza Strip to pull out and withdraw westwards across the Suez Canal. The withdrawal, the order for which reached the majority of the Egyptian commanders during the course of the afternoon, was to take place on the night of November 1st/2nd.

In the early hours of November 1st, the Israelis launched their assault on Rafah at the southern end

of the Gaza Strip. The action began with a bombardment of the Egyptian positions by the destroyers
Yafo and *Eilat;* it began at 02.00, lasted for half an
hour and was totally ineffective. The Israeli Air
Force was then called in with disastrous results. At
02.35 a Mosquito dropped a cluster of target markers
right on the Israeli positions. Before anything could
be done, half a dozen more Mosquitoes droned over
and dropped their bombs on the heads of the
Israelis. Fortunately, no one was killed and the
Mosquito crews quickly realized their mistake, but
precious time had been lost. The Mosquitoes made
a second attack, and this time dropped their bombs
in the right place. However, the bombing had not
softened up the Egyptian defences as the Israelis had
hoped, and they met with heavy fire as they began
to move forward. It was not until first light that real
progress was made, when Mustangs and Ouragans
appeared and pounded the enemy positions with
rockets and napalm. One after the other, after
vicious hand-to-hand fighting, the Egyptian defensive
positions were overrun; at 08.00 the Egyptian commander ordered his units to withdraw to El Arish,
and an hour later Rafah was in Israeli hands.

At 10.00, a battalion of AMX tanks trundled westward along the coastal road, heading for El Arish.
Two hours later they ran into a road-block, and
dispersed under heavy anti-tank fire. Fifteen minutes later, a flight of Ouragans appeared and blasted
the obstacle into smouldering wreckage, but more
Egyptians came up and still kept the Israeli tanks
pinned down. It was not until 15.00 hours, after
another air strike and an assault by Israeli infantry,
that the defences were overwhelmed and the tanks
were able to push on. In the meantime, Ouragans

and Mustangs strafed the Egyptian airfield at El
Arish, where a number of Vampires had been reported that morning; the Vampires had gone, but
the aircraft attacked three MiGs and one Mraz
Sokol, a Czech-built three-seat communications aircraft, on the ground. The MiGs were later found to
be wood and canvas dummies, but the Sokol was
real enough and the cannon-shells tore it apart.

Sporadic air attacks on the Israeli forces in the
Mitla area continued during the morning of November 1st, and one Egyptian Vampire was shot
down. By the middle of the afternoon, however,
EAF activity over Sinai had almost ceased. By this
time, the main Egyptian airfields were being subjected to heavy strafing attacks by carrier aircraft
and RAF fighter-bombers operating out of Cyprus.
A few MiG-15s and three 'long-nose' Meteor
NF.13 fighters were seen over Sinai during the early
part of the afternoon, but they stayed at high
altitude and did not offer combat. From now on, the
Israeli Air Force was free to turn its attention exclusively to ground-attack operations. Throughout
the remainder of the day, Mustangs, Mystères,
Ouragans and the old Harvards hammered Egyptian
positions that remained near Abu Ageila and at Um
Gatef. The Egyptians counter-attacked with armour
after dark in a desperate attempt to smash through
the road-blocks set up by the Israelis to the west of
Abu Ageila, but the attempt was shattered by
accurate Israeli fire.

At dawn on November 1st, the men of the 1st
Battalion of the Airborne Brigade stood-to in their
positions in the mist-shrouded Mitla Pass and waited
to repel the armoured thrust that was expected from
Bir Gifgafa. It never came. Moreover, the Egyptians

had gone from their remaining positions in the area of the Pass: they had received orders to withdraw during the night, and were making their way westwards in scattered groups.

During the late afternoon, IAF reconnaissance aircraft reported that everywhere columns of Egyptian troops and vehicles were streaming back across the desert towards the Canal. The majority of the Egyptian commanders, realizing that their position was hopeless without hope of reinforcements or air cover, had decided not to wait until nightfall to begin the withdrawal of their units. The retreating columns were harassed by the IAF until darkness fell.

By the morning of November 2nd, it was obvious that the Egyptian forces in Sinai were in full retreat. Only two major formations showed no sign of moving: the 1st Armoured Brigade, which was still at Bir Gifgafa, and a Mobile Brigade straddling the coastal road between El Arish and Kantara. The function of both these units was to keep the way open for the Egyptian withdrawal. The 1st Armoured Brigade began to move back at dawn, covering the tail-end of the withdrawal. The advancing Israelis arrived in Bir Gifgafa about noon to find the place empty; they went on with their pursuit and finally caught up with the Egyptian rearguard near Katib el Sabha, 30 miles from the Canal, at about 16.00. There was a short, sharp battle and three T-34s were left in flames by the roadside.

It was the final encounter between the Israelis and the 1st Armoured Brigade. By nightfall, the Egyptian formation had completed its crossing of the Canal; it had been under almost constant attack by the IAF since dawn and had taken a severe mauling. Altogether, the Mustangs and Ouragans had destroyed 22 T-34 tanks, 5 SU-100 self-propelled guns and at least 35 armoured personnel carriers.

The IAF was also in action from first light onwards over El Arish, towards which the Israeli 27th Armoured Brigade was advancing along the coastal road. At 05.30, Ouragans made a series of attacks on scattered Egyptian units on the western edge of the town. The units were the rearguard of the 4th Egyptian Brigade, which was in the act of withdrawing from the locality; the Israelis let them go, not wishing to have large numbers of prisoners on their hands, and did not enter the town until the afternoon. The IAF continued to strafe the Egyptian columns, now streaming westwards in a disorganized mass along the coastal road. Many of the Egyptians abandoned their vehicles as soon as the aircraft appeared overhead and continued on foot, moving through the desert in small groups. When the forward units of the 27th Armoured Brigade continued their drive along the coastal road towards El Kantara, the Israelis counted 400 Egyptian vehicles of all types and 40 tanks by the roadside. A few had been knocked out by cannon and rocket fire and were still burning, but most had been abandoned intact.

Meanwhile, at 06.00 on November 2nd, an Eastern Task Force consisting of the Israeli 9th Infantry Brigade had left Ras El Nakeb, north-west of Eilat, and begun a long advance southwards along the Gulf of Akaba towards Sharm el Sheikh. This objective, together with nearby Ras Nasrini, commanded the Straits of Tiran; both positions were strongly fortified and possessed heavy anti-aircraft defences.

The first IAF strike on Sharm el Sheikh took place at 13.00, when four Mystères attacked Egyptian gun positions with rockets. The aircraft, led by Major Benjamin Peled, ran into a heavy anti-aircraft barrage and two of them, including the leader's, were hit. Major Peled's Mystère caught fire and he ejected, landing heavily about a mile and a half from the enemy perimeter and injuring his knee. Although he was in acute pain, he managed to limp two miles across open country without being spotted and took refuge in the foothills that bordered the mountains to the west. Some time later, a Piper Cub appeared overhead. Peled waved frantically, but the pilot failed to see him and the aircraft droned away. The same Piper returned at about 17.00, and this time its crew located him. The Cub landed on a clear stretch of ground a few hundred yards away from an Egyptian guard post, whose occupants surveyed the proceedings with bored indifference. It took off again safely with Peled on board.

Peled's aircraft was the only Israeli Mystère to be lost during the entire campaign: the second damaged aircraft returned safely to base.

At 17.00, the Air Force Air Transport Command was in action once again when a flight of Dakotas dropped two companies of reserve paratroops at El Tur, on the Gulf of Suez in southern Sinai. There was a strong wind and the paratroops suffered a number of casualties, but by dark they had secured their objective, a small airfield, with hardly a struggle. A small group of IAF airfield construction and maintenance engineers had also been dropped, and now they worked to get the airstrip ready to receive the transport aircraft that were on their way with reinforcements. During the hours that followed, the Dakotas Noratlases and one requisitioned El Al Constellation made 23 flights to El Tur, bringing a battalion of the 12th Infantry Brigade and its supporting equipment. The decision to drop both paratroop companies at El Tur had been made at the last moment, in the late afternoon of November 2nd. It followed a completely erroneous report that the Egyptians had evacuated Sharm el Sheikh. One paratroop company had been on its way to drop on that objective when its Dakotas had been overtaken by a Meteor, whose pilot had signalled that Sharm el Sheikh was still occupied and strongly defended. At the last moment, the Dakota pilots had altered course and made the drop over El Tur.

November 2nd, D-Day plus Four, ended with the force at El Tur preparing to advance on Sharm el Sheikh and the 9th Infantry Brigade advancing laboriously over high ground along the Gulf of Akaba towards the same objective. One battalion of paratroops had also left the Mitla area and was advancing towards Ras Sudar on the Gulf of Suez. Its ultimate objective was to join the task force at El Tur, completing the encirclement of the Sinai Peninsula. There had been no air combats during the day and IAF operations had been confined to strafing attacks on the withdrawing Egyptians, supply missions and liaison.

At dawn the following morning, the Israeli 11th Infantry Brigade had almost completed the occupation of the Gaza Strip. Only the town of Khan Yunis held out, but when the Israelis advanced on it with their armour at 06.30 they met with only scattered small arms fire. By 08.30 the town was in their hands except for one Egyptian strongpoint, whose

defenders obstinately refused to surrender. Three
rocket-firing Ouragans were called in and reduced
it to smoking rubble.

In the south, the paratroop battalion arrived at
Ras Sudar, which it captured after a sharp fight.
The remainder of the Airborne Brigade at Mitla
assembled at the western end of the Pass and prepared to advance on Port Tewfik, but this plan was
later cancelled and the Brigade stood in readiness
to complete the occupation of southern Sinai.

The Israeli Air Force now concentrated its
main effort against the Egyptian defences at Sharm
el Sheikh and Ras Nasarini. From first light onwards,
both these positions were subjected to a series of
heavy air strikes; anti-aircraft fire was still severe
and one Ouragan was shot down. The pilot ejected
safely, but was taken prisoner.

The IAF also attacked and sank, in the Straits of
Tiran, a number of small Egyptian naval craft
reportedly bringing reinforcements to Sharm el
Sheikh. The pilots had been briefed to look out for
the frigate *Domiat*, but when they arrived over the
area they found that they were too late: the Egyptian
warship had already been sunk by the guns of the
cruiser HMS *Newfoundland*, which was patrolling
the approaches to the Gulf of Akaba with a British
naval squadron. After picking up survivors, the
British vessels entered the Straits of Tiran, where
they were spotted by the pilots of a flight of four
Mystères which had just attacked some enemy
landing-craft off Ras Nasrini. The Israelis were
unable to identify the warships, and they did not
have sufficient fuel to carry out a closer investigation; they had to be content with reporting the
position of the ships to a second flight of Mystères,

which was then on its way to the target area.

The four pilots of this flight spotted one of the warships almost immediately. It was turning away from the coast at high speed, trailing a broad crescent of wake. The pilots mistakenly identified it as the *Domiat*; in fact, the warship was the frigate HMS *Crane*, but to pilots without special training in ship recognition one warship can look very much like another. In this case the mistake was doubly excusable, for the *Domiat* had been the former HMS *Nith* and the *Crane* was built along the same classic British lines.

At any rate, the Mystères swept in to the attack. The leader let fly with a full salvo of 38 rockets; so did the other three aircraft, although the warship escaped most of them by taking violent evasive action. The rockets that did hit her were not armour-piercing and failed to cause serious damage. The last pair of Mystères attacked through a barrage of anti-aircraft fire, but fortunately the Navy gunners were not on form and the jets escaped undamaged.

The main concern of General Moshe Dayan, the Israeli Chief of Staff, was to complete the occupation of Sinai before pressure from the western powers brought a halt to the Israeli operations. In the afternoon of November 3rd, he ordered his available forces to proceed with the capture of Ras Nasrini and Sharm el Sheikh with all possible speed. At dawn the following morning, the paratroop battalion moved out of Ras Sudar and reached El Tur at noon. It was this unit that was to take part in the final pincer movement against the Egyptian stronghold. The two companies of reserve paratroops and the infantry battalion already at El Tur were evacuated in the course of the morning by Air Transport Command. A lot of their equipment, including their jeeps, had been badly damaged during the airdrop and they were consequently ill-prepared to go into action.

The 9th Infantry Brigade, meanwhile, was rapidly approaching Ras Nasrini, where the Israelis expected to meet stiff resistance. Ras Nasrini, with its battery of big naval guns commanding the Straits of Tiran and the approaches to the Gulf of Akaba, was an important strategic position. At 12.00, five Mustangs showered the Egyptian defences with napalm in what was meant to be a softening-up attack. On their return to base, the pilots reported that there had been a complete lack of anti-aircraft fire. It was hardly surprising: the 400-strong Egyptian garrison had already fallen back on Sharm el Sheikh, having spiked their naval guns. The Israeli column pressed on and arrived opposite the perimeter of Sharm el Sheikh at 14.00. The leading elements were pinned down by heavy fire from an outpost, but this gave no further trouble after two Mustangs hit it with 32 rockets an hour later.

The IAF made several more attacks on Sharm el Sheikh during the remainder of the afternoon while the Israeli ground forces regrouped and refuelled in readiness for the final assault. The air strikes had done their work well, in spite of intense and highly accurate anti-aircraft fire. By nightfall, Sharm el Sheikh was shrouded in a pall of smoke and its defenders were battered and dazed by the continual onslaught. One Ouragan was shot down; this time, the pilot parachuted into friendly territory. The strikes caused many casualties, most of whom were evacuated by sea in small craft after nightfall.

The first attack opened at 03.30 on November 5th, and was quickly pinned down by heavy Egyptian fire. The Israelis withdrew after sustaining some losses and decided to wait until daybreak before launching another assault. At 07.00 the infantry moved forward again, with Mustangs, Ouragans and Mystères strafing ahead of them in an almost continuous 'cab-rank'. Under cover of the air attacks, the first wave of Israeli infantry reached the wire on the western perimeter of Sharm el Sheikh, but they were pinned down once again by murderous machine-gun fire. For the next hour, the Israelis were involved in some of the most vicious hand-to-hand fighting of the whole campaign as they blasted their way through the wire and stormed into the Egyptian defences under cover of a smoke-screen. At 08.30 the paratroop battalion from El Tur arrived and pitched into the fray, and after that the enemy defences were quickly overrun. The shattered garrison finally surrendered at 09.00. About a third of the Egyptian force had become casualties: there were 200 dead and 300 wounded. A great many of the casualties had been sustained during the air strikes.

At 11.30, the Israeli Air Force mounted its last strike mission of the campaign when Mustangs and Ouragans attacked the small island of Sanafir, adjacent to Tiran about two miles offshore, and pounded it with rockets and napalm for 20 minutes. Soon afterwards, Israeli troops landed by LCM and made a thorough search of the island; it was deserted. The small Egyptian garrison, not more than a dozen men, had already been taken off before first light.

One more ground-attack mission was flown from

Israeli air bases, but it was carried out by the Thunderstreak pilots of the French 1e Escadre. At 15.00 hours on November 5th, some eight hours after the British 16th Parachute Brigade and the French 2nd Parachute Regiment had been dropped from an armada of transport aircraft on to their respective objectives in the Canal Zone, eight F-84Fs swept low over the Red Sea. Their target was the Egyptian air base at Luxor, where an RAF Canberra reconnaissance aircraft had previously located the 20 Ilyushin 28 jet bombers which had been flown out of Cairo West a few days earlier. The bombers, which constituted a very real threat to the Anglo-French invasion, had to be neutralized: the 1e Escadre at Lydda was the unit best situated to do the job.

The Thunderstreaks, equipped with long-range fuel tanks, entered Egyptian territory near Quseir and approached their target in a long curve from the south. The Egyptians had obviously considered Luxor to be beyond the range of Allied strike aircraft: the French pilots found the Il-28s drawn up in two neat rows. After a concentrated rocket and cannon attack lasting five minutes the F-84Fs raced away to the north-east. Behind them, an enormous mushroom of oily smoke billowed into the afternoon sky over Luxor from the piles of twisted wreckage lying in pools of blazing fuel. The Egyptian anti-aircraft gunners had not opened fire.

CHAPTER SEVEN

Modernization

The Israeli Air Force had gone into action on October 29th with the knowledge that it might have to face formidable odds. Eight days later, it had emerged from the Sinai Campaign with a clear-cut victory over its opponent. There was no escaping the fact that the Israelis had lost more aircraft than the Egyptians—eighteen, compared with eleven—but all the Egyptian aircraft had been destroyed in air combat, whereas most of the Israeli machines had fallen victim to anti-aircraft fire. Only one Israeli aircraft had been shot down by enemy fighters: a Piper Cub, which had been caught by a MiG-15 patrol and destroyed after ten minutes of violent evasive action.

The Israeli losses comprised nine Mustangs, two Harvards, three Ouragans, one Mystère and three Cubs, two of the latter having been destroyed on the ground by strafing attacks. On the other side, the Egyptians had lost four MiG-15s, one MiG-17, four Vampires and two Meteors. The Israeli pilots did not bother to claim the odd enemy liaison aircraft destroyed on the ground, such as the Mraz Sokol shot up at El Arish.

During the campaign, the Mystère IVA had shown itself to be a superlative fighting machine, and had proved capable of besting the MiG-15 on every count except rate of climb. The Israeli pilots had nothing but praise for the French aircraft and the way it handled and there is little doubt that their kill rate would have been higher if they had been more experienced on the type. As it was, the great majority of the IAF pilots went into battle with only 20 hours' flying time, and in some cases even less, on the Mystère. Compared with the MiG-15, it was a sweet aircraft: the MiG had a number of serious limitations, not the least of which was a tendency to yaw at high speed which made it a poor gun platform. It was not a novice's aircraft: because of its general instability it was prone to flick into a spin out of a maximum-rate turn and Israeli pilots saw this happen on a number of occasions. Despite this weakness, the Egyptian pilots persisted in trying to out-turn the Mystère during a dog-fight instead of making use of their superior climb-rate to get themselves out of trouble. Adding to the Egyptian pilots' difficulties was the fact that the MiG's flying controls were not power-operated, which resulted in enormous 'stick forces' during highspeed manoeuvres and made flying the aircraft extremely tiring work. Finally, the Mystère's twin 30 mm DEFA cannon had proved distinctly superior to the MiG's mixed armament of two 23 mm and one 37 mm cannon. The rate of fire of the latter was low, only about 400 rounds per minute, compared with the 1,200 of the Mystère's 30 mm weapons. On at least one occasion, a Mystère pilot had been able to avoid the heavy shells simply by cramming on power and outpacing them!

The Ouragan, too, had shown itself capable of outflying the MiG in the hands of a skilled pilot; the first MiG to be destroyed during the campaign had fallen to an Ouragan's guns. The three Ouragans that had been lost had all been shot down by anti-aircraft fire, and if it had not been for the aircraft's extremely robust construction losses would undoubtedly have been much higher. Some Ouragans had flown safely back to base after sustaining severe battle damage, one of them with a 37 mm shell lodged in one of its main fuel tanks after passing right through the main spar of the wing. The spar was shattered, but the wing had held firm. Apart from its ability to absorb tremendous punishment the Ouragan also upheld its reputation as an excellent gun platform, in this respect, it was superior to the Mystère. The Ouragan pilots evolved several methods of knocking out even heavily-armoured vehicles such as the T-34 tank with their 20 mm cannon; the aircraft's inherent stability enabled them to aim at vulnerable points like reserve fuel tanks, tracks and turret rings with great accuracy.

Of all the Israeli aircraft used during the campaign, the heaviest losses had been sustained by the piston-engined Mustang. Apart from the vulnerability of certain aspects of its design, a contributory factor in the losses suffered, which amounted to 30% of the available Mustang force, was simply that more Mustangs had been committed to the fighting than any other type of aircraft. Although the Mustang's armament had been generally ineffective against tanks, the concentrated cone of fire from its six .5 heavy-calibre Browning machine-guns had wrought havoc among soft-skinned transport and troops. Nevertheless, it was clear that the Mustang was approaching the end of its useful life in the combat role.

The same was true of the Meteor F.8, which had been the IAF's first-line interceptor up to the arrival of the Mystère IVA. The Meteors had never tangled in earnest with the MiG-15 at any time during the campaign. Shots had been exchanged during a couple of brief skirmishes on October 30th, but the outcome of these encounters had been in-decisive. The Meteor pilots had gone into action confident that they could have generally got the upper hand during any serious clash by making use of the Meteor's ability to out-turn the MiG below 15,000 feet and forcing the Egyptian pilots to fight on the 'climb-and-dive', but they never had an opportunity to put their theories into practice. The Meteors experienced some technical problems dur-ing the eight days of the campaign: one of the most frequent was difficulty in releasing the aircraft's underwing fuel tanks. Admittedly, this was some-times due to 'finger trouble' on the part of the pilot: the drop-tank release lever had to be pulled right back, and if the pilot failed to do this only one tank would come off, which sometimes resulted in the wing to which the remaining tank was attached dropping sharply and pulling the aircraft into a vicious roll. This happened to at least one pilot during the campaign, and he found himself upside down without warning at a height of less than 2,000 feet.

Because of the Meteor's obsolescence as a first-line aircraft, it was decided to turn the remaining F.8s over to the IAF's operational training units, where they were to be used for jet familiarization and gunnery instruction alongside the two-seater T.7s. Some were also used as target-tugs, and served in this role until well into the 1960s. Two other Meteor variants, however, continued to serve with first-line units for some time after the end of the

Sinai Campaign: these were the seven FR.9s bought from Flight Refuelling a couple of years earlier, and the handful of NF.13 night fighters. In addition to the three aircraft on the IAF's inventory before the start of the Sinai Campaign, three more NF.13s were delivered before March 1958 to help meet the urgent need for a fast all-weather interceptor.

From the overall assessment of the part played by the Israeli Air Force in the Sinai Campaign, three major decisions emerged. First, it was decided to phase out with all possible speed the piston-engined aircraft still in service with the IAF's first-line squadrons and replace them with the most modern jet types available. Second, it was realized that the Sinai conflict did not spell the end of the Arab threat to Israel: on the contrary, there was every likelihood that Nasser would pursue his aggressive intentions towards Israel with renewed vigour as soon as his shattered forces had been brought up to strength, despite the influence of the United Nations. It was decided, therefore, to expand the Israeli Air Force progressively until it had reached at least double its present strength. Thirdly, because of the difficulty in obtaining aircraft from other sources, French aircraft would form almost the entire basis of this expansion programme.

Because of the impressive part played by the Ouragan during the Sinai Campaign, an immediate order was placed with the French Government for an additional 45 aircraft of this type, together with three replacement aircraft to make good the losses suffered during the conflict. The additional Ouragans arrived early in 1957, four months after the order was placed; they were all aircraft that were surplus to French Air Force requirements, and they

Above: An early Israeli Spitfire 9, serial 2010. This aircraft is possibly one of the small number found in a crashed/scrap condition on RAF dumps and restored to flying condition.
Photo: IDF/AF.

Below: The sole remaining Israeli C.210—the fighter that formed the IAF's first line of air defence during the early days of the War of Independence—mounted on a plinth at an air base in the Negev.
Photo: Air-Britain Archives, via Charles W. Cain.

Above: One of the large number of Mosquitoes assembled by the Israelis from crashed/scrap aircraft. *Photo: Stephen Peltz.*

Below: A rocket-armed P-51D Mustang taxies out to take off on a mission during the Sinai Campaign of 1956. *Photo: IDF/AF.*

One of the B-17G Flying Fortresses acquired by the Israelis in 1948.
Photo: Stephen Peltz.

Above: Spitfire LF.9es running-up on a Negev airstrip during the War of Independence. *Photo: IDF/AF*

Below: Spitfire LF.9e owned and flown by Ezer Weizman. Aircraft is all black with a red spinner and IDF/AF crest on nose. Serial number is 2057, and the aircraft also has the allocated—but not used—civil registration, 17-1351. *Photo: IDF/AF.*

A pair of Mirage IIICJ fighters on patrol over the Mediterranean near the coast of Israel. *Photo: Marvin E. Newman/Camera Press.*

Above: One of the Stratocruisers converted by the Israelis for use as military transports. *Photo: Flight International.*

Below: A batch of Fouga Magisters assembled by Israel Aircraft Industries at Lod airport. *Photo: Flight International.*

Above: Gloster Meteor F.8—the IAF's first jet combat aircraft.

Photo: Stephen Peltz.

Below: Ouragan taxiing in after a training sortie.   *Photo: Stephen Peltz*

Above: Vautour IIA ground attack aircraft over the old city of Jerusalem.
Photo: Marvin E. Newman/Camera Press.

Below: Mystère 4A.   *Photo: Stephen Peltz.*

Above: Israeli airborne troops disembarking from a super Frelon assault helicopter.   *Photo: NewsPhot.*

Below: One of the S-58 helicopters acquired by the IAF from Luftwaffe surplus stocks.   *Photo: Stephen Peltz.*

Above: A MIG-17 caught by Israeli cannon fire while on quick-reaction alert at the Sinai airfield of Jebel Libni on the morning of June 5th, 1967. Note 'winged wolf' insignia on nose. *Photo: Flight International.*

Below: Israeli Air Force Commander General Mordecai Hod, who, together with Ezer Weizman, was the architect of Israel's air offensive during the Six Day War. *Photo: Marvin E. Newman/Camera Press.*

Above: A Dornier Do 27 communications aircraft at Jebel Libni—renamed IAF Station Livnat after its capture by the Israelis.

Photo: Flight International.

Below: An IAF C-47 transport at Sharm el Sheikh airstrip on the southern tip of Sinai, near the disputed Straits of Tiran over which the war began.

Photo: Flight International.

Above: A Skyhawk streaks over Sinai towards the Suez Canal for a strike on Egyptian military targets.   *Photo: Marvin E. Newman/Camera Press.*

Below: The first photograph to be released of a McDonnell Phantom in Israeli Air Force markings.   *Photo: Keystone.*

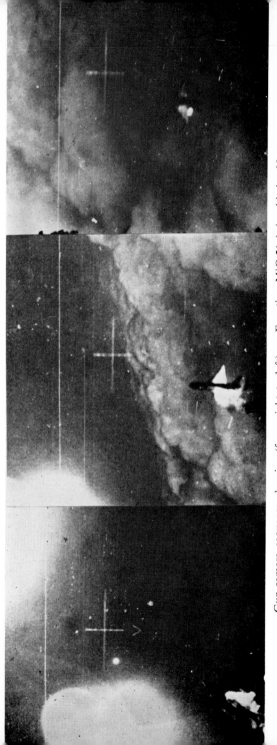

Gun-camera sequence showing (from right to left) an Egyptian MiG-21 being hit by 30 mm shells from an IAF Mirage and going down in flames over the Suez Canal during a dogfight on July 10th, 1970.

Photo: Associated Press.

Napalm canisters shower down from a flight of Mystère IVAs during a demonstration on Israeli Air Force day, July 19th, 1969. *Photo: Keystone.*

Above: Ground crewman loading the 20 mm cannon on a Skyhawk fighter bomber. *Photo: Marvin E. Newman/Camera Press.*

Below: An I.A.F. ground crewman makes a last-minute check of a Skyhawk's 2.75-inch rocket pods before a mission.

Photo: Marvin E. Newman/Camera Press.

now formed the equipment of four IAF ground-attack squadrons. During the same period more Mystère IVAs were acquired, bringing the total first-line interceptor strength up to 50 aircraft.

Finding a replacement for the versatile de Havilland Mosquito presented a problem. Up to this time, the Mosquito had been the only aircraft in the Israeli Air Force that was capable of carrying out both the light bomber and night fighter roles. The Meteor NF.13s purchased from Britain helped to bridge the gap caused by the Mosquito's retirement to a certain extent, but they were pure interceptors. The Israelis still had no jet bombers; in fact, once the Mosquitoes and the ancient B-17s had been phased out they had no bombers of any kind, and there was an urgent need for a true multi-purpose aircraft, a kind of 'jet-age Mosquito'.

There was only one aircraft that could possibly meet this requirement: the Sud-Aviation Vautour. This remarkable aircraft, the prototype of which had flown in 1952, possessed exactly the kind of capability sought by the Israeli Air Force. Powered by two SNECMA Atar 101 engines, it had a rate of climb that compared well with that of any single-engined fighter then in service; a radius of action in excess of 700 miles; and it was able to operate from short, semi-prepared runways. It was transsonic in a shallow dive, and its maximum speed of 580 knots at sea level would enable it to show a clean pair of heels to aircraft such as the MiG-17. In addition to these attributes, the Vautour was the most heavily-armed aircraft of its class in service anywhere in the world: besides its standard armament of two 30 mm cannon it could carry two Matra Type 104A rocket packs containing a total of 232 68 mm air-to-air rockets in the fuselage, or four packs each containing 19 rockets under the wings. The interceptor version was designated Vautour IIN; there were two other variants, the IIA and the IIB, which were designed for the close-support and level-bombing roles respectively. Both of them could carry up to 3,000 pounds of bombs internally, or 4,000 pounds under the wings. The Vautour IIA was a single-seater; the other variants carried a crew of two.

As well as its formidable performance and weapons-load, the Vautour had other features that appealed to the Israelis. Like the smaller Ouragan, it was simple and robust; from the outset it had been designed with ease of maintenance in mind, its designers having taken into account the possibility that the aircraft might have to operate from forward airstrips where there were limited servicing facilities.

In the summer of 1957 the Israeli Government placed an order for four Vautour IINs and 20 single seat IIAs. Towards the end of the year, a number of Israeli aircrews were attached for conversion and operational training to the Centre d'Instruction de Bombardement, where French Air Force crews were training on the Vautour IIA close-support version. Soon afterwards, more IAF personnel arrived at Tours, where the first two FAF squadrons to receive the Vautour IIN all-weather fighter, Groupes de Chasse 1/30 and 3/30, were in the process of working-up.

With the arrival of the Vautours in Israel in the spring of 1958, the IAF, for the first time in its history had the means of striking hard and effectively at any target in Egypt, or for that matter in any other hostile country in the Middle East.

In the summer of 1957, with negotiations under
way for the purchase of new aircraft such as the
Vautour to form an all-jet combat force, the Israeli
Air Staff realized that yet another revision of the
IAF's training system would be necessary if the
higher standards associated with the programme of
modernization were to be met. At that time, a
student pilot did his initial flying on the IAF's old
Boeing PT-17 Kaydet primary trainer before going
on to the Harvard for advanced training and finally
to the Meteor for jet conversion. The process was
too involved: too much time was being wasted in
piston-engined training before a pilot reached the
jet stage. What was needed was an aircraft that
would take a student pilot, after he had completed
the necessary number of hours' preliminary flying
on a piston-engined basic trainer, right through to
the operational conversion unit stage, cutting out
the Harvard and Meteor altogether. This policy of
all-jet training after the basic flying phase was now
coming into force with most major air arms, and
one, the RAF, was eliminating the piston-engine
phase altogether, training its students on jet aircraft
right from the start. The IAF Air Staff were not
in favour of the latter course; they believed that it
meant cutting too many corners, and that a thorough basic grounding on a light piston-engined aircraft was still necessary before going on to jets. As
it turned out they were right. The RAF ultimately
abandoned its 'all-through' jet training policy and
made it mandatory for students to do a certain
amount of basic flying on the Chipmunk before
going on to flying training proper on the Jet Provost.

Once again, it was the French aircraft industry

that provided a ready-made answer to the IAF's problem in the form of the Air-Fouga CM 170R Magister, a graceful lightweight jet trainer that was claimed by its makers to be the only aircraft in the world capable of carrying a pupil right through the primary, basic and advanced stages up to conversion to an operational jet type. Three hundred and fifty Magisters were already on order for the French Air Force, and several other countries were showing an interest in the type. Israeli Air Force pilots who flew the Magister at France's Mont de Marsan test centre submitted enthusiastic reports about it; apart from its qualities as a fully-aerobatic jet trainer, the aircraft had an obvious application in the close-support role because of its ability to carry air-to-ground rockets or bombs as well as two 7.5 mm machine-guns.

Following a searching assessment of the Magister's potential in both the training and ground-attack roles, a contract was signed between the Israeli Government and Air-Fouga in 1958 for the licence manufacture of the type in Israel. This was to be undertaken by Bedek Aircraft Ltd., which until this time had been almost entirely occupied in overhauling civil and military aircraft and aero-engines. The only complete airframes built by Bedek Aircraft up to now had been gliders; in 1957, the firm had undertaken the manufacture of British-designed Slingsby sailplanes for delivery to the Israeli Air Force and the Israel Flying Club. The agreement to manufacture the Magister represented a big stride forward, and immediate plans were made to expand Bedek's plant at Lod (Lydda) Airport.

The first Magister assembled in Israel was completed in June 1960, a few days after Bedek Aircraft

Ltd. had changed its name to Israel Aircraft Industries. It was followed by eleven others all assembled
from parts imported from France. The thirteenth
aircraft, which was rolled out early in 1961, was
made up of airframe parts produced entirely in
Israel, as were all subsequent machines.

In all, 60 of the 82 Magisters eventually acquired
by the IAF were built in Israel; the only imported
parts used in their construction were engines and
armament. The Israeli-built aircraft included some
refinements such as a strengthened wing to carry
a mixed load of light bombs and rockets.

Meanwhile, in 1958, the Israeli General Staff had
become seriously worried by the continued massive
build-up of Soviet equipment in the Arab countries.
Particularly disturbing was a report that the Egyptian Air Force was about to take delivery of the
first of a substantial number of supersonic MiG-19
interceptors, which were more potent than anything
the IAF then possessed. As a means of countering
this new threat, the Israeli Government placed an
order with Marcel Dassault for 24 supersonic Super
Mystère B2s. The Super Mystère was a progressive
development of the Mystère IV, and was the first
aircraft produced in quantity in Western Europe
that was capable of exceeding the speed of sound
in level flight. Equipped with two 30 mm cannon
and 55 68 mm FFAR air-to-air rockets, it could also
carry an external bomb load of up to 2,000 pounds,
and therefore had the all-important quality, as far
as the Israeli Air Force was concerned, of doubling
as an interceptor and a ground-attack aircraft. The
first IAF squadron become operational with twelve
Super Mystères in 1959, followed by a second a
little under a year later. These two units assumed

the role of first-line air defence, taking over from the Mystère IVAs.

The Super Mystère, however, was in the nature of an interim aircraft, bridging the gap between the subsonic Mystère IV and a truly supersonic aircraft capable of flying at twice the speed of sound. Such an aircraft, the Mirage III, had in fact flown in prototype form in November 1956, having been developed by Dassault as a private venture. Two months later, in the course of normal flight-testing, the Mirage had flown at Mach 1.5 at over 36,000 feet, and this speed had later been increased to Mach 1.9 with the aid of an auxiliary rocket motor. A firm order for 100 Mirage IIIC interceptors had been placed on behalf of the French Air Force in September 1958, and there was a strong possibility of more large orders, including exports, to come.

It was hardly surprising: the Mirage III was a potent weapons system and enjoyed a phenomenal performance. Possessing a speed of Mach 2.2 at 40,000 feet, it could climb to 50,000 feet in six minutes flat on the power of its Atar 9C turbojet, and with the help of its SEPR rocket motor, which carried enough fuel for an 80 second burn, the time was cut down to five minutes. Its service ceiling with the aid of the rocket motor was 75,000 feet, and the aircraft could 'zoom climb' to 100,000 feet. Its combat radius at ground level, flying subsonically and in an attack configuration, was 560 miles, and this could be extended to 750 miles by flying a high-level mission at 36,000 feet. For the interceptor role, the Mirage IIIC carried two 30 mm cannon, two Sidewinder and one Matra R.530 air-to-air missile; ground attack armament consisted of the cannon and two 1,000-pound bombs, or alternatively, two 1,000-

pounders under the wings and an AS. 30 air-to-surface missile under the fuselage. Like its predecessors, the Super Mystère IV, the Mirage III was capable of operating from short airstrips.

An evaluation of the Mirage IIIC by Israeli Air Force pilots in 1959 resulted in a firm order for 26 machines of this type, with an option on a further 14. Later, with the acquisition of Mach-Two-plus fighters such as the MiG-21 by Egypt, follow-on orders boosted the total of Mirages for the IAF to 72 aircraft. The first Israeli Mirage squadron formed early in 1963, and on May 15th that year 27 of the supersonic fighter-bombers showed their paces during a flypast over Haifa as part of the celebrations held annually to commemorate the inauguration of the State of Israel.

The Israeli aircraft were designated Mirage IIICJ, and were identical to the Mirage IIIC except for some internal electronic equipment. The total of Mirages that eventually entered service with the IAF included three Mirage IIIBJ two-seat trainers.

During the late 1950s, the Israelis also made a considerable effort to build up a small but effective helicopter force. The Sinai Campaign had shown a considerable need for machines of this kind for communications, battlefield reconnaissance and rescue duties; although the little Piper Cubs had done a magnificent job in all these connections, they had suffered from a number of limitations: the greatest being that they could not always land where they were needed most because of the nature of the terrain. Also, because of Israel's small land area, most of the IAF's training sorties were made over the sea, and although the Navy provided rescue facilities they were no longer considered adequate. To bridge the gap, two Sikorsky S-55 helicopters were purchased and formed the nucleus of an air-sea rescue flight.

The mainstay of the IAF's Helicopter Wing, however, consisted of seven Sikorsky S.58s which were obtained from the USA. In 1960, a further 24 machines of this type, ex-Luftwaffe aircraft, were bought from the Federal German Government as part of a deal that included the sale of Israeli-manufactured Uzi machine-pistols to the German Army. Although intended primarily as troop-carriers and assault helicopters, the S-58s were subsequently used for a wide variety of tasks, including the air-lifting of freight to kibbutzim along Israel's troubled frontiers.

For light observation and liaison, the Israeli Air Force obtained eight Sud-Aviation Alouette II helicopters. In 1963, also from Sud-Aviation, the IAF ordered five big three-engined heavy-duty Super Frelons; in addition to freight, these massive helicopters could carry 30 fully-equipped troops, and there was provision for anti-submarine detection and attack equipment to be fitted. Delivery of the Super Frelons began in 1965, and the helicopters provided a useful addition to the IAF's Air transport capability.

The mainstay of the IAF's Air Transport Command in the early 1960s was still the elderly C-47 Dakota, together with twelve Nord Noratlases. In addition, there was a useful weight-lifting force in the shape of five Boeing Stratocruisers, which had been converted for military use by the Israel Aircraft Industries. Apart from internal modifications, these aircraft differed from civil Stratocruisers because the whole of their tail unit could be pivoted to one side, giving easy access to the fuselage for vehicles and other large cargoes.

This, then, was the basic equipment with which the Israeli Air Force was preparing to meet the challenge of the 1960s. And at the start of the decade, it was already apparent that a threat was developing with alarming speed into potentially the greatest that Israel had ever had to face. In the half-decade since Sinai and the Suez invasion, Egypt had come a very long way in terms of equipment and aircrew proficiency. And not only that: Egypt was no longer alone. Ranged against Israel were her allies, armed with a formidable array of Russian weapons. The question that now confronted the Israeli High Command was simple: when would the Arab nations feel strong enough to attack?

CHAPTER EIGHT

The Other Side of the Fence—2

By the afternoon of November 6th, 1956, when Anglo-French strike aircraft flew their last sorties in support of the Suez landings, the Egyptian Air Force had ceased to exist. The airfields in the Canal Zone and around Cairo were littered with the burnt-out wrecks of aircraft; the shattered remnants of the powerful force with which, only a matter of days earlier, Nasser had hoped to dominate the skies of the Middle East.

In January 1957, following the withdrawal of British and French troops from the Canal Zone, the Egyptian High Command immediately laid plans to build up the country's armed forces again with all possible speed. To provide a nucleus for the reconstituted Egyptian Air Force, the 40 or so MiG-15s, MiG-17s, Il-28s and Il-14s that had escaped to Syria and Saudi Arabia were flown back to their original bases, and negotiations were started with the Soviet bloc for the supply of more aircraft to make good the crippling Egyptian losses.

The Soviet response was not long in coming. During the first week of March 1957, three Rumanian freighters unloaded fifteen MiG-17s and ten Il-28s, all of them in crates, at Alexandria. The crated aircraft were immediately transported to the Egyptian Air Force's maintenance base at Almaza, where they were assembled under the direction of Soviet technicians. During the weeks that followed, eastern

European cargo vessels continued to arrive at Alexandria in a steady stream, disgorging their loads of war material; by the end of April no fewer than 50 MiG-17s had been supplied to the EAF, and at the end of June the figure had risen to 100. Forty Ilyushin 28 bombers were also delivered. The result was that on July 23rd, 1957, the Egyptian Chief of Air Staff, Air Vice Marshal Mohamed Sidki when speaking on the anniversary of the military coup that had brought Nasser to power five years earlier, was able to claim that the first-line strength of the Egyptian Air Force had doubled since the Suez conflict.

To provide overwhelming evidence of the truth of his statement, Sidki had planned an enormous fly-past over the heads of the Cairo crowds by more than a hundred of the newly-acquired Soviet aircraft. Unfortunately, mainly as a result of the usual incompetence on the part of the Egyptian ground crews, the plan went wrong: when the big day arrived, only 52 serviceable aircraft, 21 MiG-15s, 18 MiG-17s and 13 Il-28s, could be scraped together. These aircraft duly flew over Cairo in vics of three, maintaining a large and ragged formation which indicated that the Egyptian pilots were no more proficient than they had been a year earlier. Nevertheless, the flypast was greeted enthusiastically, even hysterically, by the Egyptian crowd: at least it showed that the Egyptian Air Force was back in business.

On February 1st, 1958, the United Arab Republic came into being with the union of Egypt and Syria. The union was extended, five weeks later, to include the Yemen. Although each member state retained its own system of government, plans were drawn up

for a unified defence command. The union represented an important step forward in Nasser's ambi-
tion to achieve military domination of the Middle
East. Apart from Syria's geographical position on
Israel's northern frontier and astride the oil pipelines from the Persian Gulf, the country's
inclusion in the United Arab Republic meant that
Nasser no longer had to rely on sea traffic to bring
in arms; aircraft and other war material could be
ferried overland by the 'back door route' that ran
into Syrian territory direct from the Soviet Union,
crossing the northern tip of Iran and the mountain-
ous fringes of Turkey's south-east border.

During 1957, Syria had become the spearhead of
Soviet infiltration into the Middle East. For months
prior to the formation of the United Arab Republic,
large numbers of Soviet military personnel had been
flowing into Syria in a steady stream; they included
instructors, radar and anti-aircraft experts, and airfield construction engineers. Using locally-recruited
labour, the latter were well on the way to completing a complex of modern military airfields in
northern Syria, bases that were designed not only as
depots for the masses of Soviet aircraft arriving in
the Middle East, but intended also as a direct threat
to NATO's southern flank.

Before the arrival of the Soviet 'advisors', the
Syrian Air Force had suffered both from a lamentably low standard of training and inadequate equipment. It had been originally formed in April 1946,
following the withdrawal of Allied forces from
Syrian territory, and had occupied bases constructed
by the French during their mandate. Its first equipment had been mainly of Italian origin, in the shape
of Fiat G.55B and G.49 trainers. A few Chipmunks

and Harvards were also used for training and 'police' duties, but it was not until 1950 that the Syrian Air Force appeared to be forging a combat nucleus with the purchase of 30 de Havilland Vampire FB.5s from Italy. In fact, these aircraft were not destined for the Syrian Air Force at all : as soon as they arrived they were handed over to Egypt as part of a scheme to get round the arms embargo.

By the end of 1954, however, the Syrian Air Force had some 63 combat aircraft on its own inventory; these included 40 refurbished Spitfire 22s and 23 Meteors of various marks. The following year, like Egypt, Syria signed an arms agreement with the Soviet Bloc; this provided for the supply of 25 MiG-15s and a small number of MiG-15 UTI two-seat trainers. These aircraft were eventually delivered in crates at Alexandria along with a similar consignment for the Egyptian Air Force, and assembled by Czech technicians at Almaza. Early in 1956, a group of Syrian Air Force personnel, all with some experience on the Meteor, were sent to the EAF's advanced flying school at Abu Sueir for conversion on to the MiGs, but they proved to be even less competent than their Egyptian colleagues and they had still not completed the course by the time of the Suez invasion in October. During the Allied air strikes, the Syrians suffered the indignity of seeing their brand-new MiG-15s reduced to twisted wreckage; the only four which managed to escape were MiG-15 UTI trainers which were flown to the Syrian airfield of Hama by Czech pilots.

The Syrian Government lost no time in entering into another deal with the Russians. Towards the end of November 1956 they ordered 60 MiG-17s from the Soviet Union, and in December 20 Syrian pilots went to that country on a conversion course. Eighteen more went to Poland for the same purpose; they included Lieutenant-Colonel Musafaz Asasa, who subsequently became commander of the Syrian Air Force's fighter wing. The first twelve MiG-17s arrived in Syria in the middle of January 1957, and delivery of the other 48 had been completed by the following August. MiG-17 conversion was now carried out in Syria under the supervision of ten Soviet Air Force instructors, supported by 60 other Soviet officers, including a General, who were responsible for technical instruction.

By February 1958, when the United Arab Republic came into being, the Syrian Air Force had a total of five Russian-equipped fighter squadrons. These were based on Hama, El Rasafa, Sahl-es-Sahra and Damascus-Mezze. By this time, the British-built Gloster Meteors had been phased out of the front-line squadrons and relegated to the operational training role with the exception of a handful of NF.13s, which equipped the Syrian Air Force's only night-fighter squadron. These remained in first-line service until they were eventually replaced by all-weather versions of the MiG-17. The motley collection of western training aircraft was also progressively replaced by Russian Yak-11s and Yak-18s.

Meanwhile, the latter half of 1957 had seen a substantial drop in the number of Soviet combat aircraft being delivered to the Egyptian Air Force. The reason was not hard to find: training was going ahead at such a slow rate, largely because of the inaptitude of the Egyptian pupils, that there were still not enough qualified pilots even to fly the jet aircraft that had so far arrived in Egypt. The Russian instructors were still having to cope with the old problem of the Egyptian pilot's natural aversion to flying in bad weather or at night, as well as his general inability to put his aircraft down at the point required on the runway. When Egypt became part of the UAR, this painfully slow rate of progress proved a serious obstacle in the way of Nasser's plans for the Egyptian air component of the UAR Air Force, which called for an increase in strength to 30 operational squadrons forming six fighter/ground attack wings, three light bomber wings and one all-weather interceptor wing. Each wing was to consist of three squadrons, each with 12 aircraft. Taking the UAR Air Force as a whole, the target was an eventual strength of some 40 first-line squadrons which represented odds of approximately four to one against Israel.

Although the Russian and Czech instructors' insistence on a thorough training programme delayed Nasser's plans for expansion, it gradually began to produce the desired result. At long last, the Arab commanders were beginning to realize that quality was infinitely preferable to quantity, and from that moment on there was a general improvement in the overall standard. An enormous effort was made to improve this still further over the next two years, and by mid-1959 the Egyptian flying schools were turning out qualified pilots in sufficient numbers for the expansion scheme to be given the go-ahead.

By the end of 1960, the Egyptian component of the UAR Air Force had some 200 MiG-17 interceptors and 70 Il-28 bombers, backed up by a transport force of 50 twin-engined Il-14s and Mi-1 helicopters. The Syrian component had also acquired a dozen Il-28s and 15 Il-14s; its MiG-17 interceptor force remained at around the 60-strong mark. The influence of the Soviet instructors in both air components was now very apparent, routine training having included regular lectures on communist ideology. The whole structure of the UAR Air Force was now modelled on Soviet lines; the Egyptian component had dispensed with the old RAF-inspired system of Groups and Wings, which had been replaced by Air Divisions and Air Regiments.

Although under the terms of the Egypt-Syria unified defence agreement, both Egyptian and Syrian officers were supposed to hold positions of command on an equal basis in the UAR Armed Forces, this principle was not adhered to in practice. Ever since the Unified Command's inception, the Egyptians had made persistent efforts to see that as many responsible positions as possible were filled by Egyptian officers. There was some justification for this attitude: Nasser was well aware that many Syrian officers had been hostile towards Egypt before the creation of the United Arab Republic, and there was no reason to suppose that their outlook had changed since. His policy was, therefore, to keep as tight a hold as possible on the Syrian component, ensuring that all military decisions were made in Cairo and not Damascus. The Syrian officers who were appointed to a command were all carefully selected for their political reliability and loyalty to Nasser. To make the Syrian component even more dependent on Egypt, the flying training school at Aleppo was closed; from then on, all student pilots were trained in Egypt, where a close watch could be kept on their attitude as well as their progress.

As the quality of the available material improved,

Egyptian instructors now began to take on some of the burden of flying training, although they still remained under the supervision of Nasser's foreign 'advisors'. (These now included Poles, East Germans and Indians, as well as Russians and Czechs.) All students passed through the Central Training School at Bilbeis, where they logged a total of 300 hours on Yak-18s and Yak-11s before being commissioned and awarded their 'wings' and going on to jet conversion. There was also a Central Technical School, where officer cadets received a three-year course of instruction in engine or airframe maintenance, electronics and armaments; the instructors in this establishment were still mostly eastern Europeans.

By the middle of 1961, the Egyptian component of the UAR Air Force was well on the way to attaining a relatively high standard of efficiency. The accident rate, which only five years earlier had resulted in at least 50% of the Egyptian Air Force being unserviceable at any one time, had fallen dramatically: at long last the Egyptian pilots and ground crews were showing a sense of responsibility and, under the tuition of their Soviet Bloc instructors, appeared to have overcome their instinctive awe of anything technical.

Then, in September 1961, came the blow. Syria, torn by internal troubles, her relations with Egypt strained to breaking point, seceded from the United Arab Republic and the Egyptian military personnel who had wormed their way into high positions in Syria now found themselves on their way back to Cairo with little more than they stood up in.

For Egypt, the most serious aspect of secession was that there was no longer anyone to share the cost of the military might that had been built up laboriously over the preceding years. Syria had made a financial contribution to the UAR's military build-up out of all proportion to her actual contribution in the way of hardware; and now, with Syria gone, the whole burden fell on Egypt—with crippling effect on the country's economy.

Egypt now found herself in a position of military strength, far in excess of her actual defensive needs, from which there could be no retreat. To have attempted to reduce the size of the Egyptian armed forces at this stage would have had disastrous consequences for Nasser's government and would have resulted in a severe loss of prestige throughout the Arab world. Nasser had committed himself too far; the Egyptian forces were already beginning to receive newer and even more expensive Soviet equipment, and there was no going back on the deal even if it meant bankrupting the nation.

In the summer of 1961, the Egyptian Air Force's first-line fighter defences had been strengthened by the arrival of an initial batch of MiG-19 interceptors, the first of 80 which eventually reached the Egyptian squadrons. The MiG-19, which had entered service with the Soviet Air Force in 1955, was the first Soviet production interceptor capable of exceeding the speed of sound in level flight. Although designed as a day fighter, it had a limited all-weather capability and could be used in the ground-attack role. In Russian service, it had been largely replaced by the MiG-21 Mach Two interceptor, and in 1962 the Russians agreed to supply 40 of the latter aircraft to provide the Egyptians with a means of countering Israel's Mirage IIICJs, which were then on order. The MiG-21, however, was not in the same class as the Mirage: although it possessed an excellent rate of climb (30,000 feet per minute) it was basically a short-range, clear-weather target defence interceptor, and it had no all-weather or ground-attack potential. To meet the urgent need for a modern all-weather interceptor, Nasser had in fact asked the Russians to supply a number of twin-jet Yak-25s; but this type was not being offered for export outside the Soviet Bloc because it carried a large amount of secret electronic equipment, and the request was turned down.

In the spring of 1962, however, the Egyptians successfully completed negotiations for the supply of a Soviet aircraft type which represented the biggest threat to Israel so far. The aircraft was the Tupolev Tu-16 medium bomber, which, in addition to serving in large numbers with the Soviet Air Forces, had already been supplied to Indonesia. With a speed of over 500 knots and a range of 2,600 nautical miles with a full load, the Tu-16 could carry nine tons of bombs. Although its normal operational altitude of between 35,000 and 40,000 feet made it vulnerable to supersonic interceptors, its acquisition by the EAF would mean that Nasser would now have a formidable strategic bombing force at his disposal and a means of delivering nuclear weapons should the Egyptians succeed in building up a small stockpile of these at some future date. By the end of 1963 a total of 30 Tu-16s had been delivered to the Egyptian Air Force, forming a Strategic Bomber Regiment.

Since 1960, Nasser had been making a belated attempt to achieve a measure of independence from the Soviet Union by the formation of an indigenous Egyptian aircraft industry. The basis for this already existed; an Egyptian General Aero Organization had been founded in 1950, and its Factory 72 at Heliopolis had series-produced several versions of the wartime German Bücker Bü 181D Bestmann trainer for the Egyptian Air Force under the name of Gomhuria (Republic). Factory 72, however, had neither the plant nor the skilled personnel to undertake more advanced projects. In 1960, therefore, the Egyptians launched a drive to recruit designers and aero-engineers from western Europe, particularly from Germany, with a view to expanding the industry. Shin Bet, the Israeli Intelligence Service, was not slow to realize what was going on; in March 1960, the Israeli agents discovered that a United Arab Airlines office at No. 20 Königstrasse, Stuttgart, was being used as a recruiting agency for German scientists and that considerable numbers of the latter were already arriving in Egypt on UAA's regular Comet service between Frankfurt and Cairo. The scientists were recruited by a man calling himself Mahmud, who was quickly identified by the Israelis as Colonel Al Din Mahmud Khalil, head of Egyptian Air Force Intelligence.

Colonel Khalil had already been very active in Spain, where a large number of German aero-engineers had been employed since the war. In 1959, a mixed group of Spanish and German engineers had arrived in Egypt to assist in the construction of a second aircraft factory at Helwan, near Cairo, where the Egyptians planned to undertake licence production of the Hispano Ha-200 Saeta jet trainer, a robust little aircraft serving with the Spanish Air Force and designed by Professor Willi Messerschmitt.

The Israelis were not unduly worried by the

involvement of the Messerschmitt-Hispano team in Egyptian affairs, since this was only connected with the manufacture of training aircraft, or by the knowledge that the Egyptians were planning to continue work on a supersonic fighter project, the Ha-300, begun by Hispano several years earlier but later abandoned by the Spanish firm. They knew that the Ha-300 was little more than a status symbol, a chance for Nasser to prove that Egypt's aircraft industry was capable of turning out modern combat aircraft. Even if the Ha-300 proved successful, it would be several years before it went into production, and even then it would be inferior to the Russian-built MiG-21, not to mention Israel's Mirages.

With the arrival of more scientists from Germany in 1960 and 1961, however, the whole situation took a dramatic new turn. From the moment they arrived in Egypt, Shin Bet kept a close watch on their activities, which first of all consisted of survey work in various selected areas. What now worried the Israelis was that the man directing these surveys was Wolfgang Pilz, one of the scientists who had helped to perfect Germany's wartime V-2 rocket on von Braun's staff at Peenemünde. Also in Egypt, apparently acting as Pilz's assistant, was a specialist in radar guidance systems named Paul Goercke; a former Luftwaffe Colonel, he had worked for a time on France's Véronique rocket programme.

The presence of these two men on Nasser's payroll had a sinister significance. Following their survey, shifts of Arab workmen slaved day and night to build an extensive factory in the desert ten miles south of Cairo. Known simply as Factory 333, it was linked with Cairo's international airport by a specially-built road to afford easy transportation of materials flown in from western Europe. And on the other bank of the Nile, not far from Kilometer 62 on the main road to Alexandria, Shin Bet reported that a structure that appeared to be a rocket launching pad was beginning to take shape.

The Shin Bet agents immediately intensified their watch on the area, and in June 1962 they reported that rocket testing had begun. A few weeks later, on July 21st, Nasser himself witnessed the firings of four single-stage missiles from the Kilo 62 range. Soon afterwards, the rockets were paraded through Cairo before an hysterical crowd and the Egyptian President boasted that Egypt now had missiles capable of reaching any target south of Beirut—which meant that the whole of Israel was threatened.

The largest of the rockets shown on parade was named Al Kahir, or Conqueror. It was about 40 feet long, and its design showed a distinct V-2 influence. Its range was stated to be in the order of 370 miles, but there was no evidence of any form of guidance system. The other type of missile, named Al Zafir (Victory) was similar in design but much smaller. Its range was said to be 230 miles. A third rocket type, the Al Ared (Vanguard) made its first appearance at a military parade on July 23rd, 1963. A two-stage missile, it appeared to be a progressive development of the Al Kahir and a range of 600 miles was claimed for it.

None of the rockets appeared to be operational at this stage, but the potential threat to Israel could not be ignored. In the autumn of 1962, Shin Bet issued a high-level directive to its agents operating in Egypt and western Europe. The directive was brutally simple: the key scientists working for

Nasser were to be tracked down and killed. It was the start of a systematic terror campaign that was to continue throughout the whole of 1963: a campaign that began on the night of January 5th of that year with the attempted assassination of Dr Emil Kleinwachter, whose car was machine-gunned just outside Stuttgart as he was on his way to catch a plane for Cairo after spending a few days' leave with his family. Kleinwachter escaped with his life, but an Egyptian colleague named Hassan Khamil was not so lucky; two weeks later, while on an unspecified mission in Germany, a bomb exploded in his private aircraft and it plunged to earth in flames. Several attempts were also made on the personnel working in Factory 333; in March, a booby-trapped crate of spare parts exploded in the factory's main assembly shop, killing five technicians and injuring nine others.

As a direct result of Shin Bet's activities, life became intolerable for the 250 or so German scientists and technicians employed by Nasser. For the sake of security, they were compelled to live in a completely self-contained 'science city', with its own shops and cinema. The whole 'city' was surrounded by barbed wire and patrolled by armed guards day and night. Even so, these extra precautions failed to protect the scientists completely: three more of them died in what appeared to be accidents—but both the Egyptians and Shin Bet knew otherwise. The constant threat of liquidation, together with serious internal disputes that had arisen with the Egyptians, resulted in a group of German technicians leaving Egypt early in 1964. They were followed by more of their colleagues, in increasing numbers. By the middle of 1965, the evacuation was almost complete and the development of Nasser's rockets was brought to a complete standstill. On July 23rd, 1965 six missiles of the Al Ared type were paraded through Cairo, but it was only a gesture. They never appeared again, and Nasser's dream of launching a rain of high explosives against Israel's cities from a safe distance was shattered.

The Israeli threat, however, did not extend to the German technicians engaged in airframe and engine development, and the majority of these remained in Nasser's employment. Work on the Ha-300 prototype continued; airframe construction at Factory 36 was under the direction of an Austrian engineer named Hans Schoubaumsfeld and development of the fighter's Bristol Siddeley Orpheus turbojet was under the supervision of another Austrian, Ferdinand Brandner, at Factory 35.

Four prototypes of the Ha-300 were subsequently completed, but only two ever flew, the first on March 7th, 1964. None of the prototypes was capable of a supersonic performance, but with further development it was envisaged that the aircraft would reach a speed of Mach 2.2. Development work went ahead, although the Egyptian Air Force had already lost interest in the type, until early in 1969 when the whole project was abandoned after nine years of fluctuating fortunes. The eighty-strong German team that had been responsible for development of the Ha-300 was disbanded, and the Ministry of Military Production, whose task had been to establish an advanced aircraft technology in Egypt with the Ha-300 as its foundation, was dissolved and the Helwan factories came under the control of the Ministry of Industry.

Meanwhile, in 1965, part of Egypt's defence policy had begun to be influenced by a new factor: the Aswan High Dam. On this project hinged much of Nasser's remaining prestige among his own people, for it brought with it at least a promise of a measure of prosperity for the country's ailing economy, mortgaged as it was for years to come in order to pay for the massive supplies of Soviet arms. Trapping the Nile flood behind the dam meant that the flow of the river could now be regulated throughout the year, according to the needs of agriculture. The Aswan Dam was the one symbol of progress, the one hope for the future, that Nasser could offer his people. Protecting it against Israeli air attack in the event of hostilities became a top-priority measure, and to this end, in 1964, when the Dam was nearing completion, Nasser requested the supply of a number of Soviet SA-2 Guideline surface-to-air missiles. These were supplied as part of an air defence system that included radar and computer equipment and later models of the MiG-21 fighter. For two years an army of Soviet engineers was engaged in building missile complexes around the approaches to the dam and at other points of strategic importance; by the summer of 1966 a total of 150 SA-2s were operational with 25 six-launcher batteries scattered throughout the country's four main air defence zones. As later events were to show, however, the missiles did not provide the deterrent that Nasser had hoped: they were effective enough against aircraft flying at high and medium levels, but against low-level attackers they were generally useless.

Nevertheless, the new Soviet equipment, for all its shortcomings, had provided Egypt with the most modern air defence system in the Middle East, and by 1965 the threat to Israel's security was mounting with frightening speed. Most important of all, the Egyptian Air Force was becoming increasingly aggressive and its pilots were handling their ultra-modern equipment with a greater degree of skill. In terms of man against man, however, superiority in the air still belonged to the Israelis, which was demonstrated in a series of clashes between the two sides during the early 1960s. By 1965, the Israelis were no longer in any doubt that the Egyptians were sharpening their claws in another bid to wipe the State of Israel off the face of the earth.

CHAPTER NINE

Trial of Strength

May 25th, 1960. 10.00 hours.

Even with the perspex canopy raised, it was swelteringly hot in the narrow cockpit of the Super Mystère B.2 fighter; the strong morning sun sent waves of heat dancing over the tarmac of the operational readiness platform where the aircraft was parked. Trussed in his harness, 25 year-old Captain Aharon flicked through the pages of a paperback and tried to take his mind off the discomfort of the sweat that trickled down his back under his flying suit. It was no consolation to know that his wingman, Lieutenant Yadin, seated in the cockpit of a second Super Mystère a few yards away, was in no better shape.

The discomfort and the boredom, however, would soon be over, but not in the way Aharon and Yadin anticipated. They expected to be relieved by two more quick-reaction alert pilots in about 15 minutes' time; they were unaware that at that moment, two shimmering 'blips' creeping across the fringes of Israeli territory were being tracked by an IAF radar station in the Negev. There might be no cause for concern; Egyptian jets had made brief probing flights into Israeli territory several times during the past months, but these penetrations had never gone beyond four or five miles. This time, however, it was different: the Egyptian aircraft maintained a steady north-easterly heading, spearing deeper into the Negev.

In the Operations Room at Hatzerim, the silence was broken by the shrilling of a telephone: the 'hot line' linking the base with the IAF's Air Defence Command Control Centre. The message was brief: 'Two bogeys in Sector Five, level one five zero, heading zero four zero. Scramble your interceptors.'

Seconds later, the engines of the two Super Mystères whined into life. Aharon and Yadin wound down their cockpit canopies and the roar dwindled to a muted rumble. The ground crews pulled away the chocks and the fighters moved forward as the pilots opened the throttles slightly, checking the controls and instruments as they lined up on the runway. Without stopping, they pushed the throttles fully open and the Super Mystères accelerated rapidly under the thrust of their afterburners. A gentle backwards pressure on the control columns lifted the aircraft cleanly off the ground and they entered their battle climb, like two silver darts against the deep blue of the sky.

The Israeli pilots levelled out at 16,000 feet and headed towards the west, searching the sky for their quarry. Three minutes later, Aharon spotted two glittering dots at two o'clock, slightly lower down. The Super Mystères at once turned towards them and the distance closed rapidly, revealing the bogeys to be a pair of MiG-17s. The Egyptian pilots did not appear to be aware of the danger, continuing serenely on their course. Like a whirlwind, Aharon shot over the top of the rearmost MiG, leaving his wingman to deal with it, and closed in on the leading Egyptian aircraft. The Egyptian pilot spotted him now and broke frantically to starboard at the last moment but it was too late. A burst of shells from the Super Mystère's twin 30 mm cannon slammed into his port wing root; pieces whirled away in the slipstream and a ribbon of flame streamed back. The MiG turned over and went into a fast spin, leaving a spiral of oily smoke.

Aharon pulled his aircraft round in a tight turn, looking for the second MiG. It was heading south-westwards at high speed, closely pursued by Yadin. As Aharon watched, Yadin opened fire, the MiG took violent evasive action and the Israeli's shells went wide of the mark. They were over Egyptian territory now, and following his instructions Aharon reluctantly ordered Yadin to abandon the chase. The two pilots returned to base, elated by Aharon's success; it was the first 'kill' scored by a Super Mystère. In a second serious clash between Egyptian and Israeli aircraft over the Negev nearly a year later, Super Mystères from Hatzerim were again the victors. It happened on April 28th, 1961, when two MiG-17s penetrated Israeli air space and were engaged by the IAF fighters. One of the MiGs escaped, but the other was trapped by the two Super Mystères, whose pilots signalled the Egyptian to land. Instead, he tried to turn away towards Egyptian territory and was promptly shot down. The pilot baled out and was seen to land safely. In spite of these setbacks, the EAF continued its sporadic infiltrations and there were several more skirmishes over the next two and a half years, although these were inconclusive. In the main, the Egyptian aircraft sped for the safety of their own territory at the first sign of trouble.

However, it was not from Egypt that the threat of war now developed, but from Syria. The trouble began when, in July 1964, the Israelis completed a project known as the National Water Carrier, a conduit designed to carry water from Lake Kinneret to the arid regions of the Negev. Soon afterwards, at a summit conference, the Arab leaders agreed that force should be used to prevent the Israelis from using this water, for Lake Kinneret was fed by the River Jordan, which, according to the Arabs, Israel had no right to use. The Syrians wanted to use the project as an excuse for declaring war on Israel, but this idea was rejected by the other Arab leaders when President Nasser pointed out that the Arab nations were not yet strong enough to launch a full-scale offensive with any hope of victory.

Syria immediately began to make plans to build her own water-conduit, together with Jordan, and to divert the source of the river—effectively making Israel's project useless. In the meantime, the Syrians stepped up their military actions along the frontier, shelling border kibbutzim and firing on Israeli patrols. The worst-affected area was Tel Dan, which was overlooked by strong Syrian artillery positions at Tel Azaziyat and Nukheila. Israeli tractors and bulldozers working in the area were threatened by the Syrian guns; as a protective measure, the Israelis moved a company of tanks up to the frontier, with orders to attack the Syrian positions if they opened fire.

At noon on Tuesday, November 3rd, 1964, Syrian machine-guns opened fire on Israeli engineers engaged in levelling a dirt road opposite the Nukheila position. The Israeli Centurions immediately returned the fire, and for an hour all hell was let loose as both sides pounded each other with shells and mortar bombs. The result, however, was inconclusive: the Israeli armour had failed in its main objective, which had been to knock out the Syrian tanks and recoilless guns that were dug in on the Nukheila heights.

Ten days later, when fighting flared up again in the Nukheila sector, it was a different story. The date was Friday, November 13th. At 13.30 hours, the Syrians opened fire on an Israeli patrol and a mixed force of Israeli Shermans and Centurions moved forward to engage them. In addition to the position at Nukheila, the Syrian artillery and mortars at Tel Hamra, Banias and Tel Azaziyat also joined in, lobbing 120 mm shells on to the kibbutzim at Dan and Shaar Yishuv while a fierce tank battle raged on the wooded slopes to the north. Several Syrian tanks went up in flames, but the artillery continued to hit the kibbutzim hard and the Israeli Chief of Staff, Major-General Itzhak Rabin, decided to call in the Air Force.

At 15.00 hours a pair of Vautours appeared overhead and swooped down to pound the Syrian positions. Five minutes later they were joined by two flights of Mystère IVAs and Super Mystères, which attacked with rockets and napalm while four Mirages circled overhead to provide top cover. By 15.15, the Syrian artillery had been silenced. While the attack was still in progress three Syrian MiG-21s put in an appearance, but before they could interfere with the ground-attack aircraft they were pounced on by the Mirages and turned away, heading eastwards at high speed.

During the months that followed there were several more clashes between Israeli and Syrian armour, the most serious occurring on August 12th, 1965, when a tank battle flared up after Israeli Centurions fired across the border on Syrian tractors which were moving earth in connection with the scheme to deprive Israel of the Jordan's waters. The Israeli Air Force, however, played no direct part in these operations: the Chief of Staff was reluctant to commit his air power because of the risk of escalation, and in any case the Israeli armoured brigade had shown itself to be more than capable of dealing with the enemy without air support. The Syrian Air Force, too, stayed well clear of the battle area, which simplified the Israelis' task, but Mirage patrols were flown over the Israeli side of the border in case the MiGs showed up.

The situation deteriorated further during the first half of 1966 when the Syrians switched their earth-moving operations to the south-east of Almagor, near the Sea of Galilee. On July 14th, partly as a reprisal for a raid by saboteurs across the border in which two Israelis were killed, the Chief of Staff authorized the IAF to carry out a strike on the Syrian project. At 11.00 hours, four Mystères escorted by a flight of Mirages, attacked the Syrian plant with rockets and cannon-fire, causing considerable damage. As they were flying away from the target area, they were attacked by four Syrian MiG-21s, which dived out of the sun and hit one Mystère, damaging it slightly, on their first firing-pass. The MiGs were immediately engaged by the Mirage escort, and a dogfight developed over the Sea of Galilee. It lasted less than three minutes; one MiG was hit and the pilot ejected, and the others broke off the action and headed for home.

A month later, on August 15th, the Sea of Galilee was the scene of yet another Israeli-Syrian clash. Shortly before dawn, an Israeli patrol boat ran aground on a sandbank near the eastern shore of the lake. As soon as it was light, the Syrians opened up on the craft with small-arms fire, wounding two crew members. United Nations observers in the area made an urgent appeal to the Syrians not to interfere while the Israelis attempted to refloat the boat, but the request was ignored. Shortly afterwards, two MiG-21s appeared over the lake and made several strafing attacks on the stranded craft, causing more damage, but one of the Syrian aircraft ran into a stream of shells from the boat's solitary 20 mm cannon and crashed in the lake, throwing up an enormous fountain of spray. The second MiG turned to make another run; the pilot, intent on the boat, never saw the Mirage that came arcing down behind him until it was too late. Pulverized by 30 mm shells, the MiG went down vertically and exploded on Syrian territory to the east of the lake. The Mirage, joined by two more, then dived to attack Syrian gun positions on the heights at Massoudiye, knocking them out. The IAF's intervention, however, had come too late to save the patrol boat: riddled with bullets and shell splinters, it sank soon afterwards. After swimming around for several minutes, the crew members were picked up by rescue craft.

On November 13th, the Israelis launched their biggest reprisal raid so far against a neighbouring Arab country, and this time Jordan was the target. Soon after dawn, a two-pronged Israeli column, supported by tanks and aircraft, smashed across the border with Jordan and headed for the villages in the Mount Hebron area to the south of Jerusalem. The Israeli intention was to destroy about 40 houses in the villages of Samu, Khirbet Markaz and Khirbet Kinawa, which had been serving as bases for groups of Arab saboteurs who, during the preceding weeks, had made 13 raids into Israeli territory. As the attack developed, the Jordanians quickly rushed reinforcements to the forward areas to block the Israelis' advance, and the latter met stiff resistance around the police fort at Rajaa el Madfaa, on the Samu road.

The fort was destroyed by tanks and artillery and the column pushed on, but a few minutes later two Royal Jordanian Air Force Hunters pressed home a resolute attack through the Mirage air umbrella, causing some casualties before the Israeli jets caught up with them. One of the Hunters, flown with great expertise, escaped at ground level; the other was not so fortunate. Riddled with cannon shells and burning fiercely, it crashed with a tremendous explosion. The pilot had not baled out. The RJAF put in no further appearances during the Israeli raid, but several IAF aircraft were badly damaged by anti-aircraft fire while attacking ground targets.

After this raid the scene shifted temporarily to the Negev, where, during November, Egyptian aircraft had been becoming increasingly active. On the morning of November 29th, two Mirage III interceptors led by Captain Michael were on patrol at 25,000 feet over the southern tip of the Negev when they sighted two Egyptian MiG-19s heading into Israeli territory. The Mirage pilots immediately manoeuvred themselves into a favourable position to attack and closed in from behind. At a range of half a mile, Captain Michael launched one of his two Matra 530 air-to-air missiles at the leading MiG; like a glowing comet, the weapon homed unerringly on the Egyptian fighter's hot exhaust and exploded, tearing off the MiG's port wing. The second MiG was accounted for by Michael's wingman, using

ordinary cannon fire. It was the first time that a
Matra 530 had been fired in anger, and the Israelis
were impressed by the result. Later, however, the
missile was to prove less reliable than cannon in the
majority of air combats, it was fully effective only
when the attacking pilot was able to close with his
target from the rear.

In the spring of 1967, the action once more
shifted back to Israel's northern frontier, when there
was yet another large-scale clash between Israeli
and Syrian air and ground forces on April 7th. The
battle began when Syrian artillery shelled Israeli
tractors at work in a disputed area; the Israeli Air
Force was called in to deal with the artillery, but
the Mystères were engaged by Syrian MiG-21s and
one of them was shot down. Then the Mirages
arrived and hurled themselves into the battle, and
within a matter of minutes two MiGs had been shot
down. There was no further air action, although a
tank battle continued to rage on the ground for
several more hours. Two more MiG-21s were,
however, shot down during the next 48 hours when
they infiltrated Israeli airspace.

In the face of increasingly massive Israeli reprisals,
the position of President Nasser as leader of the
Arab world was becoming more and more difficult.
In particular, he had been strongly attacked for the
Egyptian failure to go to Jordan's aid during the
fighting in November. Then, during the first week
of May, a new factor emerged: the Soviet Union
warned both Egypt and Syria that Israel had massed
some 13 brigades along the Syrian border and was
about to launch an offensive with the ultimate
objective of toppling Syria's left-wing regime.

The report was entirely without foundation, but

Nasser, because of Egypt's defence agreement with Syria and his obligation to act in accordance with the wishes of other Arab nations, could not afford to ignore it. On May 14th, 1967, Egyptian troops and armour began to move into their forward areas in the Sinai Peninsula. One small push was all that was needed now to send the Middle East toppling into the abyss of war.

CHAPTER TEN

Preparations for War

By the spring of 1967, the total strength of the Israeli Air Force, now under the command of a new Chief of Air Staff, Brigadier Mordecai Hod stood at 350 aircraft, 250 of which were combat types. There were three interceptor/strike squadrons each with 24 Mirage IIICJs; one interceptor squadron with 18 Super Mystère B.2s; four fighter-bomber squadrons, two with 20 Mystère IVAs and two with 20 Ouragans each; one light bomber squadron with 24 Vautour IIAs and IINs; three light ground attack squadrons, each with 20 Magisters (some of which were also used in the training role and were switched to ground-attack duties when required); two transport equipped with 15 Dakotas, 18 Noratlases and five Stratocruisers; one helicopter wing with five Super Frelons, eight Alouette IIs, 30S-58s and a handful of Bell 47Gs and Hiller 12s. Two Sikorsky S-55s formed an air-sea rescue flight. Other aircraft included half a dozen Meteor T.7s and NF.13s, which were still used for target-towing and other non-operational tasks. About 25 Piper Super Cubs, which were normally used as primary trainers, could also be employed in the liaison and casualty evacuation roles when required.

Against the Israeli total, the Arab nations that menaced Israel's borders, Egypt, Syria, Jordan and the Lebanon, could muster approximately 500 combat aircraft between them; and this figure would be increased to 800 if Iraq was brought into the picture. By far the largest commitment was that of the United Arab Republic Air Force, as Egypt's air arm was still known despite the dissolution of the UAR six years earlier. In the first half of 1967, the Egyptian Air Force was built around a nucleus of three interceptor air regiments operating a total of 100 MiG-21F and -21PF fighters, operating in conjunction with the new Soviet air defence system already described in a previous chapter. Then came two fighter-bomber regiments, one with three squadrons of MiG-17Fs and the other with three squadrons of MiG-19s; there were also five fighter-bomber squadrons with the MiG-15bis and the MiG17F, operating independently of Air Force command and coming directly under the orders of Army formations.

The backbone of Egypt's bomber force was still the Ilyushin 28; there was one light bomber regiment with three squadrons of Il-28 tactical bombers and Il-28R tactical reconnaissance aircraft, making up a total of some 35 machines. Far more potent than the elderly Il-28, however, was the Tupolev Tu-16, of which 30 now equipped one strategic bomber regiment. As an insurance against these being used to attack cities and other strategic targets in Israel in the event of hostilities breaking out, the Israelis had obtained a substantial number of Raytheon Hawk missiles from the United States. Already in service with the US armed forces and on order by several other countries, the Hawk was infinitely superior to the cumbersome Soviet SA-2 missiles supplied to Egypt: it possessed a 'slant range' of 22 miles, and was capable of intercepting an aircraft flying at anything between 75 and 40,000 feet at a speed of Mach 2.5. During test-firings, Hawks had in fact successfully intercepted even supersonic missiles such as the 'Honest John'. The Israelis now had several Hawk missile batteries, deployed both on hard sites for the defence of airfields and other targets and on mobile launchers to provide effective counter-measures in the front line against low-flying close-support aircraft.

The Hawk's ability to intercept missiles as well as aircraft was a vital consideration, for in addition to their normal bomb-load the Egyptian Tu-16s were now equipped with a Soviet air-to-surface missile known by the NATO code-name of 'Kennel'. This weapon, which had a range of 30 miles, was 28 feet long and carried a high-explosive warhead. Powered by a turbojet engine and looking rather like a scaled-down MiG-15, it was primarily an anti-shipping missile, but it could conceivably be used against targets on land, enabling the launch aircraft to stay well outside the range of its target's immediate defences. Each Tu-16 carried two 'Kennels' slung under its wings.

In all, the first-line strength of the Egyptian Air Force totalled 20 squadrons and 330 combat aircraft, together with considerable reserves. Backing up the combat force were three transport squadrons, two of them equipped with Ilyushin Il-14s and one with Antonov An-12s. The transport capability was increased even more by a rapidly-expanding helicopter force, which included a few massive Mil Mi-6s.

As part of a mutual defence pact following the re-establishment of friendly relations with Syria, at least three squadrons of Egyptian aircraft were based in that country on a rotation basis to add to the

effectiveness of Syria's own 150-strong first-line force.
The request for Egyptian aid came after the Syrian
project to divert the waters of the River Jordan was
smashed by the crushing Israeli attack across the
border in July 1966, and the three squadrons of
MiG-17s and MiG-19s were transferred from Egypt
to Dumeyr, north of Damascus, about three weeks
later.

Two of the Syrian Air Force's own fighter squadrons were now equipped with MiG-21PFs, about
40 had been delivered. The other three fighter
squadrons still operated 60 MiG-17s. There was one
transport squadron, operating Il-14s, Dakotas and
Beechcraft D-18s as well as a small number of Mi-1
and Mi-4 helicopters. Syria, like Egypt had also
received a quantity of SA-2 Guideline missiles, which
had been set up in six-launcher batteries in the
vicinity of Damascus under the direct control of
the Syrian Army. The standard of efficiency of the
whole Syrian armed forces was still lamentably poor,
and attempts to raise it had been thwarted to a great
extent by the eight coups d'état suffered by the
country over the previous 17 years. Morale was at a
low ebb, especially among the officers.

By way of contrast, the morale of the armed
forces of Syria's southern neighbour—Jordan—was
perhaps the highest of any Arab country, and their
efficiency was certainly superior to that of either
Egypt or Syria. Despite continual pressure by President Nasser, King Hussein had made a tremendous
effort to remain outside the Soviet sphere of
influence and pursue a neutralist policy. The Royal
Jordanian Air Force's equipment was almost entirely
of British origin, and its first-line strength comprised
two squadrons each with 12 Hawker Siddeley

Hunter F.Mk.6s and one reserve unit with 16 Vampire FB.9s. There was also a sole Hunter FR.6 for fast photo-reconnaissance work, and a two-seater Hunter T.66B for operational conversion. For training there were four Chipmunks, three Harvards and two Vampire T.55s, but the great majority of RJAF pilots trained in Britain alongside RAF students. They were generally good pilots and achieved consistently high ratings, often excelling in certain aspects of flying such as individual aerobatics.

Although the RJAF's equipment was rapidly becoming obsolescent, King Hussein repeatedly turned down a Soviet offer to supply MiG-21s at very favourable prices. However, following an Israeli strike on Jordanian territory on November 13th, 1966, Jordan accepted a cut-price offer of 36 refurbished F-104A Starfighters from the United States. The aircraft, which were scheduled for delivery in the summer of 1967, were to equip three RJAF squadrons; pilots and ground crews had already begun training on the new type in the USA, and Jordan's Mafraq air base was being prepared to operate the Starfighters.

The fourth of Israel's immediate Arab neighbours, the Lebanon, also operated British Hunters which equipped the one and only combat squadron of the Force Aerienne Libanaise. In 1965, however, the Lebanese Government had ordered 12 Dassault Mirage IIIELs from France, together with a quantity of Matra R.530 air-to-air missiles. The Mirage IIIEL was a simpler and cheaper version of the standard Mirage IIIE; it could be used both for interception and ground-attack, and Lebanese plans envisaged the procurement of 24 to equip two squadrons. Lebanese pilots were already beginning to convert on to the new type in France in the spring of 1967, but it would be some time before deliveries began.

After Egypt, potentially the most formidable of all the Arab nations in terms of military might was Iraq, unique in that its combat units operated modern equipment obtained from both East and West. The Iraqi Air Force had been modelled closely along the lines of the Royal Air Force, and there had been a good liaison between the two Services until the summer of 1958, when a Republic was established following a bloody coup d'état and British forces withdrew from the country. As a result of the British withdrawal, the IrAF had found itself in possession of some of the best-equipped air bases in the Middle East: Habbaniya, Shaiba, Rashid, Mosul, Basra and Kirkuk.

The combat element of the IrAF had been formed around a nucleus of 15 Hunter F.Mk.6s, delivered in 1957-58. Between 1964 and 1966 more Hunters were delivered, including 18 F.Mk.59s and six two-seat T.Mk.69s. By the end of 1966, a total of 44 single-seat Hunters equipped four IrAF ground-attack squadrons, and there was also a reconnaissance flight equipped with Hunter FR.Mk.10s. The IrAF had received its first Soviet equipment towards the end of 1958, 19 MiG-15bis fighters delivered via Egypt. The following year, a large Soviet Air Force mission arrived in Iraq and virtually took control of Shaiba, Rashid and Habbaniya, where it undertook the conversion of Iraqi pilots on to the MiG-15s and a small number of Il-28 bombers which had arrived by that time.

During the early 1960s the IrAF underwent a big reorganization and modernization programme, and its squadrons took delivery of more Soviet equipment. This included 20 An-2 and eight An-12 transports, Yak-11 trainers, Mi-4 helicopters, 15 MiG-19 and twelve MiG-21F day fighters, and 20 MiG-17PF limited all-weather interceptors. A further 60 MiG-21s were delivered in the autumn of 1966, by which time a number of SA-2 Guideline missile batteries were also operational. One squadron of ten Il-28s gave the IrAF a small bombing capability, while a light ground-attack squadron was formed with 20 British-built Jet Provost T.Mk.52s. These replaced a small number of piston-engined Hawker Furies, which were operational in the counter-insurgency role until the middle of 1965. Early in 1967, the IrAF's bomber force was strengthened with the arrival of six Tu-16s, which formed a strategic bomber squadron.

This brought the first-line strength of the Iraqi Air Force to ten squadrons, operating 300 aircraft. It was a force to be reckoned with, but it suffered from the inevitable consequences of ten years of revolution and counter-revolution in that the reliability of some of its personnel was always in question. On August 16th, 1966, an Iraqi Air Force pilot flying a MiG-21F defected to Israel; the aircraft was picked up by radar while still over Syrian territory and a flight of Israeli Mirages intercepted it over the border, escorting it in for a safe landing. The IAF subsequently flew the aircraft extensively in mock combat with the Mirage and technical specialists took it apart piece by piece. Since Iraq was still technically in a state of war with Israel, repeated requests for the return of the aircraft were flatly refused. Soon afterwards, three more MiG-21 pilots defected to Jordan in rapid succession; this time, the aircraft were returned, although the pilots were granted political asylum. As a direct result of these defections the Soviet Union placed a ban on further deliveries of combat aircraft to Iraq, but this restriction was lifted in January 1967 with the delivery of a batch of MiG-21s.

The unexpected arrival of the Iraqi MiG-21 was a valuable windfall for the Israelis, for events in the Arab world and the growing belligerence of President Nasser all pointed to the possibility that an all-out conflict between Arabs and Israelis might be a matter of months away and there was a vital need to assess the potential of the enemy's equipment. The Israelis had also been keeping a close watch on operations in the Yemen, torn by civil war since the death of Imam Ahmad in 1962, where the republicans were receiving large-scale support from the Egyptians. At the beginning of 1967, some 60,000 Egyptian troops were committed to the conflict against the royalists, while Egyptian MiG-17Fs and Il-28s were operating from the Yemeni bases of Hodeida, Taiz and San'a. The Yemen's small air arm consisted entirely of Yak-11 trainers, Il-14 transports and Mi-4 helicopters supplied by the United Arab Republic, and these were also flown and maintained by Egyptian personnel. Although the five-year war had imposed a fearful drain on Egypt's already overtaxed resources, the Egyptian Air Force had undoubtedly derived some benefit from it: not only had the Egyptian pilots gained considerable experience in attacking elusive targets on the ground, but they had been able to test a variety of weapons under combat conditions. Potentially the most frightening of these was what appeared to be a gas-bomb, several of which were dropped on the royalist stronghold at Ketaf in January 1967. The bombs were said to be part of a consignment of 600 which the Egyptians had received from Communist China. Ketaf was attacked by nine Ilyushin 28s, flying in three vics and releasing 27 bombs from a height of 5,000 feet. According to eye-witness statements, the bombs shattered on impact and released a grey-green cloud of vapour that drifted slowly across the ground at a height of about three feet. When it cleared, 120 of Ketaf's inhabitants and the greater part of their livestock were dead. The only survivors were those who had been on the high ground around the village at the time of the attack; when they went down to investigate, they found that the vegetation where the bombs had impacted was strangely withered, and all food in the village was contaminated. The scene of the bombing was subsequently visited by Red Cross representatives, but their findings were inconclusive.

Nasser had been involved in the Yemen for two main reasons. First of all, he regarded Egyptian domination of the area as a preliminary step towards expansion in the direction of South Arabia following the projected British withdrawal from Aden; and second, he hoped that a republican victory in the Yemen would lead to a possible socialist revolution in neighbouring Saudi Arabia, and if that happened the richest oil fields in the Middle East would be within Egypt's reach. But Saudi Arabia had elected to provide active military support for the Yemeni royalists, with the result that the conflict had become bogged down. Moreover, Saudi Arabia had signed a deal with the British Government for the provision of an entirely new defence system, including early warning radar, telecommunications, supersonic Lightning fighters and Thunderbird surface-to-air missiles. A small force of Lightnings had already arrived in the country, and these were being flown by British pilots—employees of Airwork—while Royal Saudi Air Force personnel trained in Britain. The Lightning pilots had strict instructions not to intrude into the Yemen's airspace, but the Saudi Government had made it quite plain to Nasser that any Egyptian aircraft intruding over Saudi Arabia's frontiers would be shot down—and, with the formidable Lightning at the RSAF's disposal, it was no idle threat.

With his ambitions towards Saudi Arabia thwarted and his massive military force in the Yemen completely incapable of breaking the royalists, Nasser was experiencing a serious loss of face that endangered his leadership of the Arab world. The main challenge came from the new extreme Left-wing administration in Syria, which had come to power following yet another military coup in February 1966 and which had immediately launched a guerrilla campaign against Israel, making use of bases in Jordan.

Nasser knew that he had to act, and act quickly, if he was to have any hope of regaining the initiative. In spite of the fact that his armed forces were by no means ready to wage an all-out war, he took the decision in May 1967 to move against Israel. On May 17th, 1967, the Egyptian Government asked the United Nations security force that had been patrolling the Sinai frontier with Israel to move out in case they got caught up in the fighting in the event of war breaking out and they went within 48 hours on the orders of UN Secretary-General U Thant. During those 48 hours, both the Egyptian and Syrian forces had been put on the highest state of alert in case Israel took the opportunity to launch a pre-emptive strike against them. On May 21st, both Egypt and Israel called up their reserves; the following day, Nasser announced that the Gulf of Akaba was closed to Israeli shipping. By May 24th the blockade of the Gulf had begun in earnest, with Egyptian air and naval forces maintaining constant patrols and the shore batteries fully manned and on the alert.

Events were now approaching their climax with frightening speed. As the diplomatic war continued British and American warships moved quietly to battle stations in the Mediterranean to counter the threat of possible Soviet intervention. Israel, simultaneously with a sense of grim urgency unparalleled since the War of Independence, prepared to fight. El Al worked overtime, its aircraft evacuating tourists and other foreign nationals and bringing increasing numbers of volunteers, mostly members of Jewish youth organizations, into Israel on the return flights.

On May 24th Abba Eban, the Israeli Foreign Minister, flew to Paris, London and Washington in an abortive attempt to end the crisis by diplomatic methods rather than by force of arms. On that same day, the situation took a turn for the worse when Iraqi and Saudi Arabian troops entered Jordanian territory, the Lebanese armed forces mobilized and the Palestine Liberation Army, under Ahmed Shukeiry, received artillery and other heavy equipment from Egypt and dug in along the border of the Gaza Strip.

During this period, there was a notable lack of Israeli air activity as the IAF ground crews strove to bring the serviceability of their first-line aircraft as near to 100% as possible. Between May 21st and 27th, however, Vautours and Mirages operating mainly out of Hatzerim in the Negev made a series of reconnaissance flights over Egyptian territory to test the reaction time of the EAF's air defences. The flights ceased on the evening of the 27th; no Israeli aircraft had been lost, although there had been one or two near-interceptions.

The IAF's preparations for war went ahead smoothly. There was no need for intensive last-minute flying under simulated combat conditions, for the whole of the IAF's operational training syllabus over the past five years or more had been based on the assumption that a large-scale pre-emptive strike might one day have to be made on airfields in Egypt and the other Arab countries. Five bombing ranges in the Negev had been specially designed to simulate enemy targets, and in addition to routine practice bombing missions they were the object of a huge exercise held at least once a year and involving the greater part of the IAF's first-line force, in which the techniques that would be used in a pre-emptive strike were constantly practised and improved to keep pace with new equipment.

The Israeli Air Force's operational plan, approved in its final form during the last week of May by Brigadier Hod and Brigadier-General Ezer Weizman, the Chief of Operations, called for a surprise attack on 19 Egyptian airfields. The initial strike was to be made by the Mirages, flying a right hook out over the Mediterranean and splitting up into flights to attack Cairo West, Cairo International, El Mansura, Inchas, Abu Sueir, Fayid, Kabrit, Helwan, Beni Suef and El Minya. Of these, Cairo West was the most vital target: it was the home of the 30 Tu-16s of the EAF's Strategic Bombing Regiment. Cairo International had been left out of the original plan, but was brought in later when it was discovered that a squadron of MiG-21s had been dispersed there. The three southernmost Egyptian airfields—Hurghada, Luxor and Ras Banas—were to be attacked by the Vautour squadrons, operating at extreme range out of Hatzerim. As a result of the IAF's probing flights across the Red Sea, the Egyptian High Command considered it unlikely that if an Israeli air strike took place it would take the form of a left hook designed to hit the main bases from the south, and a mixed interceptor unit of twelve MiG-21s and eight MiG-19s had been deployed to Hurghada.

Gamil and Deversoir, in the Canal Zone, were to be hit by Mystères flying directly from their bases in Israel, while the EAF's four Sinai airfields—El Arish, Jebel Libni, Bir Gifgafa and Bir Thamada—were to be the objectives of Mystères and Ouragans. The attacks on the latter airfields called for a high degree of precision, for the Israelis wished to keep their runways as intact as possible for their own use. At the same time, since these bases were only a few minutes' flying time away from Israel's cities, the IAF pilots had to ensure the complete destruction of the Egyptian aircraft that were stationed there.

Apart from the all-important element of surprise, the Israeli plan depended on very precise timing. The three waves of Mirages selected to hit the main Egyptian bases were to be spaced out at ten-minute intervals, with a time-to-target of 20 minutes. This meant that the third wave would be forming up over its bases and setting course at the same moment that the first wave attacked its objectives, while the second wave would be on its way in over the Egyptian coast. Each wave was to spend exactly seven minutes over its target before returning to base, where the ground crews had been allotted just over seven minutes to get the aircraft refuelled and rearmed to make a second attack. On paper, therefore, the plan was as follows:

08.25: First wave sets course for target area.

08.35: Second wave sets course; first wave entering Egyptian territory.

08.45: First wave attacks objectives; second wave entering Egyptian territory; third wave setting course for target.

08.52: First wave leaves target area after 7-minute 'loiter'.

08.55: Second wave attacks targets; third wave entering Egyptian territory.

09.02: Second wave leaves target area.

09.05: Third wave attacks targets.

09.12: Third wave leaves target area; first wave arrives back at base.

09.22 (approx.): First wave (now fourth wave) sets course for target.

09.42: First (fourth) wave attacks targets for second time.

This plan meant that the Egyptian defences would have only three minutes' respite or even less between the onslaughts of the three successive waves in the first series of strikes, and because of the fast turn-round time, which allowed the first attackers to be over the target for a second time within an hour of their initial strike, there would be a delay of only 30 minutes between the last strike of the first series and the first strike of the second.

The strikes were timed to begin at 08.45 (Cairo Time) for two main reasons. First of all, Israeli Intelligence was aware that the majority of the Egyptian military and air commanders did not arrive at their offices until 09.00 and experience had shown that junior Egyptian commanders would stand by and watch their equipment reduced to burning wreckage while they waited for someone in authority to tell them what to do, rather than take the initiative themselves. Secondly, the early morning Egyptian fighter patrol habitually landed at 08.35, and a second patrol never took off until 09.10 after the pilots had had their breakfast. The Israelis believed that there would be few, if any, interceptions during the first series of strikes which were calculated to annihilate a large part of the EAF's combat force on the ground and throw the Egyptian defences into complete confusion, giving the second series a valuable advantage.

Apart from that, the main consideration that led to the choice of 08.45 as the most favourable time for an attack was one of weather conditions. In spring and early summer, the Nile Delta and the Suez Canal area are shrouded in heavy morning mist from dawn until about 08.30, when it is dispersed rapidly by the rising sun. The dispersal of the mist is followed by a period of brilliant calm accompanied by excellent visibility and an almost complete lack of turbulence—an important consideration for an accurate run-in and weapon release. The Israeli pilots were not unduly worried by the fact that they would be attacking into the sun: any disadvantage from glare would be more than offset by the certain belief on the part of the Egyptians that no pilot in his right mind would attack with the sun in his eyes, and that a morning attack must consequently develop from the east.

Enemy radar would present a considerable problem for the attackers, but not an insurmountable one. The biggest danger would come from the Jordanian Marconi 547 surveillance radar station that overlooked Israeli territory from its vantage point at Ajlun, but if the strike aircraft kept low in the shadow of the Judaean Hills they were not likely to be detected by the Ajlun station until they were about 100 miles out to sea, and even then it was likely to be some time before the Jordanians were able to identify them positively and work out their purpose. With the American Sixth Fleet on station near Cyprus, the volume of air traffic over the Eastern Mediterranean was enormous; because of the low altitude at which the Mirages would be flying the Jordanians would probably not get a true picture of the size of their formations from the radar display, and in all probability they would be taken for US Navy aircraft until they were on the point of crossing the Egyptian coast. After that, it would no longer matter: the Egyptian Air Force, past its early morning state of alert, would not have time to react before the Mirages hit their targets.

Hod and Weizman, the architects of the plan, had calculated that two separate series of strikes, one following hard on the heels of the other, should be enough to knock out at least 70% of the Egyptian Air Force's serviceable first-line strength. This allowed for the probability that under real operational conditions, the air strikes would produce results some 25% less effective than those achieved during exercises. The Israeli Air Force would, therefore, have two clear hours to dispose of most of the

EAF before switching the greater part of its commitment against Jordan, Syria and Iraq. The Israeli strategists believed that it would be long enough; taking into account the element of surprise and the inevitable confusion that would follow, the Jordanians, Syrians and Iraqis would not be in a position to make any hostile move against Israel until between two and three hours after the offensive against the Egyptian bases opened.

The two principal targets on Jordanian territory were Mafraq and Amman, where the RJAF's two Hunter squadrons were based. Five airfields—Dumeyr, Saigal, Marj Rhiyal, Damascus-Mezze and Tango Four—were targeted in Syria; the primary objectives here were Dumeyr, which housed three MiG-17 and MiG-19 squadrons of the Egyptian Air Force, and Tango Four, a staging-point for Soviet combat aircraft destined for the Middle East. In Iraq, the main objective was the former RAF airfield of Habbaniya, where the IrAF's small Tu-16 strategic bombing force was based; the other Iraqi airfield scheduled to be hit was a newly-constructed base code-named Hotel Three, which housed the greater part of the IrAF's MiG-21 interceptor force.

The object of the initial Israeli strikes against all these bases was to put the runways out of action. Not all of the strike aircraft were to carry the Matra-designed 'concrete dibber' bomb: these were in fairly short supply, and those available were mainly allotted for use against the enemy's bomber bases. There were enough to equip about 40% of the strike force; the remainder would hit their objectives with ordinary 500- or 1,000-pound bombs, dive-bombing from a height of about 5,000 feet to ensure accuracy. The damage inflicted by these weapons would be easily repairable; this was part of the plan, as the Israelis required some of the Sinai airfields for their own use with the minimum delay. The task of the first strike wave was therefore to render the runways of the enemy bases unserviceable and pin down his combat aircraft on the ground until their destruction could be completed by subsequent strike waves, attacking with cannon-fire.

By May 29th, a week after the start of Egypt's blockade of the Straits of Tiran, the Israeli Government had still not made a firm decision to go to war; the Knesset was in an uproar as 'doves' and 'hawks' fought a verbal battle over the decisions that were to mean the difference between life and death for Israel. Depending on those decisions, war might come in a day or a week; when it did come, the Israeli Defence Force would be ready for it. By midnight on the 29th, the Defence Force's commanders had succeeded in tying up the loose ends of their operational plans.

At 22.00 hours on that day, three El Al Boeing 707-320B airliners touched down at Bordeaux-Merignac airfield in France within seconds of each other. The Boeings taxied past the Marcel Dassault aircraft factories and turned on to the apron beyond, shutting down their engines. For the next six hours, the airfield was the scene of unaccustomed nocturnal activity as a convoy of lorries rolled on to the apron and drew up alongside the El Al aircraft; a small army of personnel transferred mysterious crates from the lorries into the Boeings' spacious fuselages, from which the passenger seats and other fittings had been removed to make room for up to 30 tons of freight.

Two of the Boeings left Bordeaux at dawn on May 30th; the third took off at noon. The mysterious freight that each aircraft carried included air-to-ground rocket packs and Matra 530 air-to-air missiles; the latter were partially assembled, and by the morning of June 3rd they had reached the Israeli Air Force's combat units. Eleven years earlier, on the eve of the 1956 Sinai Campaign, a last-minute delivery of French arms had enabled the IAF to go into action with an adequate stockpile of weapons behind it; now it had happened again, literally at the eleventh hour.

On the evening of Saturday, June 3rd, General Moshe Dayan, the brilliant architect of the Sinai Campaign, who had been appointed Minister of Defence two days earlier, reported to the Israeli Cabinet on the country's military preparations. The Israeli Army in the Negev was poised for a three-pronged attack on the 80,000-strong Egyptian troop concentration in Sinai, which had to be smashed as a preliminary to the complete occupation of the Peninsula. The right-hand prong would spear into the Gaza Strip, towards Khan Yunis and Rafah; the centre thrust would drive towards Abu Ageila; and the third prong would consist of an armoured race across the desert towards Jebel Libni and Bir Lahfan, with Abu Ageila also as its eventual goal. The Air Force was ready for the signal to attack, with 98% of its aircraft serviceable; and the Navy was putting the finishing touches to its plan to attack the Egyptian naval bases of Port Said and Alexandria. The weight of the Israeli offensive was to be directed against the Egyptians; Jordan was not to be attacked unless she struck first; and the Israeli forces in the north-east were to be content with containing the Syrians behind their own frontiers for the time being at least.

On Sunday, June 4th, the Israeli Cabinet gave its approval to the war plan which meant that from then on, either the Prime Minister or the Minister of Defence could order a massive Israeli attack at a moment of their own choice if the Arab threat increased. The approval came not before time: Egyptian aircraft were making probing flights into Israeli airspace, Egyptian troops were infiltrating Israeli territory, and among the Arab leaders the word invasion had turned from a dream into reality. There was no longer any doubt that Israel was going to have to fight—or die.

CHAPTER ELEVEN

Six Days of Destiny

At 08.15, Cairo Time, on the morning of June 5th, 1967, a twin-engined Ilyushin 14 droned over the Suez Canal at 10,000 feet, heading eastwards into the Sinai Peninsula. The aircraft had been fitted out as a flying command post; this morning, in addition to its normal crew, it carried three high-ranking officers: General Mohammed Sidki, the Egyptian Air Force Commander-in-Chief, General Amer, the Egyptian Chief of Staff, and a Soviet Air Force Brigadier-General who was attached to the EAF as a liaison officer.

Sidki had been a worried man for the past two days. He was convinced that the Israelis were planning a pre-emptive strike, but he didn't know when it would take place. It was logical to assume, however, that such a strike would come at first light; the Soviet advisor thought so too. Accordingly, Sidki had ordered two squadrons of MiG-17s and MiG-21s to patrol Israel's border with Sinai for the hour following sunrise at 04.00; after that, from 05.00 until 08.00, Egyptian fighters had been flying half-hourly cab-rank patrols along the length of the Suez Canal. At 08.05, a couple of minutes after the last patrol became airborne, Sidki and Amer had also taken off in their Il-14 to make an assessment of the situation; in the aircraft's fuselage four electronics operators sat huddled over their equipment, tuned in to the Israeli frequencies and listening for any tell-tale radio chatter that would indicate unusual air activity.

There was nothing: the Israeli military frequencies were silent. The Il-14 droned on over Sinai, covering the territory in a wide circle while the officers on board surveyed the Egyptian positions spread out beneath them. The crew of the Ilyushin were unaware that they were being watched by the electronic eyes of the Israeli radar, or that the aircraft's progress was causing a good deal of anxiety among the officers assembled in the Israeli war room. If the Il-14 turned north towards Gaza and lingered in the vicinity of Israel's border, its occupants would be almost certain to spot the armada of strike aircraft that would soon be skimming low over the Peninsula towards their objectives in the Canal Zone. An audible sigh of relief went up when the radar plot indicated that the Ilyushin had turned south and was heading for the Gulf of Akaba: the danger of detection was over.

08.40, Cairo time. The Ilyushin, its survey flight over, now turned north-westwards and headed for its base at Kabrit; the pilot called up Kabrit tower and informed the controller of his ETA. The controller acknowledged; there was no traffic and everything was normal. Suddenly, at 08.45, a confused babble of voices burst over the R/T. The pilot called up Kabrit again; there was no reply. The frequencies were jammed by incoherent chatter in which the words 'we are being attacked' were repeated over and over again.

The minutes ticked by. Sidki, in an agony of frustration, knew that his worst fears had materialized, but he still had no idea of the true situation. Then, with 25 miles still to go before the Ilyushin reached the Suez Canal, the crew saw the first mush-rooms of smoke.

The whole Canal Zone seemed to be on fire. The Egyptian airfields were shattered, littered with the wreckage of combat aircraft. The Il-14 pilot flew the length of the Canal, but could find nowhere to land. On Sidki's orders, he turned eastwards again and headed for the Sinai airfield of Bir Gifgafa. That, too, was in ruins; so were its neighbours, their run-ways pock-marked with bomb craters. There was nothing for it but to fly back across the Canal in the hope of finding a serviceable airstrip. Finally, the Ilyushin was cleared to land at Cairo Inter-national Airport at 10.45—two hours after the first air strikes took place. Desperately, Sidki tried to salvage something from the chaos, to organize a scratch combat force from the remnants of his squadrons. But as more situation reports reached him, he realized that it was hopeless. There was nothing left; the Egyptian Air Force had been reduced to scrap metal in the space of two hours. Sidki's carefully-laid operational plans had been torn to shreds by rockets, bombs and cannon-shells: the Israeli Air Force was mistress of the sky.

The first massive air strike had done its work well; far better than even General Hod had dared to hope. As planned, the first airfield to be hit was Cairo West, where the Mirage pilots found the 30 Tu-16s of the Strategic Bombing Regiment parked conveniently in their blast-proof revêtments. After neutralizing the runways in their first pass, the Israeli pilots turned their undivided attention to the bombers; 16 were knocked out in the space of five minutes, and the remaining 14 were destroyed by the second wave of Mirages which arrived overhead three minutes after the first wave had left.

At Cairo International, the Mirage pilots found a squadron of MiG-21s parked in a row on the far side of the airfield; all of them were destroyed. One short burst of cannon-fire disposed of four aircraft at once; the first MiG to be hit exploded violently, throwing a sheet of blazing fuel over its neighbours. At Abu Sueir, the Mirages surprised four MiG-21s which were taxiing out towards the end of the runway and pulverized the lot; it was a similar story at Inchas.

While the first wave of Mirages was on its way in to the target area, 15 twin-engined Vautours had taken off from Hatzerim in the Negev. Each aircraft carried two 500-pound bombs; the bomb-load had been cut down drastically to allow the Vautours to carry maximum fuel. The aircraft climbed steeply to 24,000 feet and then levelled off, their pilots shutting down one engine to obtain the maximum economical cruise. In two flights, one of eight and the other of seven aircraft, the Vautours headed due south, crossing the Gulf of Akaba and flying along the Red Sea after skimming over the fringes of Saudi Arabian territory. Twenty minutes later the two flights split up, heading for the Egyptian airfields of Luxor and Ras Banas, the pilots throttling right back and beginning a fast glide descent towards their objectives.

It was from these two airfields that the Egyptian Air Force's offensive against the Yemeni Royalists had been mounted; because of the Egyptian High Command's belief that they were far enough south to be safe from Israeli air attack, no attempt had been made to disperse the resident squadrons of MiG-17s and Il-28s. The Egyptian aircraft were drawn up in neat ranks, presenting an ideal target. At Ras Banas, 16 Ilyushin 28s were destroyed in a matter of minutes by two Vautours, the Israeli pilots flying low and slow along the line of parked bombers and raking them with cannon-fire. At Luxor, the Vautours also succeeded in destroying their objectives in spite of heavy anti-aircraft fire; one Vautour was hit and crashed on top of a section of four MiG-17s, which erupted in an enormous mushroom of burning fuel.

Their mission completed, the Vautours, lightened of the burden of their weapon-load and half their fuel, climbed back out over the Red Sea, levelling once more at their optimum single-engine cruise altitude of 24,000 feet. They would maintain this altitude until they reached the Israeli border, ending their flight with a glide descent to Hatzerim.

Meanwhile, as the Vautours were still on their way to Luxor and Ras Banas, the IAF's Mystères and Ouragans had launched their devastating strikes against the Egyptian airfields in Sinai. In each case the Mystères went in first, hitting the runways with their concrete dibber bombs and following up with strafing runs, then leaving the Ouragans to finish off the job.

At Jebel Libni, two Egyptian pilots were sitting in the cockpits of their MiG-17s on an operational readiness platform at the end of the main runway. The aircraft, on quick reaction alert, had their battery trollies plugged in and were all ready to go. They never had a chance to get airborne, however, a flight of four Israeli Mystères swept across the airfield at high speed in a banshee screech of engines, and two of them dropped their rocket-powered concrete dibbers on the runway intersection. While the dust and smoke of the explosions still hung in the air the second pair of Mystères racked round in a tight turn and streaked back towards the two parked MiGs. The ground crews scattered for cover, dropping everything as a stream of cannon shells churned up an avenue of dirt that raced towards them. The pilots tore at their straps, their one thought now to get away from the inferno, but it was too late. Trapped in their cockpits, they died in a nightmare of torn metal and shattered instruments as the 30 mm shells exploded around them. The MiGs' fuel tanks erupted with a thud and the fighters crumpled into the dust, blazing fiercely.

The other two Mystères, meanwhile, had dived down to strafe the 13 MiG-17s and MiG-19s parked near the airfield buildings. Several were knocked out during the fighters' first pass, and the destruction was completed when the four Mystères joined forces and made two more low-level runs. Several dummy MiGs were completely ignored; apart from the fact that IAF Intelligence had been aware of them, they were parked at the side of the runway and in other unlikely places where no genuine aircraft would have been deployed.

Attacking Jebel Libni had called for a high degree of precision: the Israeli pilots had been briefed to leave 3,500 feet of the airfield's 7,000-foot runway undamaged so that it could be used by IAF Dakota and Noratlas transports at a later stage. If the ground offensive went according to plan, the transports would be operating from the airfield the following day; in the meantime, to forestall any Egyptian attempt to repair the runways, three flights of Mystères and Ouragans hit the airfield again at dusk and showered it with small delayed-action bombs, set to detonate throughout the night.

El Arish, too, was required for use by the IAF Air Transport Command, and the Israelis took a calculated risk in leaving its runways completely intact; however, the six MiG-17s that were based there were knocked out by rocket-firing Super Mystères, and the Israelis flew continual combat patrols over the airfield until their ground forces eventually captured it.

The two remaining Sinai bases, Bir Gifgafa and Bir Thamada, were occupied mainly by transport aircraft and support helicopters. Bir Gifgafa was hit by two flights of Mystères, which knocked out the runways and destroyed four twin-engined Il-14 transports with cannon-fire. The Mystères then curved round to attack a trio of helicopters on the far side of the field; one of them, a huge Mil Mi-6, had just got airborne and was hanging in mid-air at the point of transition from hovering to forward flight when a burst of 30 mm shells from the leading Mystère scythed off its main rotor. The mighty fuselage dropped to the ground like a stone and burst into flames. The other two helicopters were destroyed on the ground. Several more Egyptian transports, including one big Antonov An-12, were knocked out during a strike by Mystères on Bir Thamada.

Meanwhile, the strikes against the Egyptian airfields across the Suez Canal continued. The first devastating strike wave had hit its targets and escaped relatively unscathed, but it alerted the Egyptian anti-aircraft defences and subsequent strikes began to suffer losses as they ran into heavy defensive fire. At about 09.30, a whirling dog-fight developed over Abu Sueir between 16 Mirages and

20 MiG-21s, the squadron from Hurghada, which had been flown south a couple of days earlier to cover the south-eastern approaches to the Canal Zone and which had now been hurriedly 'scrambled' and sent north again. The Egyptian pilots hurled themselves on the Mirages, but they were hopelessly outclassed: four of the MiGs were shot down in a matter of minutes, and the rest, probably critically low on fuel, broke off the engagement and headed west, looking for somewhere to land. A few managed to get down safely, but most smashed themselves to pieces while attempting to land on shattered runways or were destroyed when their pilots ejected after the aircraft ran out of fuel.

Apart from that, only eight MiG-21s managed to get airborne to challenge the attackers; they succeeded in destroying two Mystères, but they were pounced on almost immediately by a squadron of Mirages and all of them were shot down. All the enemy aircraft were destroyed by cannon-fire, which the Israeli pilots found more effective than their Matra air-to-air missiles.

At about 10.20, Jordanian forces launched their expected attack when Arab Legion troops took over the United Nations building in Jerusalem and began to infiltrate the Israeli half of the city. Thirty minutes later, four RJAF Hunters from Mafraq strafed the Israeli forward airstrip at Kefar Sirkin, just across the border, knocking out two Super Cubs and a few vehicles. The Hunters returned to their base to find it under attack by IAF Mirages, which the Jordanian pilots promptly engaged. One Hunter was shot down almost at once; a second, flown brilliantly by its pilot, made three determined attacks on the Mirages before shortage of fuel compelled him to break off. He was immediately trapped by two Mirages and torn apart by 30 mm shells. The two remaining Hunters were wrecked when their pilots tried to land on Mafraq's shattered runways.

The IAF pilots detailed to hit Mafraq had been briefed to make very careful positive identification of aircraft on the base before attacking, for it was known that the airfield was being used by a United States mission which was training Jordanian pilots to fly Starfighters. The mission consisted of about 100 personnel including ten flying instructors, the rest technical instructors and ground crews and was equipped with two single-seat F-104A Starfighters and three two-seat F-104Bs. Although they carried Jordanian markings, the Starfighters were still technically owned by the USAF and came under the direct control of the US Ambassador.

When the Israeli pilots arrived over Mafraq, however, they found that the Starfighters had gone; in fact, they had been flown out by the American instructors to the NATO base at Cigli, in Turkey, 36 hours previously, an indication, perhaps, that US Intelligence sources had received a tip-off that an Israeli strike was imminent. The American support crews had also been evacuated aboard two C-130 transports.

The airport at Amman was also hit by the Israelis about noon, a flight of four Mystères bombing the runways while four more strafed aircraft on the ground. Two Alouette helicopters and six light transport aircraft were destroyed; the latter included three de Havilland Doves, one of which unfortunately turned out to be the personal aircraft of the British Air Attaché. By 14.00 hours the Royal

Jordanian Air Force had ceased to exist; the Israelis had destroyed 21 of the RJAF's 24 Hunters.

The Israeli Air Force now switched its attention to Syria and Iraq. At 13.10, a lone Iraqi Tu-16 swept in low across the Plain of Sharon and bombed the Israeli town of Netanya, astride the main Haifa-Tel Aviv railway line. The bomber ran into a concentrated cone of fire from half a dozen multiple-barrelled 40 mm anti-aircraft guns and burst into flames, hitting the ground in a cloud of burning fuel and debris.

It was the only foray made into Israel by the Iraqi Air Force; an hour later, Israeli Mystères and Mirages hit Habbaniya and Hotel Three, destroying nine MiG-21s, five Hunters and a pair of Il-14 transports. The IAF's strike against the IrAF was far from devastating: the targets were distant, there was only a small number of strike aircraft available, and the Iraqis had dispersed their aircraft carefully. There is little doubt that the IrAF could have mounted a large-scale air offensive against Israel before the IAF had a chance to deal a harder blow; that it failed to do so was probably the result of Soviet pressure, the Russians no doubt realizing that the Israelis were bound to achieve complete air superiority in the long run and anxious to prevent the IrAF from suffering the same fate as its Egyptian counterpart. In any event, the Iraqi Air Force took no further part in the conflict and the Israelis left it alone.

A further important reason why Iraq was spared a crushing air strike on June 5th was that the Israeli Air Force, having destroyed the greater part of the Egyptian Air Force and the whole of the RJAF in the morning, was now preoccupied with Syria and with attacks on enemy radar stations. The Syrian air bases at Saigal, Dumeyr, Damascus, Marj Rhiyal and Tango Four were heavily attacked by Mirages and Mystères, which destroyed 32 MiG-21s, 23 MiG-17s and -15s, two Ilyushin 28s and three Mi-4 helicopters. Anti-aircraft fire was heavy and several Israeli aircraft were shot down including two Mystères, destroyed over Damascus. Another Mystère was jumped by a flight of MiG-17s, which shot it down as it was attacking targets on the ground; the MiGs were themselves shot down by Mirages a few minutes later.

The turn of the enemy radar stations came at noon on the 5th, when four Super Mystères attacked the Jordanian installation at Ajlun with rockets and reduced it to a heap of smoking rubble. Further strikes were made in the course of the afternoon against 23 Egyptian radar stations, 16 of them in Sinai; all of them were knocked out.

For 90 minutes prior to the strikes, the Israeli Air Force had carried out an intensive electronic countermeasures operation designed to throw the enemy radar into complete confusion. Among the methods used was one which proved just as effective in 1967 as it had done in 1944: three IAF Dakotas cruised backwards and forwards along the frontier of northern Sinai, dropping 'window'—strips of tinfoil cut to the exact wavelength of the Egyptian radar. The Israeli ECM caused trouble in other quarters, too: by noon on June 5th, the eastern Mediterranean area was the scene of intense radar activity as Americans, Russians and British strove to gather electronic intelligence. Thanks to the intensive Israeli jamming, it was only towards nightfall on Monday that a clear picture began to emerge.

Meanwhile, the Israeli ground offensive was beginning to get under way. At 08.15 hours, the first wave of Israeli armour had begun to roll forwards to the border of the Gaza Strip, kicking up a great cloud of yellow dust that drifted slowly across the countryside. Shells from the Egyptian artillery concentration in the Rafah defensive zone had begun to burst among the advancing tanks when for the first time, the IAF's little Fouga Magister strike/trainers went into action. The Israeli tanks' crews cheered as the Magisters swept low overhead, their wings laden with rockets. The aircraft were flown almost entirely by IAF reserve pilots, and they pounded the Egyptian gun batteries with great accuracy and without let-up for 15 minutes. The air attack was followed by an Israeli artillery barrage, which covered the advance of the tanks towards their first objectives. At 08.48 the first armoured company crossed the frontier, the Pattons moving in column because of the danger from mines. More air strikes were flown by the Magisters, this time against Egyptian artillery at Khan Yunis which was now also raining shells on the Israeli tanks.

At 10.00, after bitter fighting, the tanks broke through the enemy positions and raced across the narrow four-mile strip of land towards Khan Yunis. The first units charged through the town and sped on towards their next objectives, one defensive position to the north-west of Rafah and another on the Khan Yunis—Rafah highway. The Rafah army camp was bypassed; the infantry occupying it would be mopped up later. Fierce fighting developed around the strategic Rafah Junction, which controlled the approaches into Sinai, and the Israelis suffered considerable losses. The Magisters went into action again, once more with the Egyptian artillery as their target, but this time they failed to destroy most of the gun batteries. The gun crews simply took cover in deep bunkers when they heard the jets approaching, and emerged to take up their posts again when the Magisters had flown away.

Nevertheless, the Israeli armour succeeded in breaking through the northern wing of the junction at about 12.00, smashing through more defences at Sheikh Zuweid and driving on towards El Arish, while other armoured formations joined battle with the Egyptians on the junction's southern flank. By 16.00 the Israeli spearhead, consisting of a battalion of Centurion tanks, had reached the suburbs of El Arish after fighting its way through the strongly-fortified Jiradi Pass. 'Fighting' was perhaps not strictly accurate: the armoured dash through the Pass took the Egyptians completely by surprise, and they rallied only just in time to open fire on the tail-end of the Israeli column. When the remainder of the Israeli armour arrived, however, it was a different story: the Pass was effectively blocked and the tanks were pinned down by heavy fire. In spite of air attacks by Magisters and Ouragans, the Egyptians could not be dislodged; finally, at midnight, an armoured infantry battalion arrived and the troops stormed into the Egyptian positions, which they captured after four hours of savage fighting. Soon after the infantry's arrival, an armoured command group succeeded in punching its way through the pass and at 02.00 it linked up with the two battalions already at El Arish, where the Israelis were preparing to fight off an expected counter-attack by a large Egyptian armoured force in the vicinity.

A section of seven tanks was detailed to press on

immediately and capture El Arish airfield, possession
of which was vital to subsequent Israeli operations
in the peninsula. The force ran into opposition
in the shape of tanks and 57 mm anti-aircraft
guns, but this was overcome after a sharp battle
and the Israeli tanks captured the airfield at 04.00.
It was the soldiers' first glimpse of the devastation
wrought by the Israeli Air Force; the runways were
cratered and the airfield was littered with the
burnt-out wrecks of aircraft, but the buildings and
other installations had been left undamaged. Air
Force personnel who arrived at the base soon afterwards found large stocks of ammunition, including
80 mm air-to-ground rockets which were immediately reserved for use by the Magister squadrons. The Egyptians had apparently evacuated El
Arish in such a hurry that they had forgotten to
destroy their secret documents; searching through
the underground bunkers, Israeli Intelligence
officers later found the Egyptian Air Force's complete operational plan. Among other things, it
revealed why the EAF had been so slow to react to
the Israeli air offensive: the plans for a proposed
Egyptian strike on Israel had envisaged a delay of
at least three hours between successive strike waves
attacking their targets, and the Egyptian Air Staff
had made the fatal mistake of assuming that an
Israeli air offensive would follow the same lines,
allowing plenty of time for fighters to intercept a
second wave of strike aircraft and for Egyptian
bombers to get into the air. They had taken no
account of the infinitely higher efficiency of the
IAF's ground crews, which had reduced turn-round
time to a matter of minutes and enabled the IAF
to mount a non-stop offensive for nearly three hours.

At 06.00, the Israeli armour paused to regroup on the outskirts of El Arish. While the tanks were refuelled and rearmed, their crews—grimed with the dirt that stuck to their sweat-stained bodies, their faces blackened with smoke—snatched a few hasty mouthfuls of breakfast. At that moment, the alarm went up. Racing in from the west, heading straight for the tanks that were being refuelled, came two Sukkoi Su-7 fighter-bombers. The Israeli crews scattered, reaching for their weapons as the jets bored in relentlessly towards their target.

The Egyptian pilots never made it. Two Mirages curved down behind them, cannon twinkling. It was all over within seconds; one of the Su-7s, its wing torn off by the 30 mm shells, went into a fast upward roll before arcing over and crashing into the desert with a terrific thump. The second, burning fiercely, dropped its bombs harmlessly and turned away, losing height rapidly and finally hitting the ground, cartwheeling in a cloud of burning debris. The Mirages swept low over the tanks in a crackle of afterburners and were gone, leaving two columns of smoke rising into the morning air.

Meanwhile, as General Israel Tal's armoured division had been pushing on towards El Arish, a brigade of Centurion tanks had crossed the frontier and headed through the desert towards the vital crossroads of Bir Lahfan with the object of blocking the way to any Egyptian reinforcements coming up from Jebel Libni. The 24 tanks were in position at the crossroads by 18.45 hours on the 5th, and held on under intense but highly inaccurate Egyptian artillery fire. During the night a column of heavy Egyptian T-55s came up from the direction of Jebel Libni; the Egyptians had not expected to find any

opposition at the crossroads, and the four leading
T-55s were hit and set on fire. The remainder scattered among the surrounding sand-dunes and
returned the fire, throughout the night both sides
stayed where they were and hammered away at each
other, the Israelis being content to wait and fight a
holding action until dawn, when they could call up
air support.

During the night, one battalion of Centurions was
detached to take part in the assault on Abu Ageila,
which had begun in earnest on Monday evening,
but by the time the tanks arrived, intent on attacking the Egyptians from the rear, Abu Ageila had
already fallen. The main attack on the Egyptian
position, which was defended by a division, with
supporting artillery and about 90 tanks, had been
launched by an infantry brigade at 22.00 hours.
The infantry captured the first three lines of
defences, clearing the way for Israeli armour which
moved in through the minefield that surrounded
the perimeter and engaged the Egyptians inside
their own defences. The scales were finally tipped
in the Israelis' favour when, at 01.00 on Tuesday
morning, a squadron of Super Frelon and S-58 helicopters came clattering through the darkness and
landed a paratroop unit in the Egyptian rear. The
paratroops stormed the enemy artillery positions and
quickly silenced the guns. Nevertheless, it was not
until the following noon that the last shots were
fired in Abu Ageila, after a hectic tank battle.

It was at this point on the morning of Tuesday,
June 6th, that the Israeli Air Force, having achieved
in full its primary aim of neutralizing the enemy air
forces during the massive series of strikes the day
before, appeared over the Sinai battlefields in

strength to give its full support to the ground forces. At 06.00, four Mystères whistled down on the Egyptian armour at Bir Lahfan and attacked with rockets and cannon-fire, knocking out four tanks and setting a number of soft-skinned vehicles on fire. One Mystère was hit and went into a steep climb, trailing a thick streamer of smoke; the pilot ejected and was picked up safely by Israeli forces a few minutes later. Flights of Mystères and Ouragans kept up the air attack for four hours, until finally, at 10.00, the remnants of the Egyptian force began to stream back towards Jebel Libni, leaving the desert dotted with the hulks of burning tanks and vehicles.

Meanwhile, in the north, the Israeli assault on Gaza had begun. The town was heavily defended, and the Israeli forces found themselves engaged in savage fighting. Air strikes on the Egyptian positions made by Magisters and Ouragans followed an Israeli artillery barrage at noon; during one of these strikes, cannon shells and bomb splinters tore through the United Nations HQ, killing ten Indian UN observers. Israeli tanks broke into the town in the wake of the strikes, and a fierce battle flared up as the armour pushed on relentlessly towards Gaza's centre, which was seized by the Israelis during the afternoon. However, it was several more hours before the last pockets of resistance were mopped up. With the capture of Gaza, a large quantity of Egyptian war material fell into Israeli hands.

In the east, while the battle for Gaza continued, Israeli forces were engaged in bitter fighting against the Jordanians, with the capture of Jerusalem as the primary objective. By Tuesday afternoon the encirclement of the city had been completed following an armoured pincer movement, one arm of which seized Tel el-Pul on the Jerusalem-Ramallah road, the other by-passing Bethlehem to the south and cutting the strategic road to Jericho and Amman; while further north, Israeli paratroops were fighting hard for possession of the vital road junction at Jenin as a preliminary to advancing southwards on the town of Nablus. The Israelis used 'steamroller' tactics here: each new advance was preceded by a devastating air strike. The IAF made use of every available type of aircraft: Mirages, Mystères, Super Mystères, Vautours, Ouragans and Magisters. They were able to operate without fear of opposition in the air: the Royal Jordanian Air Force no longer existed. For 18 hours, the strike aircraft pounded the Jordanian positions that stretched along the north-south ridge overlooking the fertile plain of Israel with a non-stop rain of high explosive and napalm. At Jenin, several battalions of Jordanian Patton tanks were smashed by rocket-firing Magisters and Ouragans, opening the way for the advance on Nablus. At the same time, an Israeli thrust northwards from Ramallah drove up the heights road against stiff resistance, and before dawn on Wednesday they were in position for the final assault on the town. The defenders of Nablus, stunned by the fury of the air attacks and with no hope of relief, fought on desperately and it was not until Wednesday afternoon that the Israeli armour managed to break into the town. During the last phase of the assault, the Israeli jets swooped down to attack enemy positions only yards ahead of the advancing troops.

Jerusalem, defended by the hardy troops of the Arab Legion, proved a tough nut to crack. Here, because of the danger of destroying or damaging

buildings of priceless historical and religious value,
large-scale air strikes were out of the question;
instead, the Israelis relied on rocket, napalm and
machine-gun attacks by Magisters operating in twos
and threes, the pilots relying on their low speed and
manoeuvrability to ensure a high degree of accuracy
and faultless target identification. By attacking at a
relatively low speed the pilots exposed themselves
time and again to the worst of the intense small-
arms fire that was directed against them, and every
aircraft returned to base with some degree of battle
damage. While the Magisters saturated the Arab
Legion's defences on the Mount of Olives and elsewhere in the environs of Jerusalem, flights of Mystères and Mirages howled eastwards to attack enemy
armour and transport on the road to Amman, knocking out the Allenby Bridge across the Jordan and
so forestalling any possible enemy plans for a strong
counter-offensive. The strike aircraft roved at will
over Amman itself, lobbing their rockets into
public buildings and damaging the Royal Palace.

By nightfall on Wednesday the worst of the
fighting was over. The Israelis were mopping up the
last pockets of resistance in the eastern half of
Jerusalem; the nine Jordanian brigades which had
been deployed along the Israeli frontier at the outbreak of hostilities had been shattered, and the
Israeli forces raced on to take Jericho and Hebron.

Meanwhile, as the sun began to go down
through the dust and smoke of Tuesday's conflict,
General Israel Tal's armoured division had begun
the long chase across Sinai that would end only when
the tanks reached the Suez Canal. The plan was for
a two-pronged race across the Peninsula, one
column moving towards Kantara from El Arish and

the other from Bir Lahfan through Jebel Libni and Bir Gifgafa to Ismailia. At 19.30 the tanks moved off, the rattle of their tracks and the roar of diesels reverberating through the silence of the desert.

It was only now, at the end of the second day's fighting, that Israeli Air Force Intelligence was able to piece together a reasonably accurate picture of the havoc wrought on the Arab air forces during the preceding 48 hours. By midnight on June 6th, the Egyptian Air Force had lost 319 operational aircraft and helicopters, all but a handful of which had been destroyed on the ground. The total included all 30 Tu-16s of the Strategic Bombing Wing, knocked out in a matter of minutes during the strike on Cairo West; 27 Il-28s, most of which were destroyed during the Vautour strikes on Luxor and Ras Banas in the south; 12 brand-new Sukhoi Su-7 ground-attack jets, 15 of which had arrived in Egypt at the end of April and which at the outbreak of hostilities equipped a reserve squadron of the 12th Air Division; 95 MiG-21s; 25 MiG-19s; 85 MiG-15s and MiG-17s; 24 Ilyushin 14 and eight Antonov 12 transports; eight Mi-6, four Mi-1 and one Mi-4 helicopters. The total did not include trainer and communications aircraft, several of which had also been destroyed during the air attacks.

On the other fronts, the Syrians had lost two Il-28s, 32 MiG-21s, 23 MiG-15s and MiG-17s and three Mi-4 helicopters, while Iraq had lost one Tu-16, nine MiG-21s, five Hunters and five Il-14s. The unfortunate Royal Jordanian Air Force had been wiped out, losing 21 Hunters, five transport aircraft and three helicopters. The last

Arab air casualty was a Lebanese Hunter, which infiltrated Israeli air space late on Tuesday and which was promptly shot down by a Mirage patrol.

On the other side, Israeli Air Force losses at the end of the first two days' fighting stood at 26 aircraft: seven Mystère IVAs, five Ouragans, four Super Mystères, two Mirages, seven Magisters and one Vautour. None of the Israeli aircraft had been destroyed in air combat, although three Mystères had been 'bounced' by MiGs during strafing attacks on the enemy airfields; in all three cases it was unlikely that the Israeli pilots, intent on delivering their rockets or bombs accurately, ever saw the aircraft that shot them down. The great majority of the Israeli aircraft casualties had been caused by ground fire, which in many cases was heavy and extremely accurate and the task of the enemy anti-aircraft gunners was often made easier by the fact that most Israeli pilots made their runs at comparatively low speed to make sure of hitting their targets. On several occasions, Israeli aircraft made their firing-passes with flaps fully down and under-carriages lowered to increase the drag and cut down the speed still further.

No Israeli aircraft had fallen victim to Egyptian SA-2 Guideline surface-to-air missiles, although several were fired at the attackers in the Canal Zone. The exact sites of these missiles were known to Israeli Intelligence, and where they were particularly thick on the ground the strike aircraft were routed to avoid them. Otherwise, they were treated with something approaching contempt, the Israelis knowing the missiles' limitations against low-flying aircraft. The missile sites were not even considered to be primary targets; instead, they were listed among the targets to be attacked during later IAF missions over enemy territory, after the primaries had been destroyed. On the one or two occasions when surface-to-air missiles came anywhere near Israeli aircraft, the pilots simply manoeuvred out of the way smartly and the missiles vanished in the distance.

Five of the shot-down Israeli pilots managed to bale out over friendly territory or were picked up by rescue helicopters under the noses of the enemy; of the remaining 21 eight were killed, ten were taken prisoner and three were reported missing. Of the dead pilots, at least two, both of them hit over Syria, were known to have elected to stay in their aircraft and crash rather than bale out and fall into the hands of the Syrians. One pilot who did bale out over the Suez Canal tried to surrender to a group of Egyptian peasants; probably terror-stricken, they hacked him to death with knives and axes.

On the morning of Wednesday, June 7th, the Egyptian forces in Sinai were in full retreat, falling back towards the two positions where they might hope to block the Israeli advance: Bir Gifgafa and Mitla. The wake of their retreat was strewn with shattered vehicles as the Israeli Air Force pounded the Egyptian columns incessantly; there was no respite even during the hours of darkness, for the Vautours from Hatzerim struck at the Egyptians again and again, turning night into day with their flares and roving back and forth along the enemy columns. The armour and vehicles were silhouetted sharply against the desert, and it was impossible to miss; the desert roads quickly became choked with burnt-out carcases. On several occasions, a single bomb dropped in the right place was enough to cause a big pile-up.

Although the air strikes achieved their primary object of throwing the retreating Egyptians into confusion, they also hindered, to a certain extent, the headlong dash of the three Israeli divisions across Sinai several times, the Israeli armour had to make detours to by-pass roads that were clogged with shattered vehicles. For the Israeli commanders, these delays were extremely frustrating, since the whole idea behind the race across the desert was to reach Mitla and Bir Gifgafa before the Egyptians in order to cut off their escape route. To complicate matters still further, the divisions had to fight their way through the Egyptian rearguard and by this time, the Israeli tanks were running desperately short of fuel and ammunition. The Israelis were having to rely to a great extent on captured stocks for replenishment; the Noratlases and Dakotas of Air Transport Command did what they could throughout Wednesday, but the Israeli armour was never in one place long enough for a large supply drop to be organized.

At 18.30 on Wednesday, a neck-and-neck race for the Mitla Pass developed between the first Egyptian columns and an Israeli armoured battalion. The battalion, whose strength had been reduced from 28 Centurions to nine by shortage of fuel and breakdowns in the course of the day, got there first and set up a roadblock, but the first enemy columns managed to evade it and enter the Pass. As the disorganized mass of armour and trucks crowded into the narrow defiles, there was a whine of jet engines and two Vautours swept overhead to plant their bombs squarely on the leading Egyptian echelons.

An ammunition truck went up with a tremendous roar and the Israelis were treated to a lethal firework display as exploding shells and bullets shot in all directions, ripping through other vehicles and sparking off a chain reaction.

During the night, more Israeli units arrived at the entrance to the Pass and reinforced the blockade. A few Egyptian tanks managed to break through, but several were knocked out in the attempt. By dawn, however, the blockading Israeli forces were in a desperate situation; their fuel and ammunition were almost exhausted, and now they found themselves fighting a bitter engagement with a strong Egyptian force—including 28 T-54 tanks—that came up from the east. Fortunately, the situation was not allowed to get out of hand: three flights of Ouragans arrived overhead and swooped down to hammer the Egyptian column with napalm, rockets and cannon. Thirty minutes later, the danger was over: the enemy column had been torn apart and the survivors had scattered for cover in the surrounding desert.

Meanwhile, General Tal's armoured division had been driving rapidly towards Ismailia, advancing on Bir Gifgafa as its first objective. At 15.30 on Wednesday, as it was approaching Bir Gifgafa airfield, it ran into heavy fire from a group of T-55 tanks which was dug in on a hill near a radar station. As the two sides exchanged shells, two flights of MiG-17s whistled up and made several high-speed strafing runs over the Israeli column. The enemy jets were engaged by Mystères, but on this occasion the Israelis got the worst of the encounter: one Mystère was shot down and the MiGs got away. As a demonstration of Egyptian air power it was hardly impressive, but it showed that the Egyptian Air Force, although no longer in a position to offer any challenge to Israeli air superiority, was still capable of hitting back. The fighter-bombers that were sent into action now were mostly aircraft that had been dispersed on airfields deep inside Egypt at the time of the Israeli air strike of June 5th, but a few were Algerian machines with their national markings painted out. They had been sent to Egypt following a desperate request from Nasser for aircraft to cover the retreat of the Egyptian forces from Sinai; although the majority were flown by Egyptians, Algerian pilots also took part in the attacks on several occasions. More MiGs attacked the Israeli column that was advancing along the Ismailia axis before nightfall, but this time the Mirages were waiting for them and three of the enemy aircraft were destroyed.

Israeli air attacks kept the Egyptian forces in the vicinity of Bir Gifgafa pinned down until it grew dark, but in the course of the night the enemy counter-attacked furiously with heavy T-55 tanks, which overran an Israeli battalion equipped with light French AMX-13s and gave it a severe mauling. Reinforcements were rushed to the spot and charged into the mêlée, and two Vautours arrived in response to an urgent request for air support. They circled the area watchfully, dropping flares and diving down from time to time to attack targets that looked promising. Finally, a company of Centurions came thundering up, tipping the scales in the Israelis' favour, and the enemy withdrew. At dawn, reconnaissance patrols reported that organized enemy resistance appeared to have ceased, and General Tal decided to assign part of his force to push on towards the Canal with all possible speed.

The last hours of the Sinai campaign were characterized by some of the most vicious fighting of the
six-day war as the Egyptians hurled their last
reserves across the Canal in a desperate attempt to
stem the Israeli advance for long enough to allow
their shattered armies to escape. The Egyptian tank
crews fought back with suicidal courage; of the 950
Egyptian tanks committed to the fighting in Sinai,
only about 100 fell into Israeli hands intact; the
remainder fought on until they were destroyed by
Israeli armour and anti-tank guns or knocked out
by the air strikes. The Egyptian Air Force, too, went
on hitting the advancing Israeli columns with
sporadic hit-and-run attacks; these were pressed
home with fierce determination, and some of them
inflicted considerable damage. But the courage of
the Egyptian pilots was not enough; the prowling
Mirages and Mystères were everywhere, and the
fighter-bombers suffered appalling losses.

Nevertheless, the Israeli Air Force did not have
everything its own way during the closing phases of
the campaign. Shortly before dusk on Thursday,
following the capture of Kantara by Israeli forces,
the IAF was called in to attack several formations
of Egyptian tanks that were fighting back savagely
on the Ismailia axis. The jets inflicted heavy casualties, but they had to attack through a storm of
highly-accurate anti-aircraft fire which destroyed
three Ouragans and two Mystères.

It was the heaviest loss suffered by the IAF during
a single attack in the entire Sinai Campaign. Immediately after the air strike, the surviving Egyptian
tanks withdrew across the Canal and the Israeli
divisions moved forward to take up their positions
along the narrow waterway. From the south, another

Israeli column advanced along the Gulf of Suez towards the Canal Zone from Sharm El Sheikh, which had been captured by an airborne force from Eilat the previous day. The initial assault had been carried out by paratroops flown in by the IAF's Super Frelon helicopters; they had arrived to find the small airfield at Sharm El Sheikh practically deserted. So were the big Egyptian gun batteries that dominated the Straits of Tiran; their crews had abandoned them and scattered northwards into the desert, terrified of being cut off by the Israeli thrust. It was better than the Israelis had dared to hope; the airstrip at Sharm El Sheikh was captured almost without a shot being fired, and for the next few hours troops and equipment arrived aboard a stream of IAF Noratlases, Dakotas and Strato-cruisers.

By 03.00 on Friday morning, the fighting in Sinai was over. Along the Suez Canal, the battle-weary men of three Israeli divisions flung themselves down beside their vehicles and slept like the dead. For most of them, it was the first real rest they had enjoyed for a week. Ahead of them, across the Canal, lay a beaten Egypt; behind them, scattered through-out the Peninsula, lay the burnt-out remnants of ten Egyptian brigades and scattered groups of Egyptian soldiers, tormented by thirst as they struggled westward through the pitiless desert. Most of them would never reach the Canal; a few, the lucky ones, would be picked up by Israeli patrols, but most were condemned to die, abandoned and forgotten, their bodies bleaching in the sun.

In the early hours of Thursday, June 8th, the Egyptian delegate to the United Nations had announced that his country was willing to accept a ceasefire. Even as he spoke, Israeli forces were moving up to positions north of the Sea of Galilee in readiness to strike at the third Arab belligerent: Syria. So far, Syria's part in the war had been small; since the beginning of the week, apart from a few hastily-organized commando raids across the border, she had confined herself to shelling the Israeli kibbutzim in the north from the artillery positions on the Golan Heights. After the strike on her airfields, Syria had been in no position to challenge Israel's air superiority: the few MiGs that had made short forays into Israeli air space had been dealt with speedily by Mirages and Super Mystères.

For four days, the Israeli High Command had relied on the armed kibbutzim in the north to fend off any threat from Syria; and this they succeeded in doing, in spite of heavy and highly accurate shelling. On Friday morning, however, the Israeli Army moved in at last with the object of wiping out the Syrian artillery on the Heights, which were stiff with 135 mm long-range cannon and 120 mm mortars.

The assault began at 10.00, and almost immediately the Israeli armour came under heavy mortar fire from the Heights. The plan envisaged a breakthrough at Givat Ha'em by two Israeli task forces, which were to capture the strong Syrian defensive positions at Zaoura. It would be no easy task; the Syrians had turned the whole of the Golan Heights into a vast stronghold, sewn with minefields, gun emplacements, underground bunkers and tank traps. The Israelis would have to batter their way through a narrow sector between Tel Azaziyat and Kfar Szold, where the slope was not too steep and relatively free from boulders and other obstacles.

If they could smash the Syrian defences here, they would have access to the vital road that linked together all the major Syrian positions along the Heights.

The advance across the frontier at Givat Ha'em was preceded by a massive air bombardment on the Syrian positions, at the request of General David Elazar, the GOC Northern Command. From 09.45 onwards, waves of Mirages, Mystères, Super Mystères, Ouragans and Vautours hit the Syrian defences time after time; after 30 minutes, the murderous artillery fire had already slackened off appreciably. The first Syrian position, at Gur el Askar, had been shattered by rockets and napalm and was quickly overrun by the Israeli armour. There was some fighting around the second defensive position, Na'amush, but this was also overwhelmed within minutes and the Israeli tanks charged on, by-passing Ukda to confront the next Syrian position at Sir Adib.

It was then that the Syrians hit the Israelis with everything they had. A dozen anti-tank guns roared at once and several of the leading Shermans slewed to a stop, burning fiercely. The anti-tank guns were knocked out by the other Shermans, assisted by rocket-firing Magisters, but then mortar shells began to pour down on the attackers. A pitched battle developed, and in the dust and smoke an Israeli armoured battalion managed to crash through the Syrian positions and drive on towards the next objective, Kala'a. This was a formidable defensive position, consisting of a long, strongly-fortified ridge on which the air attacks had made little impression. Several fuel and ammunition dumps had been hit, however, and a pall of smoke from these and from burning vegetation hung over the whole area.

The road from Sir Adib to Kala'a was just over a mile long, and every foot of it was bitterly contested. It took the Israelis five hours to cover that mile, and by the time they reached the outer perimeter of Kala'a fortress their forces were almost decimated. Twenty-one tanks set out to make the final dash into the Syrian defences; of these, eleven took up positions from which they could shell Kana'a while the remaining ten fought their way slowly forward through a withering barrage. By the time they reached the outskirts of Kana'a village, the Syrian headquarters, only three were left and there were reports that Syrian armour was moving up to engage them. Desperately, the tank commander radioed for air support, only to be told that no aircraft were immediately available and that he would have to wait. But dusk was approaching fast, and after nightfall air support would be useless. Again and again, the tank commander repeated his request and each time he received the same answer. A savage battle was going on around Zaoura, and all available IAF close-support aircraft were fully committed.

At 18.00, three Syrian heavy tanks lumbered into view at the far end of Kala'a village. Grimly, certain that they would be sighted at any moment and knowing that the Syrian tanks were just the vanguard of even larger reinforcements, the Israelis prepared to fight. At that moment, three Magisters appeared over the village and circled the area several times; the Israeli tank crews realized that the pilots were unable to distinguish between friend and foe in the failing light, and signalled their own position with a coloured smoke marker. By this time, the Syrian tanks, their crews having also spotted the aircraft, were pulling out of Kala'a at top speed, but they were too late. Two of them were knocked out within seconds by 80 mm air to ground rockets, and the third was destroyed by a well-aimed shell from one of the Shermans. A few minutes later more Israeli armour arrived on the scene from Sir Adib, and several attacks by small groups of Syrian tanks were successfully beaten off during the hours of darkness that followed.

During the night, while an Israeli infantry brigade fought a savage battle for the lower Syrian defensive positions on the Golan Heights, Israeli tanks and supply columns continued to pour into the newly-captured territory through the corridors blasted open by the armoured spearhead. Several artillery positions that still fought on were subdued before dawn by paratroops, airlifted behind the Heights by Super Frelon and S-58 helicopters. After this, the helicopters began a non-stop airlift of heavy equipment, enabling the Israelis to consolidate their positions rapidly.

On Friday morning, grimy with the dust and smoke of the previous day's conflict, the Israeli troops and tanks spread out across the Syrian plateau behind the Golan Heights. A few hours later, a cease-fire with Syria came into effect. The six-day war was finally over.

On every front, victory had been attributable to three main factors. The complete dedication of the Israeli fighting men to the task in hand; their ability to adapt themselves to new situations as they arose; and, above all, overwhelming air superiority. The Israeli pilots had gone into action with the knowledge that the survival of their nation hung on their absolute precision, and they had not been in a position to afford the luxury of mistakes. The fact that not once during the six days of fighting had Israeli aircraft attacked their own troops by mistake, even though at times they had been called upon to strafe enemy positions only yards ahead of their own forces, was evidence enough of the pilots' extremely high standard of training and amazing accuracy. This was all the more noteworthy since many of the Israeli pilots were reservists, especially those who had flown the little ground-attack Magisters. Some of the reservist pilots were employees of El Al and Arkia, Israel's two airlines; five of them were killed in action.

A large part of these two airlines' effort was devoted to logistics missions in support of the armed forces during the six-day war. Arkia Inland Airlines was the most heavily committed of the two, with all its aircraft mobilized for military purposes. The airline's three Handley-Page Heralds were employed almost continuously in flying supplies to and evacuating wounded from the forward areas once Israeli air superiority had been established, a task to which they were admirably suited because of their ability to operate from unprepared airstrips. An Arkia Herald, in fact, was one of the first Israeli aircraft to land on the newly-captured El Arish airfield.

Only once, in fact, had the Israeli Air Force made a serious error—and even then, it had not prejudiced the safety of its own forces. It happened on Thursday, June 8th, when the pilot of a Magister who was carrying out a search for Egyptian patrol craft off the coast of Israel located an unidentified vessel following a southerly heading some five nautical miles off Ashod. The vessel made no attempt to identify itself, and three Israeli MTBs immediately put to sea from Haifa to intercept it. They were preceded by a pair of Mirage IIIs, which took off from Ramat David and made several low passes over the ship. The pilots reported that the vessel did not appear to carry any armament, but was festooned with an array of aerials and other electronic equipment. The ship still made no effort to identify itself and the Mirage pilots requested authority to attack it, presuming it to be hostile and possibly engaged in an electronic countermeasures mission with the object of jamming the Israeli radar.

A few moments later the Mirages attacked, hitting the hull of the ship with two air-to-ground rockets. Black smoke billowed into the air but the strange vessel held its course. Then the MTB arrived on the scene and opened fire in turn; to the amazement of the Israeli sailors, most of their light shells simply bounced off the ship's sides. There had been no sign of life on deck—but now a solitary sailor appeared and began to fire at the Israeli craft with a heavy machine-gun. The Israelis returned the fire; the sailor was hit and dropped out of sight.

After a pause, the MTBs attacked with torpedoes, scoring one hit. The vessel developed a slight list, but showed no sign of stopping. Two IAF S-55 rescue helicopters arrived overhead and the crews signalled that they were prepared to evacuate any casualties, but there was no response; instead, the ship turned away and limped slowly westwards over the horizon. The following day she reached Malta, and the full story came out at last. She was the USS *Liberty*, a 'spy ship' attached to the American Sixth Fleet, and she had paid a heavy price for coming too close to the shores of Israel; 34 of her crew were dead, with 75 more injured. It was the only time during the six-day war that Israeli aircraft attacked a vessel at sea.

When the ceasefire finally came into effect on Saturday, June 10th, the Israeli Air Force had lost a total of 40 aircraft of all types in action, and a further ten had been written off through other causes. Some of them had arrived back at base so badly damaged that they had crashed on landing, or were of no further use other than for scrap and spare parts.

The Israeli losses, both on land and in the air, lent a serious undertone to the atmosphere of quiet jubilation that gripped the whole of Israel after Saturday's final victory. For no one in Israel and least of all the Chiefs of Staff were under any illusion that the six days of conflict through which the nation had plunged headlong marked the end of the war. It was only the beginning.

CHAPTER TWELVE

War of Attrition

December 29th, 1968. 21.30 hours.

At Beirut International Airport, everything appeared normal. In front of the hangars the crash crews sat in their fire-tenders, engines ticking over, as a Pan American World Airways Boeing 707 taxied out towards the end of the main runway.

Suddenly, a dark shape came slanting down out of the darkness with a clatter of rotors and hung poised for a moment before settling on to the apron a few yards ahead of the fire tenders. In one of them, a fire officer pressed the switch of his microphone and called up the control tower, irately pointing out that the helicopter which had just landed was blocking his crash exit. He was about to request its immediate removal when the words froze in his throat; two more helicopters touched down some distance away, and this time the blue star of Israel showed up clearly on their camouflaged sides. Seconds later, steel-helmeted troops poured from the helicopters' bellies and fanned out among the civil airliners parked in front of the airport buildings.

On the Pan Am Boeing 707 that was about to take off, startled passengers peered out into the darkness as an explosion cracked out across the airfield. They were in time to see an enormous mushroom of burning fuel billowing upwards near the hangars. The hostesses did their best to reassure them, telling them that a film company was using the airport to shoot some action scenes. Then another explosion rocked the airport, and this time the passengers saw a jet airliner collapse in a cloud of burning wreckage. This was no film scene.

The eight Super Frelon helicopters of the Israeli Air Force's Helicopter Wing, each carrying 20 para-troops, had taken the Lebanese defences completely by surprise. Forty-five minutes later they re-embarked and the helicopters took off, leaving 13 aircraft burning on the airport behind them. The tally of airliners totally destroyed comprised three Comet 4Cs, two Caravelles, a Viscount, a VC-10 and a Boeing 707 of Middle East Airlines, together with two Convair 990s, a DC-4, a DC-6 and a DC-7 of Lebanese International. The lightning Israeli attack had been made as a reprisal following an Arab guerrilla raid on an El Al Boeing 707 airliner at Athens on December 26th. There was also another reason: the Israelis had taken the opportunity to demonstrate that their defence forces were still capable of carrying out an offensive operation with impunity deep inside enemy territory. The raid had been intended as a warning to the belligerent Arab nations, whose confidence had been growing steadily over the past year, as shipments of new Soviet war materials continued to make good the crippling losses suffered during the Six-Day War, that any acts of aggression would be met by swift and devastating retaliation.

Following the United Nations ceasefire that ended the Six-Day War, there had been little delay in the renewed supply of Soviet combat aircraft to the Egyptian Air Force. By the end of June 1967, some 200 machines, mostly MiG-21s and Su-7s, had been delivered, many of them via Algeria, and the new equipment was quickly pressed into service. This time, the Egyptians were not hampered by a shortage of trained personnel; since the majority of aircraft hit on the ground during the initial Israeli air strikes of June 5th had been unmanned, Egyptian aircrew losses had been small. By the beginning of June 1968, the Egyptian Air Force's inventory included 115 MiG-21s, 50 Su-7s and 15 Tu-16s; although the total strength was 20% lower than it had been immediately before the outbreak of the Six-Day War, this was offset to a great extent by the fact that much of the older equipment, MiG-15s and MiG-17s, was being progressively phased out in favour of more modern combat types.

Israel, meanwhile, was faced with a serious problem. The loss of some 20% of the Israeli Air Force's first-line strength suffered during the Six-Day War had to be made good with the utmost urgency, but there was no prospect of the United Nations' arms embargo being lifted in the foreseeable future. Immediately before the outbreak of hostilities, Israel had opened negotiations with the United States Government for the supply of a number of LTV A-7D Corsair II low-level attack aircraft, and had successfully concluded a deal for the supply of 48 refurbished ex-US Navy A-4H Skyhawks; it was some consolation when the US State Department approved the delivery of the latter aircraft, to begin in September 1967. Nevertheless, most of the IAF's hopes of maintaining air superiority in the skies of the Middle East had been pinned on the acquisition of 50 Mirage V strike aircraft; a firm order for these had been placed with Avions Marcel Dassault in December 1966, with deliveries scheduled to begin about a year later, but, despite the fact that the aircraft were complete and that most of them had already been paid for in hard American dollars, the sale was effectively blocked by the French Government after the Middle East conflict.

Then, in the spring of 1968, came the news that Iraq had completed negotiations with the French Government for the supply of 52 Mirages comprising 32 Mirage Vs, 16 Mirage IIIEs, two IIIDs and two IIIRs, together with missile armament, which consisted of 100 Matra air-to-air and 70 Nord AS 30 air-to-surface rockets. These were to equip three Iraqi Air Force squadrons, whose personnel were to be trained in France. The Israeli response was prompt; they made final payment on the Mirages still being held in France, and threatened to sue the French Government for substantial compensation through the International Court. The Israeli attitude hardened still further when the French proved that their interpretation of the Middle East arms embargo was decidedly one-sided by supplying ten Mirage IIIEL strike aircraft and two Mirage IIIBL trainers to the Lebanon, and affording facilities for the training of Lebanese pilots at Istres, near Marseille. In an attempt to placate the Israelis the French Government agreed to supply the IAF with spare parts for the Mirage IIICJs already in service, and also to provide an additional seven Sud-Aviation Super Frelon helicopters.

The main purpose of the Israeli Government now was to convince the US State Department that the French intended to maintain their ban on the supply of arms to the Israeli Defence Forces. Early in 1968, Levi Eshkol, the Israeli Prime Minister, visited the United States with a request for an additional 20 A-4H Skyhawks to supplement the 48 already being delivered, and for 50 F-4E Phantoms. The sale of the latter aircraft to Israel depended almost entirely on the French Government's continued refusal to supply the Mirage Vs, and negotiations were made more difficult by the US Government's insistence on Israel maintaining pressure on the French for delivery of the machines. The Israeli Defence Ministry, however, flatly rejected a French offer to refund the 40 million dollars which had been paid for the Mirages, and instead deposited 200 million dollars in the USA in anticipation that the sale of the 50 Phantoms would be allowed to go ahead.

In September 1968, while the sale of the Phantoms still hung in the balance, the IAF's commander, General Mordecai Hod, announced at a press conference that the combat strength of the Egyptian Air Force had climbed back to its pre-war level, and that older types of combat aircraft had now almost entirely been replaced by MiG-21s and Su-7s. In addition, about 2,000 of the 4,000 Soviet 'advisors' in Egypt were attached to the EAF, and radio transmissions picked up from Tu-16 bombers overflying the US Sixth Fleet in the Mediterranean suggested that some of these aircraft were being flown by Russian crews.

There was no doubt that this information helped to tip the scales in Israel's favour in the American negotiations. Late in September, the House of Representatives in Washington tabled a formal request before the President of the United States, asking his approval for the sale to Israel of Phantoms 'in such numbers as shall be adequate to provide a deterrent force capable of preventing future Arab aggression'. It was some weeks, however, before approval was finally given; in fact, it was not until December 27th, that the State Department indicated that agreement had been reached, and that delivery of the Phantoms was to start in 1969. Meanwhile, the Israeli Air Force had not been wasting time; early in 1969, 120 Israeli pilots arrived at George Air Force Base in California to undergo a Phantom conversion course.

The French reaction to the Israeli-American deal was curious. In spite of the French Government's attitude, Marcel Dassault were apparently still optimistic that the embargo on the supply of their Mirages to Israel would soon be lifted; the head of the firm indicated that the aircraft were almost certain to be delivered if Israel was seriously threatened. But it was wishful thinking; soon afterwards, the French extended their embargo to include spare parts, though not before spares and equipment sufficient to build 25 Magisters had been supplied during 1968 to Israel Aircraft Industries, which in any case was by this time capable of building its own components for all types of French aircraft in service with the Israeli Air Force.

Then, in July, the French Government changed its policy yet again. On July 10th, President Pompidou announced that the total ban on the shipment of war materials to Israel would be replaced by a 'selective embargo' which would allow supplies of aircraft spares to be resumed. The partial lifting of the embargo, however, was not extended to the 50 Mirage Vs.

By this time, although the Israeli Government still held out a faint hope that the Mirages would eventually be delivered, the emphasis was definitely on the purchase of American rather than French equipment. In addition to the 50 Phantoms, the initial batch of 48 Skyhawks had been supplemented by a follow-on order for 25 more; the majority of the aircraft, 70 A-4Hs and three TA-4H trainers, would be in service with the IAF's attack squadrons by April 1970. Meanwhile, during the first week in September, the first eight Phantoms arrived; they were followed by another four before the end of the month. On September 25th, while deliveries continued at an accelerated rate, the Israeli Premier, Mrs Golda Meir, laid a formal request before representatives of the US Government during discussions in Washington for the supply of an additional 24 F-4Es, together with 80 more A-4H Skyhawks and 135 Sikorsky helicopters of various types. On March 23rd, 1970, however, Israel's hopes of acquiring more American combat aircraft immediately were dashed when the US State Department announced that it would not commit itself to the sale of additional Phantoms and Skyhawks to Israel at this stage. At the same time, the US Government announced that it would sanction further sales if there was an appreciable deterioration in the Middle East situation, resulting in Israel's position becoming dangerous; for the time being, the general American feeling was that the Israeli Air Force possessed sufficient combat aircraft to meet its immediate needs.

There was some justification for this point of view, as the events of the past two and a half years had shown. Between the ceasefire of June 1967 and the end of 1969, Israeli jets flew 2,700 combat missions in the Canal Zone alone and destroyed 62 Arab aircraft on all fronts for an admitted loss of eight of their own machines. Nevertheless, there was no doubt that the general quality of the Egyptian pilots was improving steadily. This was demonstrated in September 1968, when an Egyptian pilot, Captain Husain Izzat, was caught by four Israeli Mirages during a low-level reconnaissance mission over the east bank of the Suez Canal. The Mirages fired air-to-air missiles at Izzat's Su-7, but these, presumably confused by heat emissions from the ground, all missed. The Israeli aircraft then attacked with their cannon, but after several minutes of skilful flying Izzat managed to escape and return safely to base with his photographs. For this exploit he received the Medal of Excellence, the highest Egyptian military award.

The Egyptian Air Force also emerged with a marginal victory from an air battle that took place west of El Kantara on November 3rd, 1968, when four Mirages penetrated Egyptian airspace and were intercepted by an equal number of MiG-21s. After a fierce dog-fight that lasted five minutes, during which one aircraft on either side was hit and damaged, the Mirages turned and headed for home. The damaged aircraft was unable to keep station with the others; it lost height steadily and was bracketed by heavy anti-aircraft fire, finally exploding in mid-air half a mile north of Kantara.

The Egyptian air defence system, however, was unable to cope with what was fast becoming the most prickly thorn in Nasser's flesh: the growing number of raids on selected Egyptian targets by helicopter-borne Israeli commandos. The most daring took place on the night of October 31st/November 1st, when an Israeli commando force struck deep into Egypt as a reprisal for an Egyptian artillery barrage which had killed 14 Israeli soldiers on the Suez Canal four days earlier.

On the morning of Thursday, October 31st, a small Israeli naval vessel put out from the port of Eilat on what to all intents and purposes was a routine patrol along the Gulf of Akaba. But this was no ordinary mission: nestling under a camouflage net on a platform built on the vessel's afterdeck was a Super Frelon of the IAF's Helicopter Wing, its rotor blades removed.

The vessel moved slowly down the gulf and into the Red Sea, tracked by Egyptian radar installations near Hurghada. The crew maintained a sharp lookout for Egyptian aircraft or the first signs of an artillery barrage, but nothing happened; the vessel went on its way unmolested. Shortly before dusk, the vessel rendezvoused with two others from Sharm el Sheikh at the southern tip of the Sinai Peninsula; these also carried Super Frelons. As soon as it was dark, Air Force technical personnel fitted the rotor blades to the helicopters and 18 paratroops filed aboard, six men to an aircraft. Thirty minutes later the three Super Frelons took off and headed southwestwards towards the Egyptian coastline, skimming low over the water to escape radar detection. Fifty miles inland, they touched down on the rock-strewn desert; as their rotors swung to a stop, their rear cargo doors opened and three jeeps emerged, each carrying four paratroops. While the other six paras stayed behind to guard the helicopters, the jeeps vanished into the desert darkness, heading for the Nile and their objective: the Mag Hammadi power station and the Quena Dam, 300 miles south of Cairo and only 140 miles north of the Aswan High Dam.

Both objectives were manned by a skeleton staff

of night-watchmen, who were quickly overpowered by the Israelis. Their task was made easier by the fact that they all spoke Arabic and wore Egyptian-style camouflage suits; the watchmen had no idea that they were Israelis until it was too late. The para-troops planted their charges and raced back to the waiting helicopters, which took off and reached the vessels in the Red Sea without incident. The ex-plosives caused substantial damage to the Egyptian installations and destroyed the main electricity cable leading from Aswan to Cairo.

In the spring of 1969, the Egyptian air defence system underwent a substantial re-organization and the EAF's policy began a rapid changeover from defence to attack. This was made possible by the return to Egypt of 250 pilots who had been under-going combat training in the Soviet Union for 18 months, and new dispersal plans which eliminated the fear that the bulk of the Egyptian Air Force might yet again be wiped out by an Israeli pre-emptive strike. By June 1969, the EAF had 30 modern air bases at its disposal throughout the country, many of them built with Soviet aid since the Six-Day War; in addition, a number of EAF combat units, including one squadron of Tu-16 bombers, were now dispersed on bases in Algeria.

The Egyptian Air Force's new-found aggressive spirit manifested itself during July 1969, when Egyptian and Israeli fighters were involved in the most intense period of air combat since the June War. On July 2nd, two squadrons of Egyptian and Israeli jets, MiG-21s and Mirages, tangled over the Gulf of Suez and a whirling dogfight developed in which four MiGs were shot down. Five days later, two patrolling Mirages attacked a flight of four

MiG-21s over the Sinai Peninsula near Sharm el Sheikh and destroyed two before the startled Egyptian pilots knew what had hit them; the remaining two broke away and headed for safety across the Gulf of Suez. The following day, it was Syria's turn: a squadron of seven MiG-21s made a short foray into Israeli airspace near Kuneitra and were immediately engaged by eight Mirages, which shot down all seven for the loss of one of their own number. This tremendous one-sided victory gave the Israelis new heart; it showed that although the Arab pilots were entering the lists with renewed vigour, they were still no match for Israeli teamwork and skill.

The Israeli Air Force's strike squadrons, meanwhile, had found their new Skyhawks to be a versatile and potent weapon. As the Americans had already discovered over North Vietnam, the aircraft was capable of absorbing tremendous punishment and still returning safely to base. During strikes against guerrilla targets in Jordan during 1968 and 1969, many Skyhawks had returned home with battle damage that would have knocked most other aircraft out of the sky. Not all the Skyhawks were as fortunate, however: one was shot down during a four-hour raid against guerrilla positions and concentrations of Jordanian and Iraqi artillery on December 21st, 1969. One Super Mystère was also hit and badly damaged in this attack, which was yet another reprisal for alleged frontier violations and terrorist activities.

The IAF's strike capability was dramatically increased in September 1969, with the arrival of the first Phantoms from the United States. These were immediately sent into action alongside the Skyhawks, knocking out troublesome Egyptian artillery positions on the west bank of the Suez Canal. The strike squadrons then set about the systematic destruction of Egyptian missile and radar sites in the Canal Zone, concentrating on installations spread along an 18-mile-wide defence perimeter between the Canal and Cairo, and on strategic roads in the same area. In two months, the strike aircraft destroyed twelve SA-2 surface-to-air missile sites and some 20 early-warning radar stations.

Early in December, Israeli Intelligence learned that several new Soviet low-level radar installations had arrived in Egypt, and that three were already in operation along the Gulf of Suez as part of the air defence network covering the approaches to the Aswan Dam. With their ability to detect aircraft coming in at low level, these radar stations presented a serious threat to future IAF operations over Egyptian territory—just how serious the Israelis were desperately anxious to find out. It was a job for the paratroops and the Super Frelons of the Helicopter Wing again. In a daring raid across the Gulf of Suez on December 26th, 1969, the Israeli commandos captured a complete low-level radar installation at Ras Ghareb and fought off Egyptian reinforcements while technical specialists dismantled it and loaded it on board the helicopters. The capture of the radar was a severe blow for the Soviet Union as well as for Egypt; very little information on installations of this type had so far been available to the West, and the data gathered by the IAF's experts on their examination of the radar, together with observations on possible methods of jamming it, would now almost certainly be made available to the Americans.

On Thursday, January 22nd, 1970, the Israelis captured another Egyptian radar station; a British-built one this time, which was seized when the Israelis invaded by air and sea the island of Shadwan at the entrance to the Gulf of Suez. The Israelis withdrew after stripping the island of its equipment.

Three days later, an IAF reconnaissance aircraft located a 120-ton Egyptian Navy auxiliary vessel standing off the island, having presumably ferried a new garrison from the mainland to replace the 27 Egyptian soldiers killed and 62 captured during the Israeli raid. Shortly afterwards, the craft was attacked and put out of action by rocket-firing Mystères.

Meanwhile, on January 7th, IAF Phantoms, profiting from the serious disruption of the Egyptian missile and radar defences, attacked targets in the vicinity of Cairo for the first time since the Six-Day War. It was the beginning of an intensive period of air strikes against military targets in the heart of Egypt, using both Phantoms and Mirages. On February 3rd, six Mirages and two Phantoms attacked the big Egyptian Air Force supply base at El Khanka, 13 miles north-west of Cairo, with rockets, napalm and delayed-action bombs, causing severe damage. The bomb-load of one of the Phantoms, however, became hung-up for a fraction of a second because of an electrical fault, and went down on a metal works at Abu Zabal, about a mile north of the target area. The majority of the bombs, which included napalm as well as high explosive, hit a maintenance shop and a power-generating plant, killing 70 civilians and injuring 98 others. Delayed-action bombs were still exploding several hours after the raid.

In just over a month, the Phantoms and Mirages made a total of nine deep-penetration raids into

Egypt, attacking military depots near Cairo; meanwhile, the Skyhawks and Mystères kept up a sustained air offensive against Egyptian artillery positions and missile sites in the vicinity of the Suez Canal. The IAF's orders were clear; the Egyptians, whose commandos were carrying out an increasing number of attacks on the Israeli-held east bank of the Canal, were to be allowed no respite that would enable them to move large quantities of war material up to the Canal Zone in preparation for a large-scale raid into Sinai.

The IAF succeeded in maintaining the pressure, but only at a price; in the month of intense air strikes, from January 7th until February 7th, 1970, the Israelis lost nine aircraft—bringing their total air losses since the June War to 17. The tally included two Mirages, two Skyhawks, three Mystères, a Super Mystère and a Phantom. On the credit side, the Israeli pilots had added four more Egyptian MiG-21s to their score; two of them were shot down by Mirages during a raid on the Helwan industrial complex near Cairo on February 8th. The Syrian Air Force, too, had taken some punishment in January, when three air battles took place over Syrian territory. During one of these, on January 8th, three MiG-21s were destroyed for no loss. This brought the total of Syrian aircraft shot down since the Six-Day War to 17. This figure does not include two MiG-17s which landed at the Israeli satellite airfield of Bezet near the Lebanese border in August 1968, as a result of a navigational error. The aircraft were captured and the pilots taken prisoner. Most of the Israeli losses had been sustained while attacking ground targets through heavy anti-aircraft fire; the mission against the Helwan complex of

Febrary 8th, when the two Egyptian MiGs were shot down, was the first time that the EAF's fighters had made a serious attempt to intercept one of the Israeli deep-penetration raids.

Then, towards the end of February, the situation took a new turn. Israeli pilots, returning from recon- naissance missions in the vicinity of Cairo, brought back some disturbing photographs: 15 new-type missile sites were under construction in the Egyptian capital's defence zone. Three weeks later, on March 20th, two huge Antonov An-22 transports of the Soviet Air Force landed at Cairo West and unloaded their cargoes: four SA-3 'Goa' surface-to-air missiles, together with their supporting electronic equip- ment.

By mid-April, five batteries of SA-3s were opera- tional on the approaches to Cairo. At last, the Egyptian air defences had a weapon capable of deal- ing with the IAF's marauding Phantoms and Mirages, for, like the American Hawk missile, the two-stage SA-3 was designed specifically to inter- cept fast, low-flying strike aircraft. The Egyptians planned to instal 40 SA-3 batteries; initially, until Egyptian personnel were fully trained in their use, each of these would be manned by an 80-strong Russian crew. And this was not all; a week later, Israeli and American intelligence sources revealed that some EAF fighter units, particularly those selected for the defence of vital strategic targets such as the Aswan Dam, were manned by Soviet Air Force personnel.

The fact that Soviet pilots were flying Egyptian aircraft was not new; they had been doing it since 1955, when Nasser received his first MiGs and Il-28s, although they had confined their activities to flying training. More recently, Soviet crews had been flying Algeria-based Tu-16s on surveillance missions over the United States Sixth Fleet. Now, however, faced with the possibility that IAF and Soviet pilots might meet in direct combat, and with the threat posed by the SA-3 batteries, the Israeli High Command made a difficult decision: on April 18th, deep-penetration missions beyond the Nile were suspended indefinitely.

In May, the Israeli High Command turned its attention to the northern front and the Lebanon. At dawn on May 12th, Israeli armour and infantry advanced across the Lebanese border on a 15-mile front in a 'search and destroy' operation designed to trap 1,500 Palestinian guerrillas in a massive pincer movement. The operation unfolded under a huge IAF air umbrella, with Skyhawks and Mirages mounting incessant strafing attacks on Lebanese Army positions. Suddenly, a dog-fight flared up overhead as six Syrian Air Force MiG-17s arrived and hurled themselves into the battle, completely surprising a flight of Skyhawks which was climbing away after attacking a target on the ground. Two Skyhawks were hit and crashed, the pilot of one baling out. He was picked up from Lebanese territory by an Israeli S-58 'plane guard' helicopter. But the MiGs' success was short-lived; the Syrian fighters were immediately engaged by a squadron of Mirages, and three of them went down in flames. The Israeli ground forces were attacked by MiG-17s twice during the course of the afternoon before they withdrew at the end of their operation, but all the Syrian fighters escaped. One Israeli helicopter, an S-58, was shot down by Lebanese machine-gun fire. Although the part played by the Syrian aircraft had

not been significant in terms of the damage they had
inflicted on the attackers, they had nevertheless compelled the Israeli Air Force to switch part of its
effort from ground attack to air defence, with the
result that the operation did not go entirely as
planned.

Meanwhile, the Israeli strike squadrons were
maintaining their pressure on Egyptian military
installations on the west bank of the Suez Canal.
These operations grew more intense on May 15th,
when reconnaissance photographs revealed that the
Egyptians were setting up a number of SA-3 missile
sites in the Canal Zone itself. For two days, Skyhawks and Mirages pounded a dozen new sites
spread over an area 20 miles in depth beyond the
Canal; the attacking aircraft suffered considerable
damage from Russian radar-controlled 57 mm anti-
aircraft batteries grouped around the missile sites,
although all but two returned to base. The Egyptian
Air Force, too, was up in strength to engage the
Israelis, but for some reason about half the aircraft
committed were obsolescent MiG-17s. In the space of
48 hours, the IAF claimed the destruction of seven
Egyptian aircraft.

Also on May 15th, Israeli aircraft visited Ras
Banas to attack Egyptian shipping as a reprisal for
the sinking of an Israeli fishing vessel and an attack
on Eilat harbour by Egyptian Navy frogmen. It
was the first time that Ras Banas had been the IAF's
objective since the June War. On that occasion,
Vautours had raided the airfield there; this time,
the attacking aircraft were Phantoms. The raid
involved a round trip of 1,300 miles, and was
intended partly as an indication to the Egyptians
that, in spite of their new SA-3s, the Israeli Air

Force was still capable of striking at any spot on Egyptian territory. Attacking through heavy anti-aircraft fire, the Phantoms sank a 2,575-ton Z-class destroyer and a Soviet-built Komar-class missile patrol boat. The latter was torn apart by fire from the Phantoms' internally-mounted M-61A1 multi-barrelled 20 mm cannon.

More strikes across the Suez Canal followed during the last two weeks in May, and this time road and rail communications as well as missile sites and artillery batteries were the targets. The strikes were stepped up after Egyptian commandos twice crossed the Canal in strength and ambushed Israeli patrols, causing some casualties. During the first days of June, Port Said was completely deprived of all land connections for a time after Israeli jets carried out several raids on roads, railways and water pipelines in the area, and the Egyptians had to enforce emergency measures to bring supplies of food and water into the town. This was followed, on June 7th, by a concentrated series of strikes lasting 12 hours on targets all along the west bank of the Canal, designed to frustrate a new Egyptian military build-up which was regarded by Israeli intelligence as a preparatory move towards an attempt to establish a bridgehead on Israeli-held territory. There was a lull in the air attacks when dusk descended, but they started again about midnight and continued without pause for five hours.

The air war over the Middle East, however, was showing signs of becoming less one-sided. In two ten-day periods between April 25th and June 5th, Egyptian Air Force Sukhoi Su-7s carried out a total of eleven offensive missions over the Sinai Peninsula. On each occasion a minimum of four aircraft was involved, and on two occasions the raids were carried out at squadron strength. For the first time in 22 years of hostilities, the Egyptian Air Force appeared to be working its way towards a position of real strength from which it could throw an aggressive challenge in the face of Israel's traditional air superiority.

CHAPTER THIRTEEN

The Future

The war between Israel and her Arab neighbours has now been going on for almost a quarter of a century. Any attempt to predict how much longer it is likely to last would be futile, for as Arab military strength continues to grow it is becoming more apparent that they will be satisfied with nothing less than the re-occupation of Palestine and the total surrender of Israel, if not her complete annihilation.

It is questionable whether any other nation in history has found itself in such a position of isolation as does Israel today; not even Britain during the grim days at the start of the Second World War. For Israel has no Commonwealth, no allies who are dedicated to her cause and on whose resources she may draw. Although her arms agreement with the United States gave her the means to maintain combat superiority for the time being, the reluctance on the part of the Americans to meet her full requirement for military aircraft and other equipment has been a clear indication that Israel must rely more and more on her own resources to counter a threat to her survival that shows every sign of escalating in the immediate future.

Since the June War of 1967, the amount of military equipment produced by Israel's own industries has tripled. The famous Israeli 'Uzi' machine-pistol is in service in 50 countries, equipping organizations ranging from special units of the American Green Berets to the Shah of Persia's personal bodyguard; all ammunition for the Israeli Defence Forces, with the exception of the Matra and Hawk missiles, is turned out by Israel's munitions factories, as are mortars up to a calibre of 160 mm and 30 mm quick-firing aircraft cannon. Israeli engineers are also working on the prototype of an advanced battle tank, provisionally known as the Sabra, which will be armed with surface-to-surface missiles as well as a heavy-calibre cannon.

It is in the field of combat aircraft, however, that some of the biggest strides are being made. It has been revealed that Israel Aircraft Industries are currently converting a basic Mirage IIICJ airframe into a 'Super Mirage'; the prototype could be flying in 1971, and if flight tests are successful the Israeli-built type will be produced in quantity for the IAF's combat squadrons. The aircraft will be powered by two turbojet engines—possibly American General Electric J79s or Pratt and Whitney J52s. A J79 is at present undergoing exhaustive flight tests in Israel in a Mirage III, replacing the aircraft's normal Atar 9—supplies of which have not been getting through because of the French Government's arms embargo. Later production 'Super Mirages' may be powered by a locally-built variant of the Atar, detailed blueprints of which were 'acquired' by Israeli agents from Switzerland in 1969.

The Super Mirage—and the more advanced designs that are likely to follow it—will mean that the Israeli Air Force will no longer be entirely dependent on the goodwill of foreign governments for its equipment. It means that Israel's own aircraft industry will be able to replace planes destroyed

in combat with little or no delay—a vital consideration, taking into account the fact that over the past
months aircraft losses have increased in direct proportion to the escalation in the war of attrition.

However, the availability of aircraft is no guaran-
tee that the Israeli Air Force will continue to be
effective for much longer in its primary role, which
is to keep the enemy at arm's length by striking
powerful blows at his military machine wherever
and whenever possible. Already, because of the
presence of new Soviet air defence equipment such
as the SA-3, the series of deep-penetration strikes
against Egyptian targets by the IAF has had to be
severely curtailed, and short-range attacks across the
Suez Canal may before long result in the IAF suffering prohibitive losses from defences that grow more
formidable with every passing month. Added to this,
the Israeli Air Force is being compelled to keep up
sustained pressure on more than one front; there
is no real respite for either air or ground crews.

How long the IAF can go on holding off the Arab
belligerents by the means it employs at present is a
matter of guesswork, but with its relatively limited
resources there has to be a limit somewhere. Once
that limit has been reached—and once the Arab
belligerents know that it has been reached—the
conclusion is that they will launch a full-scale attack
on Israel on all fronts. They will be able to choose
their own time, for the disaster of June 1967 will
not be repeated. With the massive pre-emptive
strike that bought her vital time, Israel played her
last trump card.

Her last conventional trump card, that is. There
is another frightening possibility. If Israel finds
that conventional warfare is no longer adequate to

ensure her survival—assuming that political endeavours to bring about a settlement in the Middle East end in failure—then she may turn to the creation of a nuclear deterrent in an attempt to find the answer.

On June 5th, 1967—the first day of the June War—a damaged IAF Mystère fighter returning from a strike in the Suez Canal area was shot down by a Hawk missile over the northern part of the Negev. The pilot was killed. He had inadvertently strayed over the most secret area in Israel—a segment of barren desert not far from the southern tip of the Dead Sea, where the sand and scrub are dominated by a tall chimney and a strange structure like a huge golf ball, surrounded by a complex of low concrete buildings.

Here, at Dimona, several hundred scientists and technicians are employed on top-secret work at Israel's nuclear reactor plant. Construction of the 'atom city' was begun with the help of French technicians shortly after the Arab-Israeli war of 1956. The secret was well kept; it was not until 1959 that the Egyptians raised the alarm and turned the spotlight on what was going on in the Negev. The US Government immediately asked the Israelis to make a statement about the purpose of the Dimona project; they replied that it was merely a textile factory.

The Central Intelligence Agency was far from satisfied. In 1960, U-2 spy planes based on Incirlik in Turkey made a series of high-altitude reconnaissance flights over the Negev. The photos they brought back were carefully studied by experts who discovered that for a textile factory, the installations were very strangely equipped indeed; in fact, the centre-piece appeared to be a 24-megawatt nuclear reactor of the EL-3 type, the kind suitable for the manufacture of fissile materials used in atomic weapons.

Israel then admitted that a nuclear reactor did exist in the Negev, but that it was being used for peaceful purposes. An American team was invited to inspect the installations, and came away having found no evidence to support the theory that fissile material was being manufactured for use in atomic bombs. What the Americans did not know, however, was that the French were producing uranium for Israel in their atomic centre at Pierrelatte. Neither did they know that nuclear research in Israel had reached the point where France's help was no longer needed; that the Israelis were buying uranium from South Africa and the Argentine, and extending their installations at Dimona to include a plant capable of producing six kilogrammes of plutonium a year. Most important of all, Israeli scientists had discovered a method of extracting uranium cheaply from waste products left over from potash workings on the shores of the Dead Sea; as much as 50 kilogrammes a year, enough to power the nuclear reactor.

In 1965, after a visit to Israel, Edward Teller—the 'father' of the hydrogen bomb—commented that Israel would have the capability to manufacture her own nuclear weapons 'at a very early date'. And one neutral member of the Atomic Energy Commission in Vienna observed that 'the Israelis have been carrying out a major scientific project in the Negev for over a decade now, and no information about what they are doing has been forthcoming. The only conclusion we can draw is that the project is of a top-secret military nature; namely, the manufacture of an atomic bomb'.

Conclusions are one thing; facts are another. The facts, as they now stand, are simply these. Israel now has the technical and physical resources to build a series of atomic bombs in the 20-kiloton range: tactical weapons of roughly the same yield as the two that were dropped on Japan. She also has a highly efficient delivery system in the shape of her Phantoms and Skyhawks, both of which are equipped to carry tactical nuclear devices, and is developing a tactical surface-to-surface rocket which can carry either a conventional or a nuclear warhead. Designated MD-660, the rocket was originally developed by Marcel Dassault of France under contract to the Israeli Government, and firing trials were carried out in the Mediterranean off Toulon in 1968. The missile can be launched from a mobile ramp and has a range in the order of 280 miles—sufficient for it to reach targets around Cairo and beyond the Nile from launching-sites in Sinai or the southern Negev.

Equipped with an advanced guidance system— which is now under development—these weapons could be used to hit targets in Egyptian territory where a high degree of precision was not required, relieving the Israeli Air Force of the burden of having to pit its men and machines against ever-stronger defences. If a conventional war is to continue in the Middle East, the increasing use of missiles as long-range artillery is a logical next step.

Missiles, however, will never replace pilots and aircraft. As long as the war continues the men and machines of the Israeli Air Force will retain their first-line role both in attack and defence. At present, with their SA-3s, the Egyptians appear to be one step ahead; but the SA-3 is not infallible, and although its presence has acted as a deterrent to Israeli plans for attacking targets deep inside Egypt for the time being—simply because the IAF has to balance the importance of a target against the losses that are likely to be incurred in attacking it— neither these missiles, nor the Russian pilots who are being attached to Egyptian combat squadrons in growing numbers, would prevent the IAF from mounting occasional deep-penetration strikes if a real need arose.

In any case, the Israeli High Command has taken what is very probably the right decision in electing to concentrate the IAF's operations on the zone immediately across the Suez Canal, over which the Egyptians must enjoy complete air superiority if they are to launch an invasion of Sinai with any hope of success. Since the beginning of 1970, repeated IAF strikes have turned the Egyptian air defence system in the Canal Zone into a shambles; A-3 sites have been pinpointed in their early stages of construction and promptly destroyed, as have the radar stations on which the missiles depend. The Egyptians realized from the outset that the installation of their new SA-3 batteries in the Canal Zone would be fraught with difficulties, and it was for this reason that—early in 1967—they approached the Russians with a request for a quantity of ultra-modern MiG-23 interceptors, all-weather fighters capable of flying at three times the speed of sound and reaching altitudes of over 100,000 feet. The Egyptian request received considerable publicity, but very little mention was made of one important fact: that the MiG-23 can only carry out its role as a high-altitude all-weather interceptor effectively if it works in conjunction with advanced early warning aircraft crammed with electronic detection gear. What happens is this: the early warning aircraft, usually the military version of the Tu-114 airliner, cruises at 40,000 feet with its radar 'eyes' on the lookout for low-flying strike aircraft. The MiG-23 cruises at 80,000 feet. Once an incoming enemy aircraft is picked up, the Tu-114's computers flash its co-ordinates to the MiG-23's advanced fire-control system. The MiG then launches 'snap-down' missiles, which streak down through the stratosphere to intercept the intruding aircraft at the point determined by the computers.

Without the Tu-114 to work with it, the MiG-23 loses much of its effectiveness. The combination of the two aircraft would have gone a long way towards solving Egypt's air defence problems, but the Russians were understandably reluctant to operate the AEW version of the Tu-114 outside the borders of the Soviet Union; the aircraft carries so much secret equipment that if it fell into the wrong hands the consequences to Russia's own defence network could be serious. Instead, the Russians agreed to supply the Egyptians with the MiG-21J, an improved version of the MiG-21PF with an all-weather capability, a higher-powered engine and an armament of both infra-red homing missiles and a 23 mm gun pod. The first MiG-21Js arrived in Egypt in April 1970, and to counter this new threat the Israelis asked McDonnell Douglas Aircraft to modify ten of the F-4E Phantoms on order as pure interceptors.

So, while the balance of power in the Middle East teeters on a knife-edge, and politicians fight their war of words, the shooting war goes on. Yet on the

Israeli Air Force's bases, the fact that the nation is in a state of war is not readily apparent; there is no sense of grim urgency, no exhausted, red-eyed pilots snatching an hour's sleep between missions, no 'Battle of Britain' atmosphere. Life on one of the big IAF bases closely resembles that of a peacetime RAF station; the married quarters are full; there is no thought that the situation might become dangerous enough to warrant the evacuation of non-military personnel. To an outsider, there is nothing to indicate whether the flight of Skyhawks or Mirages thundering down the runway is taking off on a combat mission or a routine training flight.

But the signs of war are there, if one knows where to look for them. Ground crews working in a hangar on an aircraft with wings torn by shell splinters; an empty chair in the mess; an unfinished book, turned face down on the crew-room table where its reader left it before take-off; a sad-faced girl, leaving the base for the home of relatives, with children who are fighting desperately to hold back their tears because they are old enough to sense their mother's tragedy, but too young to know that their father will not be coming back.

That is Israel's war.

APPENDIX

AIRCRAFT IN ISRAELI SERVICE SINCE 1948

Type	Remarks
Aeronca L-18	One obtained in 1950.
Airspeed Consul	Used for twin-engine conversion and navigation training during 1950s.
Auster AOP 3	Nineteen aircraft assembled from 25 crashed/scrap RAF machines purchased in 1948.
Avia C.210	Twenty-six aircraft purchased from Czechoslovakia in 1948. All were flown in aboard transport aircraft and assembled in Israel.
Avro Anson	Small number purchased in 1950-51 and used as crew trainers.
Beech A-35 Bonanza	Three aircraft obtained in 1949. Registrations 4X-AC1, 4X-AER; registration of third aircraft unknown.
Beech 18	One acquired by Arkia (Israel Inland Airlines) in 1961 for service between Tel Aviv and northern kibbutzim.

Type	Remarks
Boeing B-17G	Three obtained from USAF surplus stocks in 1948. Bombed Cairo, Gaza and El Arish on July 15th, 1948; two still in service during the Sinai Campaign of 1956.
Boeing PT-17 Kaydet	Large number obtained during 1949-50 for use as primary trainers.
Boeing Stratocruiser	Five aircraft purchased from Pan-American World Airways in 1962, and modified for use as military transports.
Boeing 707-320B and 707-420	Total of nine aircraft in service with El Al. Registrations 4X-ABA, 4X-ABB, 4X-ATA, 4X-ATB, 4X-ATC, 4X-ATR, -S, -T and -U.
Bristol Britannia Series 300	Four served with El Al from 1957, until replaced by Boeing 707s.
Convair PBY-5A Catalina	Two purchased in 1952 for service with IAF's Maritime Flight.
Curtiss C-46 Commando	Total of ten purchased from USAF surplus stocks in 1948.
Dassault Mirage IIICJ	Seventy-two delivered to IAF from 1963 onwards; equips three first-line squadrons and played a leading part in the destruction of the Arab air forces during the Six-Day War.

Type	Remarks
Dassault Mystère IVA	Sixty entered service with IAF as interceptors in 1956. Currently serves as a ground-attack aircraft and equips operational training units.
Dassault Ouragan	Seventy-five supplied during 1955-57. Still equips some secondline close support units, but now mainly relegated to the operational training role.
Dassault Super Mystère B.2	Twenty-four acquired in 1959-60. Equipped two interceptor squadrons; still in use as a close-support aircraft.
De Havilland Chipmunk	Small number used as primary trainers and communications aircraft during the 1950s.
De Havilland Mosquito	Approximately 250 Mosquitoes of all types purchased by the IAF during 1950-54, mostly in a crashed/scrap condition. Forty—mostly FB.6s and T.3s—assembled from cannibalized parts, and a few more (about ten) purchased direct from Britain.
De Havilland Rapide	Two purchased in 1949, and one more captured intact on an Egyptian airstrip in Sinai.

Type	Remarks
De Havilland Tiger Moth	A few bought from Canada in 1949-50 and used as basic trainers.
Douglas C-47 Dakota	Thirty Dakotas acquired between 1948 and 1956. Still used by Air Transport Command.
Douglas C-54	Three C-54s used by the IAF during the 1950s. The first entered service in 1948 and was used to ferry Avia C.210s from Czechoslovakia.
Douglas DC-5	One aircraft acquired in 1950.
Fairchild F-24R Argus	One aircraft acquired in 1949.
Fokker S.11	Forty-one purchased from the Netherlands during the early 1950s for use as primary trainers. Many turned over to flying clubs, including 4X-ADM, 4X-ADR, 4X-ANA, 4X-ANB and 4X-AWC.
Gloster Meteor T.7	Four T.7s purchased from Britain in 1953 (serials 2162-2165) and two from Belgium.

Type	Remarks
Gloster Meteor F.8	Eleven F.8.s acquired from Britain between August 1933 and January 1954 (serials 2166-2169 and 2172-2178).
Gloster Meteor F.R.9	Seven FR.9s purchased from Flight Refuelling Ltd. Serials 211-217.
Gloster Meteor NF.13	Six NF.13s delivered between September 1956 and March 1958. Serials 4X-FNA, -B, -C, -D, -E, and -F.
Grumman Widgeon	Two Widgeons acquired in 1950.
Handley-Page Herald	Three Heralds purchased in 1961-2 by Arkia Airlines.
Hiller 360	Small number in use for training, light transport and communications with IAF Helicopter Wing.
Lockheed Hudson	One aircraft purchased from Canada in 1950.
Lockheed Constellation	Three acquired from USAF surplus stocks in 1948.
McDonnell Douglas F.4E Phantom	Fifty F-4E strike aircraft and six RF-4E reconnaissance machines ordered. Over 30 delivered by May 1970, and further deliveries proceeding at the rate of four aircraft per month. Equips three IAF strike/interceptor squadrons.

251

Type	Remarks
McDonnel Douglas A-4H Skyhawk	Seventy A-4Hs and three TA-4Hs ordered, A-4Hs as an Ouragan replacement. Deliveries completed early in 1970; type equips three ground-attack squadrons.
Miles Aerovan	One delivered to Israel in June 1948 by Mayfair Air Services Ltd, via South Africa.
Miles Gemini	One acquired in 1949.
Noorduyn Norseman	Six purchased from Canada in 1948; three lost en route to Israel. One of the surviving three (4X-ARS) withdrawn from the Israel Civil Register in 1962.
Nord Noratlas	Twelve purchased from France in 1956; later supplemented by eight ex-Luftwaffe machines. Still in service with Air Transport Command.
Nord Norecrin	Two purchased from France in 1950. Registrations 4X-ADT and 4X-ADY.
North American Harvard	Twenty-five purchased between 1950 and 1953. Used as advanced trainers and light ground-attack aircraft.